The Algorithm
Design Manual

Steven S. Skiena

The Algorithm
Design Manual

With 72 Figures

 Includes CD-ROM

Steven S. Skiena
Department of Computer Science
State University of New York
Stony Brook, NY 11794-4400
USA

Library of Congress Cataloging-in-Publication Data
Skiena, Steven S.
     The algorithm design manual / Steven S. Skiena
          p.   cm.
     Includes bibliographical references and index.
     ISBN 0-387-94860-0 (hardcover : alk. paper)
     1. Computer algorithms.   I. Title.
QA76.9.A43S55   1997
005.1—dc21   97-20712

Printed on acid-free paper.

Production managed by Anthony Guardiola; manufacturing supervised by Johanna Tschebull.
Photocomposed pages prepared using the author's electronic files by Integre Technical Publishing Company, Inc., Albuquerque, NM.
Printed and bounded by Hamilton Printing Co., Rensselaer, NY.
Printed in the United States of America.

9 8 7 6 5 4 3

ISBN 0-387-94860-0 Springer-Verlag New York Berlin Heidelberg   SPIN 10725301

THE
ELECTRONIC
LIBRARY
OF
SCIENCE

TELOS, The Electronic Library of Science, is an imprint of Springer-Verlag New York. Its publishing program encompasses the natural and physical sciences, computer science, mathematics, economics, and engineering. All TELOS publications have a computational orientation to them, as TELOS' primary publishing strategy is to wed the traditional print medium with the emerging new electronic media in order to provide the reader with a truly interactive multimedia information environment. To achieve this, every TELOS publication delivered on paper has an associated electronic component. This can take the form of book/diskette combinations, book/CD-ROM packages, books delivered via networks, electronic journals, newsletters, plus a multitude of other exciting possibilities. Since TELOS is not committed to any one technology, any delivery medium can be considered. We also do not foresee the imminent demise of the paper book, or journal, as we know them. Instead we believe paper and electronic media can coexist side-by-side, since both offer valuable means by which to convey information to consumers.

The range of TELOS publications extends from research level reference works to textbook materials for the higher education audience, practical handbooks for working professionals, and broadly accessible science, computer science, and high technology general interest publications. Many TELOS publications are interdisciplinary in nature, and most are targeted for the individual buyer, which dictates that TELOS publications be affordably priced.

Of the numerous definitions of the Greek word "telos," the one most representative of our publishing philosophy is "to turn," or "turning point." We perceive the establishment of the TELOS publishing program to be a significant step forward towards attaining a new plateau of high quality information packaging and dissemination in the interactive learning environment of the future. TELOS welcomes you to join us in the exploration and development of this exciting frontier as a reader and user, an author, editor, consultant, strategic partner, or in whatever other capacity one might imagine.

TELOS, The Electronic Library of Science
Springer-Verlag New York, Inc.

THE ELECTRONIC LIBRARY OF SCIENCE

TELOS Diskettes

Unless otherwise designated, computer diskettes packaged with TELOS publications are 3.5″ high-density DOS-formatted diskettes. They may be read by any IBM-compatible computer running DOS or Windows. They may also be read by computers running NEXTSTEP, by most UNIX machines, and by Macintosh computers using a file exchange utility.

In those cases where the diskettes require the availability of specific software programs in order to run them, or to take full advantage of their capabilities, then the specific requirements regarding these software packages will be indicated.

TELOS CD-ROM Discs

For buyers of TELOS publications containing CD-ROM discs, or in those cases where the product is a stand-alone CD-ROM, it is always indicated on which specific platform, or platforms, the disc is designed to run. For example, Macintosh only; Windows only; cross-platform, and so forth.

TELOSpub.com (Online)

Interact with TELOS online via the Internet by setting your World-Wide-Web browser to the URL: *http://www.telospub.com*.

The TELOS Web site features new product information and updates, an online catalog and ordering, samples from our publications, information about TELOS, data-files related to and enhancements of our products, and a broad selection of other unique features. Presented in hypertext format with rich graphics, it's your best way to discover what's new at TELOS.

TELOS also maintains these additional Internet resources:

gopher://gopher.telospub.com
ftp://ftp.telospub.com

For up-to-date information regarding TELOS online services, send the one-line e-mail message:

send info to: info@TELOSpub.com.

Preface

Most of the professional programmers that I've encountered are not well prepared to tackle algorithm design problems. This is a pity, because the techniques of algorithm design form one of the core practical *technologies* of computer science. Designing correct, efficient, and implementable algorithms for real-world problems is a tricky business, because the successful algorithm designer needs access to two distinct bodies of knowledge:

- *Techniques* – Good algorithm designers understand several fundamental algorithm design techniques, including data structures, dynamic programming, depth-first search, backtracking, and heuristics. Perhaps the single most important design technique is *modeling*, the art of abstracting a messy real-world application into a clean problem suitable for algorithmic attack.

- *Resources* – Good algorithm designers stand on the shoulders of giants. Rather than laboring from scratch to produce a new algorithm for every task, they know how to find out what is known about a particular problem. Rather than reimplementing popular algorithms from scratch, they know where to seek existing implementations to serve as a starting point. They are familiar with a large set of basic algorithmic problems, which provides sufficient source material to model most any application.

This book is intended as a manual on algorithm design, providing access to both aspects of combinatorial algorithms technology for computer professionals and students. Thus this book looks considerably different from other books on algorithms. Why?

- We reduce the algorithm design process to a sequence of questions to ask about the problem at hand. This provides a concrete path to take the nonexpert from an initial problem statement to a reasonable solution.

- Since the practical person is usually looking for a program more than an algorithm, we provide pointers to solid implementations whenever they are available. We have collected these implementations on the enclosed CD-ROM and at one central WWW site for easy retrieval. Further, we provide recommendations to make it easier to identify the correct code for the job. With these implementations available, the critical issue in algorithm design becomes properly modeling your application, more so than becoming intimate with the details of the actual algorithm. This focus permeates the entire book.

- Since finding out what is known about a problem can be a difficult task, we provide a catalog of important algorithmic problems as a major component of this book. By browsing through this catalog, the reader can quickly identify what their problem is called, what is known about it, and how they should proceed to solve it. To aid in problem identification, we include a pair of "before" and "after" pictures for each problem, illustrating the required input and output specifications.

- For each problem in the catalog, we provide an honest and convincing motivation, showing how it arises in practice. If we could not find such an application, then the problem doesn't appear in this book.

- In practice, algorithm problems do not arise at the beginning of a large project. Rather, they typically arise as subproblems when it suddenly becomes clear that the programmer does not know how to proceed or that the current program is inadequate. To provide a better perspective on how algorithm problems arise in the real world, we include a collection of "war stories," tales from our experience on real problems. The moral of these stories is that algorithm design and analysis is not just theory, but an important tool to be pulled out and used as needed.

Equally important is what we do not do in this book. We do not stress the mathematical analysis of algorithms, leaving most of the analysis as informal arguments. You will not find a single theorem anywhere in this book. Further, we do not try to be encyclopedic in our descriptions of algorithms, but only in our pointers to descriptions of algorithms. When more details are needed, the reader should follow the given references or study the cited programs. The goal of this manual is to get you going in the right direction as quickly as possible.

But what is a manual without software? This book comes with a substantial electronic supplement, an ISO-9660 compatible, multiplatform CD-ROM, which can be viewed using Netscape, Microsoft Explorer, or any other WWW browser. This CD-ROM contains:

- A complete hypertext version of the full printed book. Indeed, the extensive cross-references within the book are best followed using the hypertext version.

- The source code and URLs for all cited implementations, mirroring the Stony Brook Algorithm Repository WWW site. Programs in C, C++, Fortran, and Pascal are included, providing an average of four different implementations for each algorithmic problem. Over 50 megabytes of software is included.

- More than thirty hours of *audio* lectures on the design and analysis of algorithms are provided, all keyed to the on-line lecture notes. Following these lectures provides another approach to learning algorithm design techniques.

Listening to all the audio is equivalent to taking a one-semester college course on algorithms!

This book is divided into two parts, *techniques* and *resources*. The former is a general guide to techniques for the design and analysis of computer algorithms. The resources section is intended for browsing and reference, and comprises the catalog of algorithmic resources, implementations, and an extensive bibliography.

Altogether, this book covers material sufficient for a standard *Introduction to Algorithms* course, albeit one stressing design over analysis. We assume the reader has completed the equivalent of a second programming course, typically titled *Data Structures* or *Computer Science II*. Textbook-oriented features include:

- Beyond standard pen-and-paper exercises, this book includes "implementation challenges" suitable for teams or individual students. These projects and the applied focus of the text can be used to provide a new laboratory focus to the traditional algorithms course. More difficult exercises are marked by (*) or (**).

- "Take-home lessons" at the beginning of each chapter emphasize the concepts to be gained from the chapter.

- This book stresses design over analysis. It is suitable for both traditional lecture courses and the new "active learning" method, where the professor does not lecture but instead guides student groups to solve real problems. The "war stories" provide an appropriate introduction to the active learning method.

- A full set of lecture slides for teaching this course is available on the CD-ROM and via the World Wide Web, both keyed to unique on-line audio lectures covering a full-semester algorithm course. Further, a complete set of my videotaped lectures using these slides is available for interested parties. See http://www.cs.sunysb.edu/~algorith for details.

Acknowledgments

I would like to thank several people for their concrete contributions to this project. Ricky Bradley built up the substantial infrastructure required for both the WWW site and CD-ROM in a logical and extensible manner. Zhong Li did a spectacular job drawing most of the catalog figures using xfig and entering the lecture notes that served as the foundation of Part I of this book. Frank Ruscica, Kenneth McNicholas and Dario Vlah came up big in the pinch, redoing all the audio and helping as the completion deadline approached. Filip Bujanic, David Ecker, David Gerstl, Jim Klosowski, Ted Lin, Kostis Sagonas, Kirsten Starcher, Brian Tria, and Lei Zhao all made contributions at various stages of the project.

Richard Crandall, Ron Danielson, Takis Metaxas, Dave Miller, Giri Narasimhan, and Joe Zachary all reviewed preliminary versions of the manuscript and/or CD-ROM; their thoughtful feedback helped to shape what you see here. Thanks also to Allan Wylde, the editor of my previous book as well as this one, and Keisha Sherbecoe and Anthony Guardiola of Springer-Verlag. I am grateful for the support given me by the National Science Foundation and the Office of Naval Research.

I learned much of what I know about algorithms along with my graduate students Yaw-Ling Lin, Sundaram Gopalakrishnan, Ting Chen, Francine Evans, Harald Rau, Ricky Bradley, and Dimitris Margaritis. They are the real heroes of many of the war stories related within. Much of the rest I have learned with my Stony Brook friends and colleagues Estie Arkin and Joe Mitchell, who have always been a pleasure to work and be with.

Finally, I'd like to send personal thanks to several people. Mom, Dad, Len, and Rob all provided moral support. Michael Brochstein took charge of organizing my social life, thus freeing time for me to actually write the book. Through his good offices I met Renee. Her love and patience since then have made it all worthwhile.

Caveat

It is traditional for the author to magnanimously accept the blame for whatever deficiencies remain. I don't. Any errors, deficiencies, or problems in this book are somebody else's fault, but I would appreciate knowing about them so as to determine who is to blame.

Steven S. Skiena
Department of Computer Science
State University of New York
Stony Brook, NY 11794-4400
http://www.cs.sunysb.edu/~skiena
September 1997

Contents

II RESOURCES

P
A
R
T

I

Techniques

Introduction to Algorithms

What is an algorithm? An algorithm is a procedure to accomplish a specific task. It is the idea behind any computer program.

To be interesting, an algorithm has to solve a general, well-specified *problem*. An algorithmic problem is specified by describing the complete set of *instances* it must work on and what properties the output must have as a result of running on one of these instances. This distinction between a problem and an instance of a problem is fundamental. For example, the algorithmic problem known as *sorting* is defined as follows:

Input: A sequence of n keys $a_1, \ldots, a_n$.
Output: The permutation (reordering) of the input sequence such that $a_1' \leq a_2', \cdots, \leq a_n'$.

An instance of sorting might be an array of names, such as {*Mike, Bob, Sally, Jill, Jan*}, or a list of numbers like {*154, 245, 568, 324, 654, 324*}. Determining whether you in fact have a general problem to deal with, as opposed to an instance of a problem, is your first step towards solving it. This is true in algorithms as it is in life.

An *algorithm* is a procedure that takes any of the possible input instances and transforms it to the desired output. There are many different algorithms for solving the problem of sorting. For example, one method for sorting starts with a single element (thus forming a trivially sorted list) and then incrementally inserts the remaining elements so that the list stays sorted. This algorithm, *insertion sort*, is described below:

```
InsertionSort(A)
      for i = 1 to n − 1 do
            for j = i + 1 downto 2 do
                  if (A[j] < A[j − 1]) then swap(A[j], A[j − 1])
```

Note the generality of this algorithm. It works equally well on names as it does on numbers, given the appropriate $<$ comparison operation to test which of the two keys should appear first in sorted order. Given our definition of the sorting problem, it can be readily verified that this algorithm correctly orders every possible input instance.

In this chapter, we introduce the desirable properties that good algorithms have, as well as how to measure whether a given algorithm achieves these goals. Assessing algorithmic performance requires a modest amount of mathematical notation, which we also present. Although initially intimidating, this notation proves essential for us to compare algorithms and design more efficient ones.

While the hopelessly "practical" person may blanch at the notion of theoretical analysis, we present this material because it is useful. In particular, this chapter offers the following "take-home" lessons:

- Reasonable-looking algorithms can easily be incorrect. Algorithm correctness is a property that must be carefully demonstrated.

- Algorithms can be understood and studied in a machine independent way.

- The "big Oh" notation and worst-case analysis are tools that greatly simplify our ability to compare the efficiency of algorithms.

- We seek algorithms whose running times grow logarithmically, because $\log_b(n)$ grows very slowly with increasing n.

- Modeling your application in terms of well-defined structures and algorithms is the most important single step towards a solution.

1.1   Correctness and Efficiency

Throughout this book we will seek algorithms that are *correct* and *efficient*, while being *easy to implement*. All three goals are obviously desirable, but they may not be simultaneously achievable. For this reason, one or more of them are often ignored. Theoretical algorithm designers have traditionally been unconcerned with implementation complexity, since they often do not program their algorithms. Instead, theoreticians focus on efficiency and correctness. Conversely, quick-and-dirty is typically the rule of thumb in industrial settings. Any program that seems to give good enough answers without slowing the application down is acceptable, regardless of whether a better algorithm exists. The issue of finding the best possible answer or achieving maximum efficiency usually does not arise in industry until serious troubles do.

Here, we stress the importance of recognizing the difference between *algorithms*, which always produce a correct result, and *heuristics*, which often do a good job without providing any guarantee. We also emphasize the potential efficiency gains resulting from using faster algorithms.

1.1.1 Correctness

It is seldom obvious whether a given algorithm correctly solves a given problem. This is why correct algorithms usually come with a proof of correctness, which is an explanation of *why* we know that the algorithm correctly takes all instances of the problem to the desired result. In this book, we will not stress formal proofs of correctness, primarily because they take substantial mathematical maturity to properly appreciate and construct. However, before we go further it is important to demonstrate why *"it's obvious"* never suffices as a proof of correctness and usually is flat-out wrong.

To illustrate, let us consider a problem that often arises in manufacturing, transportation, and testing applications. Suppose we are given a robot arm equipped with a tool, say a soldering iron. In manufacturing circuit boards, all the chips and other components must be fastened onto the substrate. More specifically, each chip has a set of contact points (or wires) that must be soldered to the board. To program the robot arm to perform this soldering job, we must first construct an ordering of the contact points, so that the robot visits (and solders) the first contact point, then visits the second point, third, and so forth until the job is done. The robot arm must then proceed back to the first contact point to prepare for the next board, thus turning the tool-path into a closed tour, or cycle.

Since robots are expensive devices, we want to find the tour that minimizes the time it takes to assemble the circuit board. A reasonable assumption is that the robot arm moves with fixed speed, so that the time it takes to travel between two points is the same as its distance. In short, we must solve the following algorithm problem:

Input: A set S of n points in the plane.
Output: What is the shortest cycle tour that visits each point in the set S?

You are given the job of programming the robot arm. Stop right now and think about an algorithm to solve this problem. I'll be happy to wait until you find one.

Several possible algorithms might come to mind to solve this problem. Perhaps the most popular idea is the *nearest-neighbor* heuristic. Starting from some point p_0, we walk first to its nearest neighbor p_1. From p_1, we walk to its nearest unvisited neighbor, thus excluding only p_0 as a candidate. We now repeat this process until we run out of unvisited points, at which time we return to p_0 to close off the tour. In pseudocode, the nearest-neighbor heuristic looks like this:

```
NearestNeighborTSP(P)
    Pick and visit an initial point p0 from P
    p = p0
```

$$i = 0$$

While there are still unvisited points

$$i = i + 1$$

Select p_i to be the closest unvisited point to p_{i-1}

Visit p_i

Return to p_0 from p_{n-1}

This algorithm has a lot to recommend it. It is simple to understand and implement. It makes sense to visit nearby points before we visit faraway points if we want to minimize the total travel time. The algorithm works perfectly on the example in Figure 1-1. The nearest neighbor rule is very efficient, for it looks at each pair of points (p_i, p_j) at most twice, once when adding p_i to the tour, the other when adding p_j. Against all these positives there is only one problem. This algorithm is completely wrong.

Wrong? How can it be wrong? The algorithm always finds a tour, but the trouble is that it doesn't necessarily find the shortest possible tour. It doesn't necessarily even come close. Consider the set of points in Figure 1-2, all of which lie spaced along a line. The numbers describe the distance that each point lies to the left or right of the point labeled '0'. When we start from the point '0' and repeatedly walk to the nearest unvisited neighbor, you will see that we keep jumping left-right-left-right over '0'. A much better (indeed optimal) tour for these points starts from the leftmost point and visits each point as we walk right before returning at the rightmost point.

Try now to imagine your boss's delight as she watches a demo of your robot arm hopscotching left-right-left-right during the assembly of such a simple circuit board.

"But wait," you might be saying. "The problem was in starting at point '0'. Instead, why don't we always start the nearest-neighbor rule using the leftmost point as the starting point p_0? By doing this, we will find the optimal solution on this example."

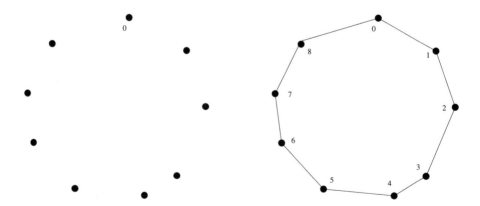

Figure 1–1. A good example for the nearest neighbor heuristic

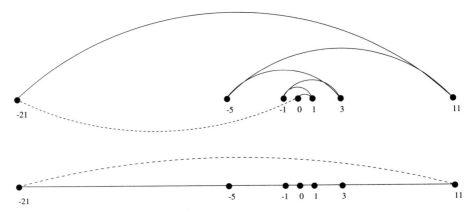

Figure 1-2. A bad example for the nearest neighbor heuristic

That is 100% true, at least until we rotate our example 90 degrees. Now all points are equally leftmost. If the point '0' were moved just slightly to the left, it would be picked as the starting point. Now the robot arm will hopscotch up-down-up-down instead of left-right-left-right, but the travel time will be just as bad as before. No matter what you do to pick the first point, the nearest neighbor rule is doomed not to work correctly on certain point sets.

Maybe what we need is a different approach. Always walking to the closest point is too restrictive, since it seems to trap us into making moves we didn't want. A different idea would be to repeatedly connect the closest pair of points whose connection will not cause a cycle or a three-way branch to be formed. Eventually, we will end up with a single chain containing all the points in it. At any moment during the execution of this *closest-pair heuristic*, we will have a set of partial paths to merge, where each vertex begins as its own partial path. Connecting the final two endpoints gives us a cycle. In pseudocode:

ClosestPairTSP(P)
    Let n be the number of points in set P.
    $d = \infty$
    For $i = 1$ to $n - 1$ do
        For each pair of endpoints (s, t) of distinct partial paths
            if $dist(s, t) \leq d$ then $s_m = s$, $t_m = t$, and $d = dist(s, t)$
        Connect (s_m, t_m) by an edge
    Connect the two endpoints by an edge

This closest-pair rule does the right thing on the example in Figure 1-2. It starts by connecting '0' to its immediate neighbors, the points 1 and -1. Subsequently, the next closest pair will alternate left-right, growing the central path by one link at a time. The closest-pair heuristic is somewhat more complicated and less efficient than the previous one, but at least it gives the right answer. On this example.

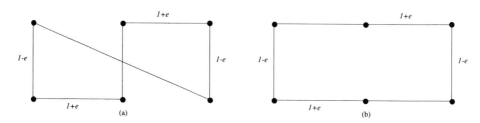

Figure 1–3. A bad example for the closest-pair heuristic.

But not all examples. Consider what the algorithm does on the point set in Figure 1-3(a). It consists of two rows of equally spaced points, with the rows slightly closer together (distance $1 - e$) than the neighboring points are spaced within each row (distance $1 + e$). Thus the closest pairs of points stretch across the gap, not around the boundary. After we pair off these points, the closest remaining pairs will connect these pairs alternately around the boundary. The total path length of the closest-pair tour is $3(1-e)+2(1+e)+\sqrt{(1-e)^2 + (2+e)^2}$. Compared to the tour shown in Figure 1-3(b), we travel over 20% farther than necessary when $e \approx 0$. Examples exist where the penalty is considerably worse than this.

Thus this second algorithm is also wrong. Which one of these algorithms performs better? You can't tell just by looking at them. Clearly, both heuristics can end up with very bad tours on very innocent-looking input.

At this point, you might wonder what a correct algorithm for the problem looks like. Well, we could try all enumerating *all* possible orderings of the set of points, and then select the ordering that minimizes the total length:

> OptimalTSP(P)
>> $d = \infty$
>> For each of the $n!$ permutations P_i of points P
>>> If $(cost(P_i) \leq d)$ then $d = cost(P_i)$ and $P_{min} = P_i$
>> Return P_{min}

Since all possible orderings are considered, we are guaranteed to end up with the shortest possible tour. This algorithm is correct, since we pick the best of all the possibilities. The problem is that it is also extremely slow. The fastest computer in the world couldn't hope to enumerate all $20! = 2{,}432{,}902{,}008{,}176{,}640{,}000$ orderings of 20 points within a day. For real circuit boards, where $n \approx 1{,}000$, forget about it. All of the world's computers working full time wouldn't come close to finishing your problem before the end of the universe, at which point it presumably becomes moot.

The quest for an efficient algorithm to solve this so-called *traveling salesman problem* will take us through much of this book. If you need to know how the story ends, check out the catalog entry for the traveling salesman problem in Section 8.5.4.

Hopefully, this example has opened your eyes to some of the subtleties of algorithm correctness. Ensuring the optimal answer on all possible inputs is a difficult but often achievable goal. Seeking counterexamples that break pretender algorithms is a tricky business, but it is an important part of the algorithm design process.

1.1.2 Efficiency

The skilled algorithm designer is an efficiency expert. Through cunning and experience we seek to get jobs done as quickly as possible. The benefits of finding an efficient algorithm for a computationally expensive job can be mind-boggling, as illustrated by several of the war stories scattered through this book. Section 2.7 shows how clever algorithms yielded a program that was 30,000 times faster than our initial attempt!

Upon realizing that a given application runs too slowly, the practical person usually asks the boss to buy them a faster machine. Sometimes this is the cheapest option. However, the potential win from faster hardware is typically limited to a factor of ten or so, due to technical constraints (they don't make faster machines) or economic ones (the boss won't pay for it).

To realize larger performance improvements, we must seek better algorithms. As we will show, a faster algorithm running on a slower computer will *always* win for sufficiently large instances. *Always*. Usually, problems don't have to get very large before the faster algorithm wins.

Be aware that there are situations where finding the most efficient algorithm for a job is a complete waste of programmer effort. In any program, there is usually one bottleneck that takes the majority of computing time. The typical claim is that 90% of the run time of any program is spent in 10% of the code. Optimizing the other 90% of the program will have little impact on the total run time. Further, many programs are written with only one or two special instances in mind. If the program will be run just a few times, or if the job can easily be run overnight, it probably does not pay to make the programmer work harder in order to reduce cycle consumption.

1.2 Expressing Algorithms

Describing algorithms requires a notation for expressing a sequence of steps to be performed. The three most common options are (1) English, (2) pseudocode, or (3) a real programming language. Pseudocode is perhaps the most mysterious of the bunch, but it is best defined as a programming language that never complains about syntax errors. All three methods can be useful in certain circumstances, since there is a natural tradeoff between greater ease of expression and precision. English is the most natural but least precise language, while C and Pascal are precise but difficult to write and understand. Pseudocode is useful because it is a happy medium.

The choice of which notation is best depends upon which of the three methods you are most comfortable with. I prefer to describe the *ideas* of an algorithm in English, moving onto a more formal, programming-language-like pseudocode to clarify sufficiently tricky details of the algorithm. A common mistake among my students is to use pseudocode to take an ill-defined idea and dress it up so that it looks more formal. In the real world, you only fool yourself when you pull this kind of stunt.

The *implementation complexity* of an algorithm is usually why the fastest algorithm known for a problem may not be the most appropriate for a given application. An algorithm's implementation complexity is often a function of how it has been described. Fast algorithms often make use of very complicated data structures, or use other complicated algorithms as subroutines. Turning these algorithms into programs requires building implementations of every substructure. Each catalog entry in Section 8 points out available implementations of algorithms to solve the given problem. Hopefully, these can be used as building blocks to reduce the implementation complexity of algorithms that use them, thus making more complicated algorithms worthy of consideration in practice.

1.3   Keeping Score

Algorithms are the most important, durable, and original part of computer science *because* they can be studied in a language- and machine-independent way. This means that we need techniques that enable us to compare algorithms without implementing them. Our two most important tools are (1) the RAM model of computation and (2) asymptotic analysis of worst-case complexity.

This method of keeping score will be the most mathematically demanding part of this book. However, it is important to understand why we need both of these tools to analyze the performance of algorithms. Once you understand the intuition behind these ideas, the formalism becomes a lot easier to deal with.

1.3.1   The RAM Model of Computation

Machine-independent algorithm design depends upon a hypothetical computer called the *Random Access Machine* or RAM. Under this model of computation, we are confronted with a computer where:

- Each "simple" operation (+, *, −, =, if, call) takes exactly 1 time step.
- Loops and subroutines are *not* considered simple operations. Instead, they are the composition of many single-step operations. It makes no sense for "sort" to be a single-step operation, since sorting 1,000,000 items will take much longer than sorting 10 items. The time it takes to run through a loop or execute a subprogram depends upon the number of loop iterations or the specific nature of the subprogram.

- Each memory access takes exactly one time step, and we have as much memory as we need. The RAM model takes no notice of whether an item is in cache or on the disk, which simplifies the analysis.

Under the RAM model, we measure the run time of an algorithm by counting up the number of steps it takes on a given problem instance. By assuming that our RAM executes a given number of steps per second, the operation count converts easily to the actual run time.

The RAM is a simple model of how computers perform. A common complaint is that it is too simple, that these assumptions make the conclusions and analysis too coarse to believe in practice. For example, multiplying two numbers takes more time than adding two numbers on most processors, which violates the first assumption of the model. Memory access times differ greatly depending on whether data sits in cache or on the disk, thus violating the third assumption. Despite these complaints, the RAM is an excellent model for understanding how an algorithm will perform on a real computer. It strikes a fine balance by capturing the essential behavior of computers while being simple to work with. We use the RAM model because it is useful in practice.

Every model has a size range over which it is useful. Take, for example, the model that the earth is flat. You might argue that this is a bad model, since the earth is not flat. However, when laying the foundation of a house, the flat earth model is sufficiently accurate that it can be reliably used. Further, it is so much easier to manipulate a flat-earth model that it is inconceivable that you would try to think spherically when you don't have to.

The same situation is true with the RAM model of computation. We make an abstraction that in general is very useful. It is quite difficult to design an algorithm such that the RAM model will give you substantially misleading results, by performing either much better or much worse in practice than the model suggests. The robustness of the RAM enables us to analyze algorithms in a machine-independent way.

1.3.2 Best, Worst, and Average-Case Complexity

Using the RAM model of computation, we can count how many steps our algorithm will take on any given input instance by simply executing it on the given input. However, to really understand how good or bad an algorithm is, we must know how it works over *all* instances.

To understand the notions of the best, worst, and average-case complexity, one must think about running an algorithm on all possible instances of data that can be fed to it. For the problem of sorting, the set of possible input instances consists of all the possible arrangements of all the possible numbers of keys. We can represent every input instance as a point on a graph, where the x-axis is the size of the problem (for sorting, the number of items to sort) and the y-axis is the number of steps taken by the algorithm on this instance. Here we assume, quite reasonably, that it doesn't matter what the values of the keys are, just how many of them there are and how they are ordered. It should not take longer to sort 1,000 English names than it does to sort 1,000 French names, for example.

As shown in Figure 1-4, these points naturally align themselves into columns, because only integers represent possible input sizes. After all, it makes no sense to ask how long it takes to sort 10.57 items. Once we have these points, we can define three different functions over them:

- The *worst-case complexity* of the algorithm is the function defined by the maximum number of steps taken on any instance of size n. It represents the curve passing through the highest point of each column.
- The *best-case complexity* of the algorithm is the function defined by the minimum number of steps taken on any instance of size n. It represents the curve passing through the lowest point of each column.
- Finally, the *average-case complexity* of the algorithm is the function defined by the average number of steps taken on any instance of size n.

In practice, the most useful of these three measures proves to be the worst-case complexity, which many people find counterintuitive. To illustrate why worst-case analysis is important, consider trying to project what will happen to you if you bring n dollars to gamble in a casino. The best case, that you walk out owning the place, is possible but so unlikely that you should place no credence in it. The worst case, that you lose all n dollars, is easy to calculate and distressingly likely to happen. The average case, that the typical person loses

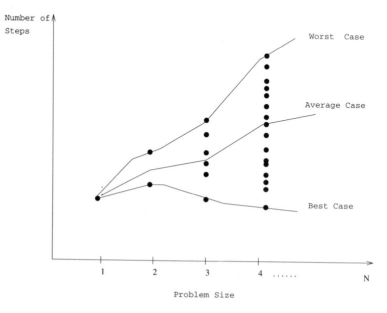

Figure 1–4. Best, worst, and average-case complexity

87.32% of the money that they bring to the casino, is difficult to compute and its meaning is subject to debate. What exactly does *average* mean? Stupid people lose more than smart people, so are you smarter or dumber than the average person, and by how much? People who play craps lose more money than those playing the nickel slots. Card counters at blackjack do better on average than customers who accept three or more free drinks. We avoid all these complexities and obtain a very useful result by just considering the worst case.

The important thing to realize is that each of these time complexities defines a numerical function, representing time versus problem size. These functions are as well-defined as any other numerical function, be it $y = x^2 - 2x + 1$ or the price of General Motors stock as a function of time. Time complexities are complicated functions, however. In order to simplify our work with such messy functions, we will need the big Oh notation.

1.4 The Big Oh Notation

We have agreed that the best, worst, and average-case complexities of a given algorithm are numerical functions of the size of the instances. However, it is difficult to work with these functions exactly because they are often very complicated, with many little up and down bumps, as shown in Figure 1-5. Thus it is usually cleaner and easier to talk about upper and lower bounds of such functions. This is where the big Oh notation comes into the picture.

We seek this smoothing in order to ignore levels of detail that do not impact our comparison of algorithms. Since running our algorithm on a machine that is twice as fast will cut the running times of all algorithms by a multiplicative constant of two, such constant factors would be washed away in upgrading machines. On the RAM we ignore such constant factors anyway and there-

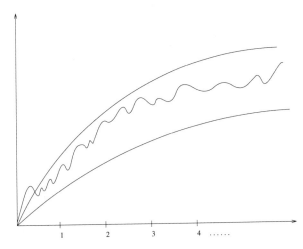

Figure 1-5. Upper and lower bounds smooth out the behavior of complex functions

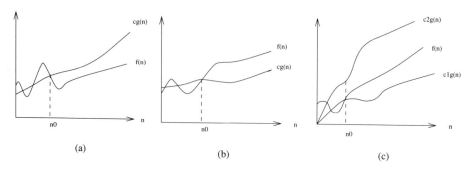

Figure 1-6. Illustrating the (a) O, (b) Ω, and (c) Θ notations

fore might as well ignore them when comparing two algorithms. We use the following notation:

- $f(n) = O(g(n))$ means $c \cdot g(n)$ is an *upper bound* on $f(n)$. Thus there exists some constant c such that $f(n)$ is always $\leq c \cdot g(n)$, for large enough n.
- $f(n) = \Omega(g(n))$ means $c \cdot g(n)$ is a *lower bound* on $f(n)$. Thus there exists some constant c such that $f(n)$ is always $\geq c \cdot g(n)$, for large enough n.
- $f(n) = \Theta(g(n))$ means $c_1 \cdot g(n)$ is an upper bound on $f(n)$ and $c_2 \cdot g(n)$ is a lower bound on $f(n)$, for large enough n. Thus there exists constants c_1 and c_2 such that $f(n) \leq c_1 \cdot g(n)$ and $f(n) \geq c_2 \cdot g(n)$. This means that $g(n)$ is a nice, tight bound on $f(n)$.

Got it? These definitions are illustrated in Figure 1-6. All of these definitions imply a constant n_0 beyond which they are always satisfied. We are not concerned about small values of n, i.e. anything to the left of n_0. After all, do you really care whether one sorting algorithm sorts six items faster than another algorithm, or which one is faster when sorting 1,000 or 1,000,000 items? The big Oh notation enables us to ignore details and focus on the big picture.

Make sure you understand this notation by working through the following examples. We choose certain constants in the explanations because they work and make a point, but you are free to choose any constant that maintains the same inequality:

$$3n^2 - 100n + 6 = O(n^2) \text{ because } 3n^2 > 3n^2 - 100n + 6;$$

$$3n^2 - 100n + 6 = O(n^3) \text{ because } .00001n^3 > 3n^2 - 100n + 6;$$

$$3n^2 - 100n + 6 \neq O(n) \text{ because } c \times n < 3n^2 \text{ when } n > c;$$

$$3n^2 - 100n + 6 = \Omega(n^2) \text{ because } 2.99n^2 < 3n^2 - 100n + 6;$$

$$3n^2 - 100n + 6 \neq \Omega(n^3) \text{ because } 3n^2 - 100n + 6 < n^3;$$

$$3n^2 - 100n + 6 = \Omega(n) \text{ because } 10^{10^{10}} n < 3n^2 - 100 + 6;$$

$$3n^2 - 100n + 6 = \Theta(n^2) \text{ because both } O \text{ and } \Omega;$$

$$3n^2 - 100n + 6 \neq \Theta(n^3) \text{ because } O \text{ only};$$

$$3n^2 - 100n + 6 \neq \Theta(n) \text{ because } \Omega \text{ only}.$$

The big Oh notation provides for a rough notion of equality when comparing functions. It is somewhat jarring to see an expression like $n^2 = O(n^3)$, but its meaning can always be resolved by going back to the definitions in terms of upper and lower bounds. It is perhaps most instructive to read '=' as meaning *one of the functions that are*. Clearly, n^2 is one of functions that are $O(n^3)$.

1.5 Growth Rates

In working with the big Oh notation, we cavalierly discard the multiplicative constants. The functions $f(n) = 0.001n^2$ and $g(n) = 1000n^2$ are treated identically, even though $g(n)$ is a million times larger than $f(n)$ for all values of n. The method behind this madness is illustrated by Figure 1-7, which tabulates the growth rate of several functions arising in algorithm analysis, on problem instances of reasonable size. Specifically, Figure 1-7 shows how long it takes for algorithms that use $f(n)$ operations to complete on a fast computer where each operation takes one nanosecond (10^{-9} seconds). Study the table for a few minutes and the following conclusions become apparent:

- All of these algorithms take about the same amount of time for $n = 10$.
- The algorithm whose running time is $n!$ becomes useless well before $n = 20$.
- The algorithm whose running time is 2^n has a greater operating range, but it becomes impractical for $n > 40$.

n	$f(n) = \lg n$	$f(n) = n$	$f(n) = n \lg n$	$f(n) = n^2$	$f(n) = 2^n$	$f(n) = n!$
10	0.003 μs	0.01 μs	0.033 μs	0.1 μs	1 μs	3.63 ms
20	0.004 μs	0.02 μs	0.086 μs	0.4 μs	1 ms	77.1 years
30	0.005 μs	0.03 μs	0.147 μs	0.9 μs	1 sec	8.4×10^{15} years
40	0.005 μs	0.04 μs	0.213 μs	1.6 μs	18.3 min	
50	0.006 μs	0.05 μs	0.282 μs	2.5 μs	13 days	
100	0.007 μs	0.1 μs	0.644 μs	10 μs	4×10^{13} years	
1,000	0.010 μs	1.00 μs	9.966 μs	1 ms		
10,000	0.013 μs	10 μs	130 μs	100 ms		
100,000	0.017 μs	0.10 ms	1.67 ms	10 sec		
1,000,000	0.020 μs	1 ms	19.93 ms	16.7 min		
10,000,000	0.023 μs	0.01 sec	0.23 sec	1.16 days		
100,000,000	0.027 μs	0.10 sec	2.66 sec	115.7 days		
1,000,000,000	0.030 μs	1 sec	29.90 sec	31.7 years		

Figure 1-7. Growth rates of common functions measured in nanoseconds

- The algorithm whose running time is n^2 is perfectly reasonable up to about $n = 100$, but it quickly deteriorates with larger inputs. For $n > 1,000,000$ it likely to be hopeless.

- Both the n and $n \lg n$ algorithms remain practical on inputs of up to one billion items.

- You can't hope to find a real problem where an $O(\lg n)$ algorithm is going to be too slow in practice.

The bottom line is that even by ignoring constant factors, we can get an excellent idea of whether a given algorithm will be able to run in a reasonable amount of time on a problem of a given size. An algorithm whose running time is $f(n) = n^3$ seconds will beat one whose running time is $g(n) = 1,000,000 \cdot n^2$ seconds only so long as $n < 1,000,000$. Such enormous constant factor differences between algorithms occur in practice far less frequently than such large problems do.

1.6 Logarithms

Logarithm is an anagram of algorithm, but that's not why the wily algorist needs to know what logarithms are and where they come from. You've seen the button on your calculator but have likely forgotten why it is there. A *logarithm* is simply an inverse exponential function. Saying $b^x = y$ is equivalent to saying that $x = \log_b y$. Exponential functions are functions that grow at a distressingly fast rate, as anyone who has ever tried to pay off a mortgage or bank loan understands. Thus inverse exponential functions, i.e. logarithms, grow refreshingly slowly. If you have an algorithm that runs in $O(\lg n)$ time, take it and run. As shown by Figure 1-7, this will be blindingly fast even on very large problem instances.

Binary search is an example of an algorithm that takes $O(\lg n)$ time. In a telephone book with n names, you start by comparing the name that you are looking for with the middle, or $(n/2)$nd name, say *Monroe, Marilyn*. Regardless of whether you are looking someone before this middle name (*Dean, James*) or after it (*Presley, Elvis*), after only one such comparison you can forever disregard one half of all the names in the book. The number of steps the algorithm takes equals the number of times we can halve n before only one name is left. By definition, this is exactly $\log_2 n$. Thus twenty comparisons suffice to find any name in the million-name Manhattan phone book! The power of binary search and logarithms is one of the most fundamental ideas in the analysis of algorithms. This power becomes apparent if we could imagine living in a world with only unsorted telephone books.

Figure 1-8 is another example of logarithms in action. This table appears in the Federal Sentencing Guidelines, used by courts throughout the United States. These guidelines are an attempt to standardize criminal sentences, so that a felon convicted of a crime before one judge receives the same sentence as they would

Loss (apply the greatest)	Increase in level
(A) $2,000 or less	no increase
(B) More than $2,000	add 1
(C) More than $5,000	add 2
(D) More than $10,000	add 3
(E) More than $20,000	add 4
(F) More than $40,000	add 5
(G) More than $70,000	add 6
(H) More than $120,000	add 7
(I) More than $200,000	add 8
(J) More than $350,000	add 9
(K) More than $500,000	add 10
(L) More than $800,000	add 11
(M) More than $1,500,000	add 12
(N) More than $2,500,000	add 13
(O) More than $5,000,000	add 14
(P) More than $10,000,000	add 15
(Q) More than $20,000,000	add 16
(R) More than $40,000,000	add 17
(Q) More than $80,000,000	add 18

Figure 1-8. The Federal Sentencing Guidelines for Fraud

before a different judge. To accomplish this, the judges have prepared an intricate point function to score the depravity of each crime and map it to time-to-serve.

Figure 1-8 gives the actual point function for fraud, a table mapping dollars stolen to points. Notice that the punishment increases by one level each time the amount of money stolen roughly doubles. That means that the level of punishment (which maps roughly linearly to the amount of time served) grows logarithmically with the amount of money stolen. Think for a moment about the consequences of this. Michael Milken sure did. It means that the total sentence grows *extremely* slowly with the amount of money you steal. Knocking off five liquor stores for $10,000 each will get you far more time than embezzling $100,000 once. The corresponding benefit of stealing really large amounts of money is even greater. The moral of logarithmic growth is clear: *"If you are gonna do the crime, make it worth the time!"*

Two mathematical properties of logarithms are important to understand:

- *The base of the logarithm has relatively little impact on the growth rate* – Compare the following three values: $\log_2(1,000,000) = 19.9316$, $\log_3(1,000,000) = 12.5754$, and $\log_{100}(1,000,000) = 3$. A big change in the base of the logarithm produces relatively little difference in the value of the log. This is a consequence of the formula for changing the base of the logarithm:

$$\log_a b = \frac{\log_c b}{\log_c a}$$

Changing the base of the log from a to c involves multiplying or dividing by $\log_c a$. This will be lost to the big Oh notation whenever a and c are constants, as is typical. Thus we are usually justified in ignoring the base of the logarithm when analyzing algorithms.

When performing binary search in a telephone book, how important is it that each query split the book exactly in half? Not much. Suppose we did such a sloppy job of picking our queries such that each time we split the book 1/3 to 2/3 instead of 1/2 to 1/2. For the Manhattan telephone book, we now use $\log_{3/2}(1,000,000) \approx 35$ queries in the worst case, not a significant change from $\log_2(1,000,000) \approx 20$. The power of binary search comes from its logarithmic complexity, not the base of the log.

- *Logarithms cut any function down to size* – The growth rate of the logarithm of any polynomial function is $O(\lg n)$. This follows because

$$\log_a n^b = b \cdot \log_a n$$

The power of binary search on a wide range of problems is a consequence of this observation. For example, note that doing a binary search on a sorted array of n^2 things requires only twice as many comparisons as a binary search on n things. Logarithms efficiently cut any function down to size.

1.7 Modeling the Problem

Modeling is the art of formulating your application in terms of precisely described, well-understood problems. Proper modeling is the key to applying algorithmic design techniques to any real-world problem. Indeed, proper modeling can eliminate the need to design or even implement algorithms by relating your application to what has been done before. Proper modeling is the key to effectively using the problem catalog in Part II of this book.

Real-world applications involve real-world objects. You might be working on a system to route traffic in a network, to find the best way to schedule classrooms in a university, or to search for patterns in a corporate database. Most algorithms, however, are designed to work on rigorously defined abstract structures such as permutations, graphs, and sets. After all, if you can't define what you want to do, you can't hope to compute it. You must first describe your problem abstractly, in terms of fundamental structures and properties.

Whatever your application is, odds are very good that others before you have stumbled upon the same algorithmic problem, perhaps in substantially different contexts. Therefore, to find out what is known about your particular "widget optimization problem," you can't hope to look in a book under *widget*. You have to formulate widget optimization in terms of computing properties of abstract structures such as:

- *Permutations*, which are arrangements, or orderings, of items. For example, $\{1, 4, 3, 2\}$ and $\{4, 3, 2, 1\}$ are two distinct permutations of the same set of four integers. Permutations are likely the object in question whenever your problem seeks an "arrangement," "tour," "ordering,", or "sequence."

- *Subsets*, which represent selections from a set of items. For example, $\{1, 3, 4\}$ and $\{2\}$ are two distinct subsets of the first four integers. Order does not matter in subsets the way it does with permutations, so the subsets $\{1, 3, 4\}$ and $\{4, 3, 1\}$ would be considered identical. Subsets are likely the object in question whenever your problem seeks a "cluster," "collection," "committee," "group," "packaging," or "selection."

- *Trees*, which represent hierarchical relationships between items. Figure 1-9(a) illustrates a portion of the family tree of the Skiena clan. Trees are likely the object in question whenever your problem seeks a "hierarchy," "dominance relationship," "ancestor/decendant relationship," or "taxonomy."

- *Graphs*, which represent relationships between arbitrary pairs of objects. Figure 1-9(b) models a network of roads as a graph, where the vertices are cities and the edges are roads connecting pairs of cities. Graphs are likely the object in question whenever you seek a "network," "circuit," "web," or "relationship."

- *Points*, which represent locations in some geometric space. For example, the locations of McDonald's restaurants can be described by points on a map/plane. Points are likely the object in question whenever your problems work on "sites," "positions," "data records," or "locations."

- *Polygons*, which represent regions in some geometric space. For example, the borders of a country can be described by a polygon on a map/plane. Polygons and polyhedra are likely the object in question whenever you are working on "shapes," "regions," "configurations," or "boundaries."

- *Strings*, which represent sequences of characters or patterns. For example, the names of students in a class can be represented by strings. Strings are likely

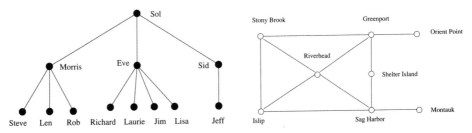

Figure 1-9. Modeling real-world structures with trees and graphs

the object in question whenever you are dealing with "text," "characters," "patterns," or "labels."

These fundamental structures all have associated problems and properties, which are presented in the catalog of Part II. Familiarity with all of these problems is important, because they provide the language we use to model applications. To become fluent in this vocabulary, browse through the catalog and study the *input* and *output* pictures for each problem. Understanding all or most of these problems, even at a cartoon/definition level, will enable you to know where to look later when the problem arises in your application.

Examples of successful application modeling will be presented in the war stories spaced throughout this book. However, some words of caution are in order. The act of modeling reduces your application to one of a small number of existing problems and structures. Such a process is inherently constraining, and certain details might not fit easily into the given model. Also, certain problems can be modeled in several different ways, some much better than others.

Modeling is only the first step in designing an algorithm for a problem. Be alert for how the details of your applications differ from a candidate model. But don't be too quick to say that your problem is unique and special. Temporarily ignoring details that don't fit can free the mind to ask whether they were fundamental in the first place.

1.8 About the War Stories

The best way to become convinced that careful algorithm design can have a significant impact on performance is to look at case studies. By carefully studying other people's experiences, we learn how they might apply to our work.

Jon Bentley's *Programming Pearls* columns are probably the most interesting collection of algorithmic "war stories" currently available. Originally published in the *Communications of the ACM*, they have been collected in two books [Ben86, Ben90]. Fredrick Brooks's *The Mythical Man Month* [Bro74] is another wonderful collection of war stories, focused more on software engineering than algorithm design, but they remain a source of considerable wisdom. Every programmer should read all three of these books, for pleasure as well as insight.

Scattered throughout this text are several of our own war stories, presenting our successful (and occasionally unsuccessful) algorithm design efforts on real applications. We hope that the reader will be able to internalize these experiences so that they will serve as models for their own attacks on problems.

Every one of the war stories is true. Of course, the stories improve somewhat in the retelling, and the dialogue has been punched up to make it more interesting to read. However, I have tried to honestly trace the process of going from a raw problem to a solution, so you can watch how the process unfolded.

The *Oxford English Dictionary* defines an *algorist* as "one skillful in reckonings or figuring." In these stories, I have tried to capture some of the attitude, or mindset, of the algorist in action as they attack a problem.

The various war stories usually involve at least one and often several problems from the problem catalog in Part II. The appropriate section of the catalog is referenced wherever it occurs. This emphasizes the benefits of modeling your application in terms of standard algorithm problems. By using the catalog, you will be able to pull out what is known about any problem whenever it is needed.

1.9 War Story: Psychic Modeling

The call came for me out of the blue as I sat in my office.

"Professor Skiena, I hope you can help me. I'm the President of Lotto Systems Group Inc., and we need an algorithm for a problem arising in our latest product."

"Sure," I replied. After all, the dean of my engineering school is always eager for our faculty to interact with industry.

"At Lotto Systems Group, we market a program designed to improve our customers' psychic ability to predict winning lottery numbers.[1] In a standard lottery, each ticket consists of 6 numbers selected from, say, 1 to 44. Thus any given ticket has only a very small chance of winning. However, after proper training, our clients can visualize a set of, say, 15 numbers out of the 44 and be certain that at least 4 of them will be on the winning ticket. Are you with me so far?"

"Probably not," I replied. But then I recalled how my dean wants us to interact with industry.

"Our problem is this. After the psychic has narrowed the choices down to 15 numbers and is certain that at least 4 of them will be on the winning ticket, we must find the most efficient way to exploit this information. Suppose that a cash prize is awarded whenever you pick at least three of the correct numbers on your ticket. We need an algorithm to construct the smallest set of tickets that we must buy in order to guarantee that we win at least one prize."

"Assuming the psychic is correct?"

"Yes, assuming the psychic is correct. We need a program that prints out a list of all the tickets that the psychic should buy in order to minimize their investment. Can you help us?"

I thought about the problem for a minute. Maybe they did have psychic ability, for they had clearly come to the right place. Identifying the best subset of tickets to buy was very much a combinatorial algorithm problem. It was going to be some type of covering problem, where each ticket we buy was going to "cover" some of the possible 4-element subsets of the psychic's set. Finding the absolute smallest set of tickets to cover everything was a special instance of

---

[1]Yes, this is a true story.

the NP-complete problem *set cover* (discussed in Section 8.7.1) and presumably computationally intractable.

It was indeed a special instance of set cover, completely specified by only four numbers: the size n of the candidate set S (typically $n \approx 15$), the number of k slots for numbers on each ticket (typically $k \approx 6$), the number of psychically promised correct numbers j from S (say $j = 4$), and finally, the number of numbers l matching the winning ticket necessary to win a prize (say $l = 3$). Figure 1-10 illustrates a covering of a smaller instance, where $n = 5$, $j = k = 3$, and $l = 2$.

"Although it will be hard to find the *exact* minimum set of tickets to buy, with heuristics I should be able to get you pretty close to the cheapest covering ticket set," I told him. "Will that be good enough?"

"So long as it generates better ticket sets than my competitor's program, that will be fine. His system doesn't always provide enough coverage to guarantee a win. I really appreciate your help on this, Professor Skiena."

"One last thing. If your program can train people to pick lottery winners, why don't you use it to win the lottery yourself?"

"I look forward to talking to you again real soon, Professor Skiena. Thanks for the help."

I hung up the phone and got back to thinking. It seemed like a fun problem, and the perfect project to give to a bright undergraduate. After it was modeled in terms of sets and subsets, the basic components of a solution seemed fairly straightforward:

- We needed the ability to generate all subsets of k numbers from the candidate set S. Algorithms for generating and ranking/unranking subsets of sets are presented in Section 8.3.5.

- We needed the right formulation of what it meant to have a covering set of purchased tickets. The obvious criteria would be to pick a small set of tickets

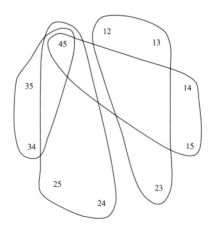

Figure 1–10. Covering all pairs of $\{1, 2, 3, 4, 5\}$ with tickets $\{1, 2, 3\}$, $\{1, 4, 5\}$, $\{2, 4, 5\}$, $\{3, 4, 5\}$

such that for each of the $\binom{n}{l}$ l-subsets of S that might pay off with the prize, we have purchased at least one ticket that contains it.

- For each new ticket we decided to buy, we would need to keep track of which prize combinations we have thus far covered. Presumably, we would want to buy tickets to cover as many thus-far-uncovered prize combinations as possible. The currently covered combinations are a subset of all possible combinations, so we would need a data structure to represent the subset. Such data structures are discussed in Section 8.1.5. The best candidate seemed to be a bit vector, which would permit a constant time query of "is this combination already covered?" in concert with a way to *rank* the subset, i.e. hash it to a unique integer.

- We would need a search mechanism to decide which ticket to buy next. For small enough set sizes, we could do an exhaustive search over all possible subsets of tickets and pick the smallest one. For larger sizes, a randomized search process like simulated annealing (see Section 5.5.1) would select tickets-to-buy to cover as many uncovered combinations as possible. By repeating this randomized procedure several times and picking the best solution, we would be likely to come up with a very good set of tickets.

Excluding the search mechanism, the pseudocode for the bookkeeping looked something like this:

LottoTicketSet(n, k, l)

    Initialize the $\binom{n}{l}$-element bit-vector V to all false

    While there exists a false entry of V

        Select a k-subset T of $\{1, \ldots, n\}$ as the next ticket to buy

        For each of the l-subsets T_i of T, $V[rank(T_i)] =$ true

    Report the set of tickets bought

The bright undergraduate, Fayyaz Younas, rose to the challenge. Based on this framework, he implemented a brute-force search algorithm and found optimal solutions for problems with $n \leq 5$ in a reasonable time. To solve larger problems, he implemented a random search procedure, tweaking it for a while before settling on the best variant. Finally, the day arrived when we could call Lotto Systems Group and announce that we had solved the problem.

"See, our program found that optimal solution for $n = 15, k = 6, j = 6, l = 3$ meant buying 28 tickets."

"Twenty-eight tickets!" complained the president. "You must have a bug. Look, these five tickets will suffice to cover everything *twice* over: $\{2, 4, 8, 10, 13, 14\}$, $\{4, 5, 7, 8, 12, 15\}$, $\{1, 2, 3, 6, 11, 13\}$, $\{3, 5, 6, 9, 10, 15\}$, $\{1, 7, 9, 11, 12, 14\}$."

We fiddled with this example for a while before admitting that he was right. *We hadn't modeled the problem correctly!* In fact, we didn't need to explicitly cover all possible winning combinations. Figure 1-11 illustrates the principle by giv-

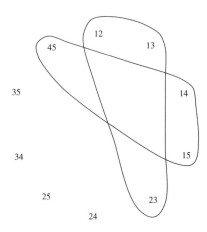

Figure 1–11. Guaranteeing a winning ticket from $\{1,2,3,4,5\}$ using only tickets $\{1,2,3\}$ and $\{1,4,5\}$

ing a better solution to the previous small example. As a different example, suppose we needed three numbers on a five element ticket to win, and the psychic had promised us that four numbers would be from $\{1,2,3,4,5,6\}$. Suppose that we had bought tickets that covered the subsets $\{1,2,3\}$, $\{1,2,4\}$, and $\{1,2,5\}$. There would be no need to buy a ticket that explicitly covered $\{1,2,6\}$, because *if* 1 and 2 and two other psychic numbers were on a winning ticket, then at least one of the other two psychic numbers would have to be either 3, 4, or 5. If 1 and 2 were not on the winning ticket, a ticket covering $\{1,2,6\}$ would do us absolutely no good. Thus we were trying to cover too many combinations, and the penny-pinching psychics were unwilling to pay for such extravagance.

Fortunately, this story has a happy ending, as reported in [YS96]. The general outline of our search-based solution described above still holds for the real problem. All that is needed to fix it is to change how many uncovered subsets we credit for covering with a given set of tickets. After this modification, we obtained the kind of results they were hoping for. Lotto Systems Group gratefully accepted our program to incorporate into their products, and we hope they hit the jackpot with it.

The moral of this story is to make sure that you correctly model the problem before trying to solve it. In our case, we came up with a reasonable model for the problem but didn't work hard enough to validate it before we started to program. Our misinterpretation would have become obvious had we worked out a small but non-trivial example by hand and bounced it off our sponsor before beginning work. Our success in recovering from the error is a tribute to the basic correctness of our initial formulation and our use of well-defined routines and abstractions for such tasks as (1) ranking/unranking k-subsets, (2) the set data structure, and (3) combinatorial search.

1.10 Exercises

1-1. Let P be a problem. The worst-case time complexity of P is $O(n^2)$. The worst-case time complexity of P is also $\Omega(n \lg n)$. Let A be an algorithm that solves P. Which subset of the following statements are consistent with this information about the complexity of P?

- A has worst-case time complexity $O(n^2)$.
- A has worst-case time complexity $O(n^{3/2})$.
- A has worst-case time complexity $O(n)$.
- A has worst-case time complexity $\Theta(n^2)$.
- A has worst-case time complexity $\Theta(n^3)$.

1-2. Suppose that an algorithm A runs in worst-case time $f(n)$ and that algorithm B runs in worst-case time $g(n)$. For each of the following questions, answer either *yes*, *no*, or *can't tell* and explain why.
(a) Is A faster than B for all n greater than some n_0 if $g(n) = \Omega(f(n) \log n)$?
(b) Is A faster than B for all n greater than some n_0 if $g(n) = O(f(n) \log n)$?
(c) Is A faster than B for all n greater than some n_0 if $g(n) = \Theta(f(n) \log n)$?
(d) Is B faster than A for all n greater than some n_0 if $g(n) = \Omega(f(n) \log n)$?
(e) Is B faster than A for all n greater than some n_0 if $g(n) = O(f(n) \log n)$?
(f) Is B faster than A for all n greater than some n_0 if $g(n) = \Theta(f(n) \log n)$?

1-3. For each of these questions, briefly explain your answer.
(a) If I prove that an algorithm takes $O(n^2)$ worst-case time, is it possible that it takes $O(n)$ on some inputs?
(b) If I prove that an algorithm takes $O(n^2)$ worst-case time, is it possible that it takes $O(n)$ on all inputs?
(c) If I prove that an algorithm takes $\Theta(n^2)$ worst-case time, is it possible that it takes $O(n)$ on some inputs?
(d) If I prove that an algorithm takes $\Theta(n^2)$ worst-case time, is it possible that it takes $O(n)$ on all inputs?
(e) Is the function $f(n) = \Theta(n^2)$, where $f(n) = 100n^2$ for even n and $f(n) = 20n^2 - n \log_2 n$ for odd n?

1-4. For each of the following, answer *yes*, *no*, or *can't tell*. Explain your reasoning!
(a) Is $3^n = O(2^n)$?
(b) Is $\log 3^n = O(\log 2^n)$?
(c) Is $3^n = \Omega(2^n)$?
(d) Is $\log 3^n = \Omega(\log 2^n)$?

1-5. (*) Give a proof or counterexample to the following claim: if $f(n) = O(F(n))$ and $g(n) = O(G(n))$, then $f(n)/g(n) = O(F(n)/G(n))$.

1-6. (*) Does $f(n) = O(g(n))$ imply that $2^{f(n)} = O(2^{g(n)})$? Explain your reasoning!

1-7. (*) Give a proof or counterexample to the following claim: for all functions $f(n)$ and $g(n)$, either $f(n) = O(g(n))$ or $g(n) = O(f(n))$.

1-8. (*) When you first learned to multiply numbers, you were told that $x \times y$ means add x a total of y times, so $5 \times 4 = 5 + 5 + 5 + 5 = 20$. What is the time complexity of multiplying two n-digit numbers in base b (people work in base 10, of course, while computers work in base 2) using the repeated addition method, as a function of n and b. Assume that single-digit by single-digit addition or multiplication takes $O(1)$ time. (hint: how big can y be as a function of n and b?)

1-9. (*) In grade school, you learned to multiply long numbers on a digit by digit basis, so that $127 \times 211 = 127 \times 1 + 127 \times 10 + 127 \times 200 = 26,397$. Analyze the time complexity of multiplying two n-digit numbers with this method as a function of n (assume constant base size). Assume that single-digit by single-digit addition or multiplication takes $O(1)$ time.

Implementation Challenges

1-1. (*) Implement the two TSP heuristics of Section 1.1.1. Which of them gives better-quality solutions in practice? Can you devise a heuristic that works better than both of them?

1-2. (*) Describe exactly how to test whether a given set of tickets proves minimum coverage in the Lotto problem of Section 1.9. Write a program to find good ticket sets.

2

Data Structures and Sorting

When things go right, changing a data structure in a slow program works the same way an organ transplant does in a sick patient. For several classes of *abstract data types*, such as containers, dictionaries, and priority queues, there exist many different but functionally equivalent *data structures* that implement the given data type. Changing the data structure does not change the correctness of the program, since we presumably replace a correct implementation with a different correct implementation. However, because the new implementation of the data type realizes different tradeoffs in the time to execute various operations, the total performance of an application can improve dramatically. Like a patient in need of a transplant, only one part might need to be replaced in order to fix the problem.

It is obviously better to be born with a good heart than have to wait for a replacement. Similarly, the maximum benefit from good data structures results from designing your program around them in the first place. Still, it is important to build your programs so that alternative implementations can be tried. This involves separating the internals of the data structure (be it a tree, a hash table, or a sorted array) from its interface (operations like search, insert, delete). Such *data abstraction* is an important part of producing clean, readable, and modifiable programs. We will not dwell on such software engineering issues here, but such a design is critical if you are to experiment with the impact of different implementations on performance.

In this chapter we will also discuss sorting, stressing how sorting can be applied to solve other problems more than the details of specific sorting algorithms. In this sense, sorting behaves more like a data structure than a problem in its own right. Sorting is also represented by a significant entry in the problem catalog; namely Section 8.3.1.

The key take-home lessons of this chapter are:

- Building algorithms around data structures such as dictionaries and priority queues leads to both clean structure and good performance.

- Picking the wrong data structure for the job can be disastrous in terms of performance. Picking the very best data structure is often not as critical, for there are typically several choices that perform similarly.
- Sorting lies at the heart of many different algorithms. Sorting the data is one of the first things any algorithm designer should try in the quest for efficiency.
- Sorting can be used to illustrate most algorithm design paradigms. Data structure techniques, divide-and-conquer, randomization, and incremental construction all lead to popular sorting algorithms.

2.1 Fundamental Data Types

An abstract data type is a collection of well-defined operations that can be performed on a particular structure. The operations define what the data type does, but not how it works. Abstract data types are black boxes that we dare not open when we design algorithms that use them.

For each of the most important abstract data types, several competing implementations, or *data structures*, are available. Often, alternative data structures realize different design tradeoffs that make certain operations (say, insertion) faster at the cost of other operations (say, search). In some applications, certain data structures yield simpler code than other implementations of a given abstract data type, or have some other specialized advantages.

We assume that the reader has had some previous exposure to elementary data structures and some fluency in pointer manipulation. Therefore, we do not discuss these topics here. The reader who wants to review elementary data structures is referred to any of the books in Section 8.1. Instead, we focus on three fundamental abstract data types: containers, dictionaries, and priority queues. Detailed discussion of the tradeoffs between implementations of these data types is deferred to the relevant catalog entry.

2.1.1 Containers

Containers are abstract data types that hold stuff. They don't do much more than hold it so that it can be retrieved later. Still, they are critical to the functioning of society. We will use the term *container* to denote a data structure that permits storage and retrieval of data items *independently of content*. The two fundamental operations of any container are:

- *Put(C,x)* – Insert a new data item x into the container C.
- *Get(C)* – Retrieve the next item from the container C. Different types of containers support different retrieval orders, based on insertion order or position.

Containers are typically most useful when they will contain only a limited number of items and when the retrieval order is predefined or irrelevant. The most popular type of containers are:

- *Stacks* – Supports retrieval in last in, first out order (LIFO). Stacks are simple to implement, and very efficient. Indeed, stacks are probably the right container to use when the retrieval order doesn't matter at all, as when processing batch jobs. The *put* and *get* operations for stacks are usually called *push* and *pop*.

- *Queues* – Supports retrieval in first in, first out order (FIFO). FIFO may seem the fairest way to control waiting times. However, for many applications, data items have infinite patience. Queues are trickier to implement than stacks and are appropriate only for applications (like certain simulations) where the order is important. The *put* and *get* operations for queues are usually called *enqueue* and *dequeue*.

- *Tables* – Supports retrieval by position, so that *put* and *get* each accept an index as an argument. Tables are naturally implemented using arrays.

Each of these containers can be implemented using either arrays or linked lists. With the exception of tables, the choice of lists versus tables probably doesn't matter very much. The key issue is whether an upper bound on the size of the container is known in advance, thus permitting a statically allocated array. Using arrays, *put* and *get* can be implemented in constant time per operation for each of the containers.

2.1.2 Dictionaries

Dictionaries are a form of container that permits access to data items by content. You put a word into a dictionary because you know you can look it up when you need it.

The primary operations dictionaries support are:

- *Search(D,k)* – Given a search key k, return a pointer to the element in dictionary D whose key value is k, if one exists.

- *Insert(D,x)* – Given a data item x, add it to the set of items in the dictionary D.

- *Delete(D,x)* – Given a pointer to a given data item x in the dictionary D, remove it from D.

Perhaps the simplest possible dictionary implementation maintains an unsorted linked list as a data structure. Insertion and deletion are supported in constant time, although a query requires potentially traversing the entire linked list. Basing an implementation on a sorted array speeds up the query operation to $O(\lg n)$ by binary search. Making room for a new item or filling a hole left by a deletion may require moving arbitrarily many items, so insertion and deletion become linear-time operations.

Many other dictionary implementations are available. Binary search trees are discussed in some detail in the next section. Hash tables are another attractive option in practice. A complete discussion of different dictionary data structures is presented catalog Section 8.1.1. We encourage the reader to browse through the data structures section of the catalog in order to learn what your options are.

Certain dictionary data structures also efficiently support the following useful operations:

- *Max(D)* or *Min(D)* – Retrieve the item with the largest (or smallest) key from *D*. This enables the dictionary to serve as a priority, as discussed below.
- *Predecessor(D,k)* or *Successor(D,k)* – Retrieve the item from *D* whose key is immediately before (or after) *k* in sorted order. By starting from the first item *Min(D)* and repeatedly calling *Successor* until we obtain *Max(D)*, we traverse all elements in sorted order.

2.1.3   Binary Search Trees

Fast support of all dictionary operations is realized by binary search trees. A *binary tree* is a rooted tree where each node contains at most two children. Each child can be identified as either a left or right child. As shown in Figure 2-1, a binary tree can be implemented where each node has *left* and *right* pointer fields, an (optional) *parent* pointer, and a data field.

A binary *search* tree labels each node in a binary tree with a single key such that for any node labeled x, all nodes in the left subtree of x have keys $< x$ while all nodes in the right subtree of x have keys $> x$. The search tree labeling enables us to find where any key is. Start at the root. If it does not contain the key we are searching for, proceed either left or right depending upon whether what we want occurs before or after the root key. This algorithm works because both the left and right subtrees of a binary search tree *are* binary search trees; the recursive structure yields a recursive algorithm. Accordingly, the dictionary *Query* operation can be performed in $O(h)$ time, where h is the height of the tree.

BinaryTreeQuery(x, k)
      if ($x = NIL$) *or* ($k = key[x]$)
          then return x
      if ($k < key[x]$)
          then return BinaryTreeQuery($left[x], k$)
          else return BinaryTreeQuery($right[x], k$)

To insert a new item x with key k into a binary search tree T, it is important to place it where it can be later be found. There is only one such location in any binary search tree, namely by replacing the nil pointer found in T after an unsuccessful query for k. Replacing this nil pointer with a pointer to x is a simple, constant-time operation after the search has been performed in $O(h)$ time.

Deletion is somewhat more tricky than insertion, because the node selected to die may not be a leaf. Leaf nodes may be deleted without mercy, by clearing the pointer to the given node. However, internal nodes have children that must remain accessible after the deletion. By restructuring or relabeling the tree, however, the item to delete can always be made into a leaf and then removed. Details appear in any data structures text.

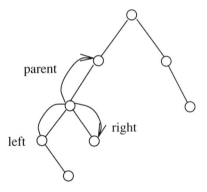

Figure 2–1. Relationships in a binary search tree

When implemented using binary search trees, all three dictionary operations take $O(h)$ time, where h is the height of the tree. The smallest height we could hope for occurs when the tree is perfectly balanced, where $h = \lceil \lg n \rceil$. In fact, if we insert the keys in random order, with high probability the tree will have $O(\lg n)$ height. However, if we get unlucky with our order of insertion or deletion, we can end up with a linear-height tree in the worst case. The worst case is a serious potential problem. Indeed, it occurs whenever the keys are inserted in sorted order.

To avoid such worst-case performance, more sophisticated *balanced* binary search tree data structures have been developed that guarantee the height of the tree always to be $O(\lg n)$. Therefore, all dictionary operations (insert, delete, query) take $O(\lg n)$ time each. Implementations of such balanced tree data structures as red-black trees are discussed in Section 8.1.1.

From an algorithm design viewpoint, it is most important to know that these trees exist and that they can be used as black boxes to provide an efficient dictionary implementation. When figuring the costs of dictionary operations for algorithm analysis, assume the worst-case complexities of balanced binary trees to be a fair measure.

2.1.4 Priority Queues

Many algorithms process items according to a particular order. For example, suppose you have to schedule a list of jobs given the deadline by which each job must be performed or else its importance relative to the other jobs. Scheduling jobs requires sorting them by time or importance, and then performing them in this sorted order.

Priority queues provide extra flexibility over sorting, which is required because jobs often enter the system at arbitrary intervals. It is much more cost-effective to insert a new job into a priority queue than to re-sort everything. Also, the need to perform certain jobs may vanish before they are executed, meaning that they must be removed from the queue.

The basic priority queue supports three primary operations:

- *Insert(Q,x)* – Given an item x with key k, insert it into the priority queue Q.
- *Find-Minimum(Q)* or *Find-Maximum(Q)* – Return a pointer to the item whose key value is smaller (larger) than any other key in the priority queue Q.
- *Delete-Minimum(Q)* or *Delete-Maximum(Q)* – Remove the item from the priority queue Q whose key is minimum (maximum).

All three of these priority queue operations can be implemented in $O(\lg n)$ time by representing the heap with a binary search tree. Implementing the *find-minimum* operation requires knowing where the minimum element in the tree is. By definition, the smallest key must reside in the left subtree of the root, since all keys in the left subtree have values less than that of the root. Therefore, as shown in Figure 2-2, the minimum element must be the leftmost decendent of the root. Similarly, the maximum element must be the rightmost decendent of the root.

Find-Maximum(x)
    while *right*[x] ≠ *NIL*
       do x = *right*[x]
    return x

Find-Minimum(x)
    while *left*[x] ≠ *NIL*
       do x = *left*[x]
    return x

Repeatedly traversing left (or right) pointers until we hit a leaf takes time proportional to the height of the tree, or $O(\lg n)$ if the tree is balanced. The insert operation can be implemented exactly as binary tree insertion. *Delete-Min* can be implemented by finding the minimum element and then using standard binary tree deletion. It follows that each of the operations can be performed in $O(\lg n)$ time.

Priority queues are very useful data structures. Indeed, they are the hero of the war story described in Section 2.6. A complete set of priority queue implementations is presented in Section 8.1.2.

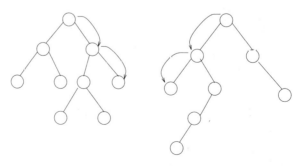

Figure 2-2. The maximum and minimum element in a binary search tree

2.2 Specialized Data Structures

The basic data structures thus far described all represent an unstructured set of items so as to facilitate retrieval operations. These data structures are well known to most programmers. Not as well known are high-powered data structures for representing more structured or specialized kinds of objects, such as points in space, strings, and graphs.

The design principles of these data structures are the same as for basic objects. There exists a set of basic operations we need to perform repeatedly. We seek a data structure that supports these operations very efficiently. These efficient, specialized data structures are as important for efficient graph, string, and geometric algorithms as lists and arrays are for basic algorithms, so one should be aware of their existence. Details appear throughout the catalog.

- *String data structures* – Character strings are typically represented by arrays of characters, with perhaps a special character to mark the end of the string. Suffix trees/arrays are special data structures that preprocess strings to make pattern matching operations faster. See Section 8.1.3 for details.

- *Geometric data structures* – Geometric data typically consists of collections of data points and regions. Regions in the plane are usually described by polygons, where the boundary of the polygon is given by a chain of line segments. Polygons can be represented using an array of points $(v_1, \ldots, v_n, v_1)$, such that (v_i, v_{i+1}) is a segment of the boundary. Spatial data structures such as *kd*-trees organize points and regions by geometric location to support fast search. See Section 8.1.6 for details.

- *Graph data structures* – Graphs are typically represented by either adjacency matrices or adjacency lists. The choice of representation can have a substantial impact on the design of the resulting graph algorithms, as discussed in Chapter 4. Implementation aspects of graph data structures are presented in Section 8.1.4.

- *Set data structures* – Subsets of items are typically represented using a dictionary, to support fast membership queries. A variety of data structures for manipulating sets is presented in Section 8.1.5.

2.3 Sorting

By the time they graduate, computer science students are likely to have studied the basic sorting algorithms in their introductory programming class, then in their data structures class, and finally in their algorithms class. Why is sorting worth so much attention? There are several reasons:

- Sorting is the basic building block around which many other algorithms are built. By understanding sorting, we obtain an amazing amount of power to solve other problems.

- Historically, computers have spent more time sorting than doing anything else. A quarter of all mainframe cycles are spent sorting data [Knu73b]. Although it is unclear whether this remains true on smaller computers, sorting remains the most ubiquitous combinatorial algorithm problem in practice.

- Sorting is the most throughly studied problem in computer science. Literally dozens of different algorithms are known, most of which possess some advantage over all other algorithms in certain situations. To become convinced of this, the reader is encouraged to browse through [Knu73b], with hundreds of pages of interesting sorting algorithms and analysis.

- Most of the interesting ideas used in the design of algorithms appear in the context of sorting, such as divide-and-conquer, data structures, and randomized algorithms.

2.4 Applications of Sorting

An important key to algorithm design is to use sorting as a basic building block, because once a set of items is sorted, many other problems become easy. Consider the following applications:

- *Searching* – Binary search enables you to test whether an item is in a dictionary in $O(\lg n)$ time, once the keys are all sorted. Search preprocessing is perhaps the single most important application of sorting.

- *Closest pair* – Given a set of n numbers, how do you find the pair of numbers that have the smallest difference between them? After the numbers are sorted, the closest pair of numbers will lie next to each other somewhere in sorted order. Thus a linear-time scan through them completes the job, for a total of $O(n \lg n)$ time including the sorting.

- *Element uniqueness* – Are there any duplicates in a given a set of n items? The most efficient algorithm is to sort them and then do a linear scan though them checking all adjacent pairs. This is a special case of the closest-pair problem above, where we ask if there is a pair separated by a gap of zero.

- *Frequency distribution* – Given a set of n items, which element occurs the largest number of times in the set? If the items are sorted, we can sweep from left to right and count them, since all identical items will be lumped together during sorting. To find out how often an arbitrary element k occurs, start by looking up k using binary search in a sorted array of keys. By walking to the left of this point until the element is not k and then walking to the right, we can find this count in $O(\lg n + c)$ time, where c is the number of occurrences of k. The number of instances of k can be found in $O(\lg n)$ time by using binary search to look for the positions of both $k - \epsilon$ and $k + \epsilon$, where ϵ is arbitrarily small, and then taking the difference of these positions.

- *Selection* – What is the kth largest item in the set? If the keys are placed in sorted order in an array, the kth largest can be found in constant time by simply looking at the kth position of the array. In particular, the median element (see Section 8.3.3) appears in the $(n/2)$nd position in sorted order.

- *Convex hulls* – Given n points in two dimensions, what is the polygon of smallest area that contains them all? The convex hull is like a rubber band stretched over the points in the plane and then released. It compresses to just cover the points, as shown in Figure 2-3. The convex hull gives a nice representation of the shape of the points and is the most important building block for more sophisticated geometric algorithms, as discussed in catalog Section 8.6.2.

  But how can we use sorting to construct the convex hull? Once you have the points sorted by x-coordinate, the points can be inserted from left to right into the hull. Since the rightmost point is always on the boundary, we know that it will be inserted into the hull. Adding this new rightmost point might cause others to be deleted, but we can quickly identify these points because they lie inside the polygon formed by adding the new point. These points to delete will be neighbors of the previous point we inserted, so they will be easy to find. The total time is linear after the sorting has been done.

While some of these problems (particularly median and selection) can be solved in linear time using more sophisticated algorithms, sorting provides quick and easy solutions to all of these problems. It is a rare application whose time complexity is such that sorting proves to be the bottleneck, especially a bottleneck that could have otherwise been removed using more clever algorithmics. Don't ever be afraid to spend time sorting whenever you use an efficient sorting routine.

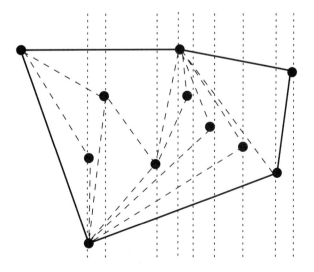

Figure 2-3. Constructing the convex hull of points in the plane

2.5 Approaches to Sorting

Sorting is a natural laboratory for studying basic algorithm design paradigms, since many useful techniques lead to interesting sorting algorithms. Indeed, we introduce several techniques here that will be described further in subsequent chapters. Consider the algorithms below as case studies for each of the relevant techniques.

2.5.1 Data Structures

Perhaps the most dramatic algorithmic improvement made possible by using appropriate data structures occurs in sorting. Selection sort is a simple-to-code algorithm that repeatedly extracts the smallest remaining element from the unsorted part of the set:

> SelectionSort(A)
> > For $i = 1$ to n do
> > > $Sort[i]$ = Find-Minimum from A
> > > Delete-Minimum from A
> > Return($Sort$)

Selection sort is typically implemented by partitioning the input array into sorted and unsorted regions. To find the smallest item, we perform a linear sweep through the unsorted portion of the array. The smallest item is then swapped with the ith item in the array before moving on the next iteration. Selection sort performs n iterations, where the average iteration takes $n/2$ steps, for a total of $O(n^2)$ time.

But what if we improve the data structure? It takes $O(1)$ time to remove a particular item from an unsorted array once it has been located, but $O(n)$ time to find the smallest item. These two are exactly the operations supported by priority queues. So what happens if we replace the data structure with a better priority queue implementation, either a heap or a balanced binary tree. Operations within the loop now take $O(\lg n)$ time each, instead of $O(n)$. By using such a priority queue implementation, selection sort is sped up to $O(n \lg n)$ from $O(n^2)$. The name typically given to this algorithm, *heapsort*, obscures the relationship between them, but heapsort is nothing but an implementation of selection sort with the right data structure.

2.5.2 Incremental Insertion

Now consider a different approach to sorting that grows the sorted set one element at a time. Select an arbitrary element from the unsorted set, and put it in the proper position in the sorted set.

InsertionSort(A)
    $A[0] = -\infty$
    for $i = 1$ to $n - 1$ do
        $j = i$
        while ($A[j] > A[j - 1]$) do swap($A[j], A[j - 1]$)

Although insertion sort takes $O(n^2)$ in the worst case, it will perform considerably better if the data is almost sorted, since few iterations of the inner loop will suffice to sift it into the proper position. Insertion sort is perhaps the simplest example of the *incremental insertion* technique, where we build up a complicated structure on n items by first building it on $n - 1$ items and then making the necessary changes to fix things in adding the last item. Incremental insertion proves a particularly useful technique in geometric algorithms.

2.5.3   Divide and Conquer

Suppose we take the n elements to sort and split them into piles S and T, each with half the elements. After sorting both piles, it is easy to combine the two sorted piles. To merge $A = \{5, 7, 12, 19\}$ and $B = \{4, 6, 13, 15\}$, note that the smallest item must sit at the top of one of the two lists. Once identified, the smallest element can be removed, and the second smallest item will again be atop one of the two lists. Repeating this operation merges the two sorted lists in $O(n)$ time. This provides the basis for a recursive algorithm.

Mergesort($A[1, n]$)
    Merge(MergeSort($A[1, \lfloor n/2 \rfloor]$), MergeSort($A[\lfloor n/2 \rfloor + 1, n]$))

Because the recursion goes $\lg n$ levels deep and a linear amount of work is done per level, Mergesort takes $O(n \lg n)$ time in the worst case.

Mergesort is a classic divide-and-conquer algorithm. Whenever we can break one large problem into two smaller problems, we are ahead of the game because the smaller problems are easier. The trick is taking advantage of the two partial solutions to put together a solution of the full problem. The merge operation takes care of the reassembly in mergesort, but not all problems can be so neatly decomposed.

2.5.4   Randomization

Suppose we select, at random, one of the n items we seek to sort. Call this element p. In *quicksort*, we separate the $n - 1$ other items into two piles: a low pile containing all the elements that will appear before p in sorted order and a high pile containing all the elements that will appear after p in sorted order. After sorting the two piles and arranging the piles correctly, we have succeeded in sorting the n items.

```
Quicksort(A, low, high)
    if (low < high)
            ploc = Partition(A, low, high)
            Quicksort(A, low, ploc − 1)
            Quicksort(A, ploc + 1, high)
Partition(A, low, high)
        swap(A[low], A[random(low . . . high)])
        pivot = A[low]
        leftwall = low
        for i = low+1 to high
                if (A[i] < pivot) then
                        leftwall = leftwall + 1
                        swap(A[i], A[leftwall])
        swap(A[low], A[leftwall])
```

Mergesort ran in $O(n \lg n)$ time because we split the keys into two equal halves, sorted them recursively, and then merged the halves in linear time. Thus whenever our pivot element is near the center of the sorted array (i.e. the pivot is close to the median element), we get a good split and realize the same performance as mergesort. Such good pivot elements will often be the case. After all, half the elements lie closer to the middle than one of the ends. On average, Quicksort runs in $O(n \lg n)$ time. If we are extremely unlucky and our randomly selected elements always are among the largest or smallest element in the array, Quicksort turn into selection sort and runs in $O(n^2)$. However, the odds against this are vanishingly small.

Randomization is a powerful, general tool to improve algorithms with bad worst-case but good average-case complexity. The worst case examples still exist, but they depend only upon how unlucky we are, not on the order that the input data is presented to us. For randomly chosen pivots, we can say that

"Randomized quicksort runs in $\Theta(n \lg n)$ time on any input, with high probability."

If instead, we used some deterministic rule for selecting pivots (like always selecting $A[(low + high)/2]$ as pivot), we can make no claim stronger than

"Quicksort runs in $\Theta(n \lg n)$ time if you give me random input data, with high probability."

Randomization can also be used to drive search techniques such as simulated annealing, which are discussed in Section 5.5.1.

2.5.5   Bucketing Techniques

If we were sorting names for the telephone book, we could start by partitioning the names according to the first letter of the last name. That will create 26 different piles, or buckets, of names. Observe that any name in the *J* pile must

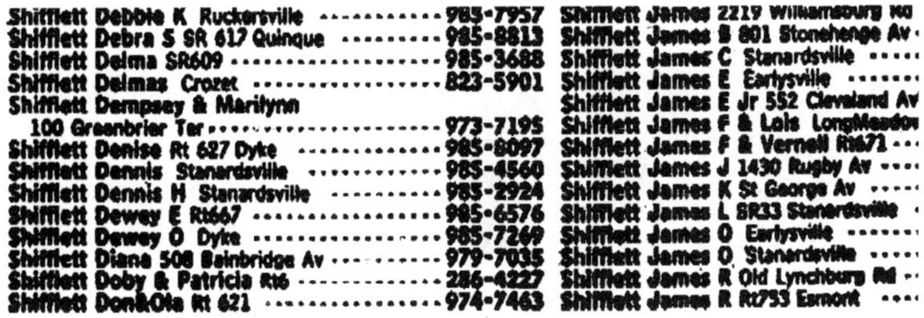

```
Shifflett Debbie K  Ruckersville ···········985-7957   Shifflett James 2219 Williamsburg Rd
Shifflett Debra S SR 617 Quinque ··········985-8813   Shifflett James B 801 Stonehenge Av ·
Shifflett Delma SR609 ·····················985-3688   Shifflett James C Stanardsville ·····
Shifflett Delmas  Crozet ··················823-5901   Shifflett James E Earlysville ······
Shifflett Dempsey & Marilynn                          Shifflett James E Jr 552 Cleveland Av
   100 Greenbrier Ter ····················973-7195   Shifflett James F & Lois Longmeadow
Shifflett Denise Rt 627 Dyke  ·············985-8097   Shifflett James F & Vernell Rt671 ··
Shifflett Dennis  Stanardsville ···········985-4560   Shifflett James J 1430 Rugby Av ····
Shifflett Dennis H Stanardsville ··········985-2924   Shifflett James K St George Av ·····
Shifflett Dewey E Rt667 ···················985-6576   Shifflett James L BR33 Stanardsville ·
Shifflett Dewey O Dyke ····················985-7269   Shifflett James O Earlysville ······
Shifflett Diana 508 Bainbridge Av ········979-7035   Shifflett James O Stanardsville ····
Shifflett Doby & Patricia Rt6 ············286-4227   Shifflett James R Old Lynchburg Rd ·
Shifflett Don&Ola Rt 621 ·················974-7463   Shifflett James R Rt753 Esmont ····
```

Figure 2-4. A small subset of Charlottesville Shiffletts

occur after every name in the *I* pile but before any name in the *K* pile. Therefore, we can proceed to sort each pile individually and just concatenate the bunch of piles together.

If the names are distributed fairly evenly among the buckets, as we might expect, the resulting 26 sorting problems should each be substantially smaller than the original problem. Further, by now partitioning each pile based on the second letter of each name, we generate smaller and smaller piles. The names will be sorted as soon as each bucket contains only a single name. The resulting algorithm is commonly called bucketsort or distribution sort.

Bucketing is a very effective idea whenever we are confident that the distribution of data will be roughly uniform. It is the idea that underlies hash tables, *kd*-trees, and a variety of other practical data structures. The downside of such techniques is that the performance can be terrible whenever the data distribution is not what we expected. Although data structures such as binary trees offer guaranteed worst-case behavior for any input distribution, no such promise exists for heuristic data structures on unexpected input distributions.

To show that nonuniform distributions occur in real life, consider Americans with the uncommon last name of Shifflett. The 1997 Manhattan telephone directory, with over one million names, contains exactly five Shiffletts. So how many Shiffletts should there be in a small city of 50,000 people? Figure 2-4 shows a small portion of the *two and a half pages* of Shiffletts in the Charlottesville, Virginia telephone book. The Shifflett clan is a fixture of the region, but it would play havoc with any distribution sort program, as refining buckets from *S* to *Sh* to *Shi* to *Shif* to ... to *Shifflett* results in no significant partitioning.

2.6 War Story: Stripping Triangulations

The most common type of geometric model used for computer graphics describes the geometry of the object as a triangulated surface, as shown in Figure 2-5a. High-performance rendering engines have special hardware for rendering and shading triangles. This hardware is so fast that the bottleneck of rendering is simply feeding the structure of the triangulation into the hardware engine.

Although each triangle can be described by specifying its three endpoints and any associated shading/normal information, an alternative representation

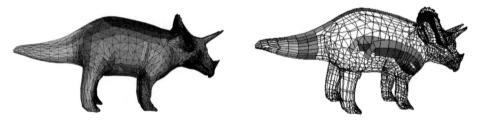

Figure 2–5. (a) A triangulated model of a dinosaur (b) Several triangle strips in the model

is more efficient. Instead of specifying each triangle in isolation, suppose that we partition the triangles into *strips* of adjacent triangles and walk along the strip, as shown in Figure 2-5(b). Since each triangle shares two vertices in common with its neighbors, we save the cost of retransmitting the two extra vertices and any associated normals. To make the description of the triangles unambiguous, the Silicon Graphics triangular-mesh renderer *OpenGL* assumes that all turns alternate from left to right (as shown in Figure 2-6).

The problem of finding a small number of strips that cover each triangle in a mesh can be thought of as a graph problem, where this graph has a vertex for every *triangle* of the mesh, and there is an edge between every pair of vertices representing adjacent triangles. This *dual graph* representation of the planar subdivision representing the triangulation (see Section 8.4.12) captures all the information about the triangulation needed to partition it into triangle strips. Section 4.3 describes our experiences constructing the graph from the triangulation.

Once we had the dual graph available, the project could begin in earnest. We sought to partition the vertices of the dual graph into as few paths or strips as possible. Partitioning it into one path implied that we had discovered a Hamiltonian path, which by definition visits each vertex exactly once. Since finding a Hamiltonian path was NP-complete (see Section 8.5.5), we knew not to look for an optimal algorithm, but to concentrate instead on heuristics.

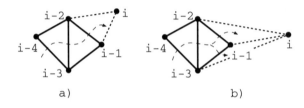

Figure 2–6. Partitioning a triangular mesh into strips: (a) with left-right turns (b) with the flexibility of arbitrary turns

It is always best to start with simple heuristics before trying more complicated ones, because simple might well suffice for the job. The most natural heuristic for strip cover would be to start from an arbitrary triangle and then do a left-right walk from there until the walk ends, either by hitting the boundary of the object or a previously visited triangle. This heuristic had the advantage that it would be fast and simple, although there could be no reason to suspect that it should find the smallest possible set of left-right strips for a given triangulation.

A heuristic more likely to result in a small number of strips would be *greedy*. Greedy heuristics always try to grab the best possible thing first. In the case of the triangulation, the natural greedy heuristic would find the starting triangle that yields the longest left-right strip, and peel that one off first.

Being greedy also does not guarantee you the best possible solution, since the first strip you peel off might break apart a lot of potential strips we would have wanted to use later. Still, being greedy is a good rule of thumb if you want to get rich. Since removing the longest strip would leave the fewest number of triangles for later strips, it seemed reasonable that the greedy heuristic would out-perform the naive heuristic.

But how much time does it take to find the largest strip to peel off next? Let k be the length of the walk possible from an average vertex. Using the simplest possible implementation, we could walk from each of the n vertices per iteration in order to find the largest remaining strip to report in $O(k \cdot n)$ time. With the total number of strips roughly equal to n/k, this yields an $O(n^2)$-time implementation, which would be hopelessly slow on a typical model of 20,000 triangles.

How could we speed this up? It seems wasteful to rewalk from each triangle after deleting a single strip. We could maintain the lengths of all the possible future strips in a data structure. However, whenever we peel off a strip, we would have to update the lengths of all the other strips that will be affected. These strips will be shortened because they walked through a triangle that now no longer exists. There are two aspects of such a data structure:

- *Priority Queue* – Since we were repeatedly interested in identifying the longest possible next strip, we needed a priority queue to store the strips, ordered according to length. The next strip to peel would always be the top of the queue. Our priority queue had to permit reducing the priority of arbitrary elements of the queue whenever we updated the strip lengths to reflect what triangles were peeled away. Because all of the strip lengths were bounded by a fairly small integer (hardware constraints prevent any strip from having more than 256 vertices), we used a bounded height priority queue (shown in Figure 2-7 and described in Section 8.1.2). An ordinary heap would also have worked just fine.

  To update a queue entry associated with a triangle, we needed to be able to quickly find where it was. This meant that we also needed a . . .

- *Dictionary* – For each triangle in the mesh, we needed a way to find where it was in the queue. This meant storing a pointer for each triangle. Since each triangle was defined by three integer vertex numbers, either a hash table or an array of lists of triangles would suffice. By integrating this dictionary

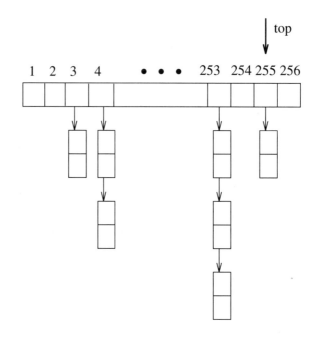

Figure 2–7. A bounded height priority queue for triangle strips

with the priority queue, we built a data structure capable of a wider range of operations.

Although there were various other complications, such as quickly recalculating the length of the strips affected by the peeling, the key idea needed to obtain better performance was to use the priority queue. Run time improved by several orders of magnitude after employing these data structures.

How much better did the greedy heuristic do than the naive heuristic? Consider the table in Figure 2-8. In all cases, the greedy heuristic led to a set of strips that cost less, as measured by the total size of the strips. The savings ranged from about 10% to 50%, quite remarkable, since the greatest possible improvement (going from three vertices per triangle down to one) could yield a savings of only 66.6%.

After implementing the greedy heuristic with our priority queue data structure, our complete algorithm ran in $O(n \cdot k)$ time, where n is the number of triangles and k is the length of the average strip. Thus the torus, which consisted of a small number of very long strips, took longer than the jaw, even though the latter contained over three times as many triangles.

There are several lessons to be gleaned from this story. First, whenever we are working with a large enough data set, only linear or close to linear algorithms (say $O(n \lg n)$) are likely to be fast enough. Second, choosing the right data structure is often the key to getting the time complexity down to this point. Finally,

Model name	Triangle count	Naive cost	Greedy cost	Greedy time
Diver	3,798	8,460	4,650	6.4 sec
Heads	4,157	10,588	4,749	9.9 sec
Framework	5,602	9,274	7,210	9.7 sec
Bart Simpson	9,654	24,934	11,676	20.5 sec
Enterprise	12,710	29,016	13,738	26.2 sec
Torus	20,000	40,000	20,200	272.7 sec
Jaw	75,842	104,203	95,020	136.2 sec

Figure 2–8. A comparison of the naive versus greedy heuristics for several triangular meshes

using a greedy or somewhat smarter heuristic over the naive approach is likely to significantly improve the quality of the results. How much the improvement is likely to be can be determined only by experimentation. Our final, optimized triangle strip program is described in [ESV96].

2.7 War Story: Mystery of the Pyramids

That look in his eyes should have warned me even before he started talking.

"We want to use a parallel supercomputer for a numerical calculation up to 1,000,000,000, but we need a faster algorithm to do it."

I'd seen that distant look before. Eyes dulled from too much exposure to the raw horsepower of supercomputers. Machines so fast that brute force seemed to eliminate the need for clever algorithms. So it always seemed, at least until the problems got hard enough.

"I am working with a Nobel prize winner to use a computer on a famous problem in number theory. Are you familiar with Waring's problem?"

I knew some number theory [NZ80]. "Sure. Waring's problem asks whether every integer can be expressed at least one way as the sum of at most four integer squares. For example, $78 = 8^2 + 3^2 + 2^2 + 1^2 = 7^2 + 5^2 + 2^2$. It's a cute problem. I remember proving that four squares suffice in my undergraduate number theory class. Yes, it's a famous problem, sure, but one that got solved about 200 years ago."

"No, we are interested in a different version of Waring's problem. A *pyramidal number* is a number of the form $(m^3 - m)/6$, for $m \geq 2$. Thus the first several pyramidal numbers are 1, 4, 10, 20, 35, 56, 84, 120, and 165. The conjecture since 1928 is that every integer can be represented by the sum of at most five such pyramidal numbers. We want to use a supercomputer to prove this conjecture on all numbers from 1 to 1,000,000,000."

"Doing a billion of anything will take a serious amount of time," I warned. "The time you spend to compute the minimum representation of each number will be critical, because you are going to do it one billion times. Have you thought about what kind of an algorithm you are going to use?"

"We have already written our program and run it on a parallel supercomputer. It works very fast on smaller numbers. Still, it take much too much time as soon as we get to 100,000 or so."

"Terrific," I thought. Our supercomputer junkie had discovered asymptotic growth. No doubt his algorithm ran in something like quadratic time, and he got burned as soon as n got large.

"We need a faster program in order to get to one billion. Can you help us? Of course, we can run it on our parallel supercomputer when you are ready."

I'll confess that I am a sucker for this kind of challenge, finding better algorithms to speed up programs. I agreed to think about it and got down to work.

I started by looking at the program that the other guy had written. He had built an array of all the $\Theta(n^{1/3})$ pyramidal numbers from 1 to n. To test each number k in this range, he did a brute force test to establish whether it was the sum of two pyramidal numbers. If not, the program tested whether it was the sum of three of them, then four, and finally five of them, until it first got an answer. About 45% of the integers are expressible as the sum of three pyramidal numbers, while most of the remaining 55% require the sum of four; usually each can be represented many different ways. Only 241 integers are known to require the sum of five pyramidal numbers, the largest one being 343,867. For about half of the n numbers, this algorithm presumably went through all of the three-tests and at least some of the four-tests before terminating. Thus the total time for this algorithm would be at least $O(n \times (n^{1/3})^3) = O(n^2)$ time, where $n = 1,000,000,000$. No wonder his program cried uncle.

Anything that was going to do significantly better on a problem this large had to avoid explicitly testing all triples. For each value of k, we were seeking the smallest number of pyramidal numbers that sum exactly to k. This problem has a name, the *knapsack problem*, and it is discussed in Section 8.2.10. In our case of interest, the weights are the set of pyramidal numbers no greater than n, with an additional constraint that the knapsack holds exactly k items.

The standard approach to solving knapsack precomputes the sum of smaller subsets of the items for use in computing larger subsets. In other words, if we want to know whether k is expressible as the sum of three numbers, and we have a table of all sums of two numbers, all we have to do is ask whether our number is expressible as the sum of a single number plus a number in this two-table.

Thus what I needed was a table of all integers less than n that could be expressed as the sum of two of the 1,818 pyramidal numbers less than 1,000,000,000. There could be at most $1,818^2 = 3,305,124$ of them. Actually, there would be only about half this many because we could eliminate duplicates, or any sum bigger than our target. Building a data structure, such as a sorted array, to hold these numbers would be no big deal. Call this data structure the *two*-table.

To find the minimum decomposition for a given k, I would start out by checking whether it was one of the 1,818 pyramidal numbers. If not, I would then search to see whether k was in the sorted table of the sums of two pyramidal numbers. If it wasn't, to see whether k was expressible as the sum of three such numbers, all I had to do was check whether $k - p[i]$ was in the *two*-table for $1 \leq i \leq 1,818$. This could be done quickly using binary search. To see whether k was expressible as the sum of four pyramidal numbers, I had to check whether $k - two[i]$ was in the two-table for all $1 \leq i \leq |two|$. However, since almost every k was expressible in many ways as the sum of four pyramidal numbers, this latter test would terminate quickly, and the time taken would be dominated by the cost of the threes. Testing whether k was the sum of three pyramidal numbers would take $O(n^{1/3} \lg n)$. Running this on each of the n integers gives an $O(n^{4/3} \lg n)$ algorithm for the complete job. Comparing this to his $O(n^2)$ algorithm for $n = 1,000,000,000$ suggested that my algorithm was a cool *30,000* times faster than the original!

My first attempt to code this up solved up to $n = 1,000,000$ on my crummy Sparc ELC in about 20 minutes. From here, I experimented with different data structures to represent the sets of numbers and different algorithms to search these tables. I tried using hash tables and bit vectors instead of sorted arrays and experimented with variants of binary search such as interpolation search (see Section 8.3.2). My reward for this work was solving up to n =1,000,000 in under three minutes, a factor of six improvement over my original program.

With the real thinking done, I worked to tweak a little more performance out of the program. I avoided doing a sum-of-four computation on any k when $k - 1$ was the sum-of-three, since 1 is a pyramidal number, saving about 10% of the total run time on this trick alone. Finally, I got out my profiler and tried some low-level tricks to squeeze a little more performance out of the code. For example, I saved another 10% by replacing a single procedure call with in-line code.

At this point, I turned the code over to the supercomputer guy. What he did with it is a depressing tale, which is reported in Section 5.8.

In writing up this war story, I went back to rerun our program almost five years later. On my desktop Sparc 5, getting to 1,000,000 now took 167.5 seconds using the cc compiler without turning on any compiler optimization. With level 2 optimization, the job ran in only 81.8 seconds, quite a tribute to the quality of the optimizer. The gcc compiler with optimization did even better, taking only 74.9 seconds to get to 1,000,000. The run time on my desktop machine had improved by a factor of about three over a four-year period. This is probably typical for most desktops.

The primary importance of this war story is to illustrate the enormous potential for algorithmic speedups, as opposed to the fairly limited speedup obtainable via more expensive hardware. I sped his program up by about 30,000 times. His million-dollar computer had 16 processors, each reportedly five times faster on integer computations than the $3,000 machine on my desk. That gave a maximum potential speedup of less than 100 times. Clearly, the algorithmic improvement was the big winner here, as it is certain to be in any sufficiently large computation.

2.8 War Story: String 'em Up

Biologists are hard at work on a fifteen-year project to sequence the human genome. This pattern of nucleotides encodes all the information necessary to build the proteins that we are built of. This project has already had an enormous impact on medicine and molecular biology.

Algorists have become interested in the human genome project as well, for several reasons:

- DNA sequences can be accurately represented as strings of characters on the four-letter alphabet (A,C,T,G). Biologist's needs have sparked new interest in old algorithmic problems (such as string matching – see Section 8.7.3) as well as creating new problems of substantial interest (such as shortest common superstring – see Section 8.7.9).

- DNA sequences are very *long* strings. The human genome is approximately three billion base pairs (or characters) long. Thus sophisticated computational techniques are necessary to deal with them. Such large problem sizes means that asymptotic (big-Oh) complexity analysis is usually fully justified on biological problems.

- Enough money is being invested in the human genome project for computer scientists to want to claim their piece of the action.

My particular interest in computational biology has revolved around a recently proposed but algorithmically intensive technique for DNA sequencing called sequencing by hybridization (SBH) [CK94, PL94]. The traditional sequencing by hybridization procedure attaches a set of probes to an array, forming a *sequencing chip*. Each of these probes determines whether or not the probe string occurs as a substring of the DNA target. The target DNA can now be sequenced based on the constraints of which strings are and are not substrings of the target.

One problem with SBH is that enormous arrays (say $4^8 = 65,536$ strings) are necessary to sequence relatively short pieces of DNA (typically about 200 base pairs long). The major reason is that all of these 4^8 probes are made at the same time. If you want to look up a name in the telephone book but are only allowed to consult the book once, you must copy down every single name from the book at that time. But if you are allowed to ask "is the name before Mendoza?" and wait for the answer before asking your next question, you can use binary search to greatly reduce your total effort.

We were convinced that using several small arrays would be more efficient than using one big array. We even had theory to justify our technique, but biologists aren't very inclined to believe theory. They demand experiments for proof. Hence we had to implement our algorithms and use simulation to prove that they worked.

So much for motivation. The rest of this tale will demonstrate the impact that clever data structures can have on a string processing application.

Our technique involved identifying all the strings of length $2k$ that are possible substrings of an unknown string S, given that we know all length k substrings of S. For example, suppose we know that AC, CA, and CC are the only length-2 substrings of S. It is certainly possible that $ACCA$ is a substring of S, since the center substring is one of our possibilities. However, $CAAC$ *cannot* be a substring of S, since we know that AA is not a substring of S. We needed to find a fast algorithm to construct all the consistent length-$2k$ strings, since S could be very long.

The simplest algorithm to build the $2k$ strings would be to concatenate all $O(n^2)$ pairs of k-strings together, and then for each pair to make sure that all $(k-1)$ length-k substrings spanning the boundary of the concatenation were in fact substrings, as shown in Figure 2-9. For example, the nine possible concatenations of AC, CA, and CC are $ACAC$, $ACCA$, $ACCC$, $CAAC$, $CACA$, $CACC$, $CCAC$, $CCCA$, and $CCCC$. Only $CAAC$ can be eliminated because of the absence of AA.

We needed a fast way of testing whether each of the $k-1$ substrings straddling the concatenation was a member of our dictionary of permissible k-strings. The time it takes to do this depends upon which kind of data structure we use to maintain this dictionary. With a binary search tree, we could find the correct string within $O(\lg n)$ comparisons, where each comparison involved testing which of two length-k strings appeared first in alphabetical order. Since each such comparison could require testing k pairs of characters, the total time using a binary search tree would be $O(k \log n)$.

That seemed pretty good. So my graduate student Dimitris Margaritis implemented a binary search tree data structure for our implementation. It worked great up until the moment we ran it.

"I've tried the fastest computer in our department, but our program is too slow," Dimitris complained. "It takes forever on strings of length only 2,000 characters. We will never get up to 50,000."

For interactive SBH to be competitive as a sequencing method, we had to be able to sequence long fragments of DNA, ideally over 50 kilobases in length. If we couldn't speed up the program, we would be in the embarrassing position of having a biological technique invented by computer scientists fail because the computations took too long.

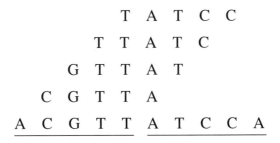

Figure 2-9. The concatenation of 2 fragments can be in S only if all subfragments are

We profiled our program and discovered that almost all the time was spent searching in this data structure, which was no surprise. For each of the $O(n^2)$ possible concatenations, we did this $k - 1$ times. We needed a faster dictionary data structure, since search was the innermost operation in such a deep loop.

"What about using a hash table?" I suggested. "If we do it right, it should take $O(k)$ time to hash a k-character string and look it up in our table. That should knock off a factor of $O(\log n)$, which will mean something when $n \approx 2,000$."

Dimitris went back and implemented a hash table implementation for our dictionary. Again, it worked great up until the moment we ran it.

"Our program is still too slow," Dimitris complained. "Sure, it is now about ten times faster on strings of length 2,000. So now we can get up to about 4,000 characters. Big deal. We will never get up to 50,000."

"We should have expected only a factor ten speedup," I mused. "After all, $\lg_2(2,000) \approx 11$. We need a faster data structure to search in our dictionary of strings."

"But what can be faster than a hash table?" Dimitris countered. "To look up a k-character string, you must read all k characters. Our hash table already does $O(k)$ searching."

"Sure, it takes k comparisons to test the first substring. But maybe we can do better on the second test. Remember where our dictionary queries are coming from. When we concatenate $ABCD$ with $EFGH$, we are first testing whether $BCDE$ is in the dictionary, then $CDEF$. These strings differ from each other by only one character. We should be able to exploit this so that each subsequent test takes constant time to perform...."

"We can't do that with a hash table," Dimitris observed. "The second key is not going to be anywhere near the first in the table. A binary search tree won't help, either. Since the keys $ABCD$ and $BCDE$ differ according to the first character, the two strings will be in different parts of the tree."

"But we can use a suffix tree to do this," I countered. "A suffix tree is a trie containing all the suffixes of a given set of strings. For example, the suffixes of $ACAC$ are $\{ACAC, CAC, AC, C\}$. Coupled with suffixes of string $CACT$, we get the suffix tree of Figure 2-10. By following a pointer from $ACAC$ to its longest proper suffix CAC, we get to the right place to test whether $CACT$ is in our set of strings. One character comparison is all we need to do from there."

Suffix trees are amazing data structures, discussed in considerably more detail in Section 8.1.3. Dimitris did some reading about them, then built a nice suffix tree implementation for our dictionary. Once again, it worked great up until the moment we ran it.

"Now our program is faster, but it runs out of memory," Dimitris complained. "And this on a 128 megabyte machine with 400 megabyte virtual memory! The suffix tree builds a path of length k for each suffix of length k, so all told there can be $\sum_{i=1}^{n} i = \Theta(n^2)$ nodes in the tree. It crashes when we go beyond 2,000 characters. We will never get up to strings with 50,000 characters."

I wasn't yet ready to give up. "There is a way around the space problem, by using compressed suffix trees," I recalled. "Instead of explicitly representing long

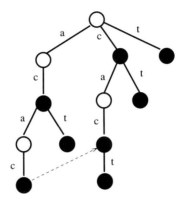

Figure 2-10. Suffix tree on *ACAC* and *CACT*, with the pointer to the suffix of *ACAC*

paths of character nodes, we can refer back to the original string." Compressed suffix trees always take linear space, as described in Section 8.1.3.

Dimitris went back one last time and implemented the compressed suffix tree data structure. *Now* it worked great! As shown in Figure 2-11, we ran our simulation for strings of length $n = 65,536$ on a SPARCstation 20 without incident. Our results, reported in [MS95a], showed that interactive SBH could be a very efficient sequencing technique. Based on these simulations, we were able to arouse interest in our technique from biologists. Making the actual wet lab-

String Length	Binary Tree	Hash Table	Suffix Tree	Compressed Suffix
8	0.0	0.0	0.0	0.0
16	0.0	0.0	0.0	0.0
32	0.1	0.0	0.0	0.0
64	0.3	0.4	0.3	0.0
128	2.4	1.1	0.5	0.0
256	17.1	9.4	3.8	0.2
512	31.6	67.0	6.9	1.3
1,024	1,828.9	96.6	31.5	2.7
2,048	11,441.7	941.7	553.6	39.0
4,096	> 2 days	5,246.7	out of	45.4
8,192		> 2 days	memory	642.0
16,384				1,614.0
32,768				13,657.8
65,536				39,776.9

Figure 2-11. Run times (in seconds) for the SBH simulation using various data structures

oratory experiments feasible provided another computational challenge, which is reported in Section 5.6.

The take home lessons for programmers from Figure 2-11 should be apparent. We isolated a single operation (dictionary string search) that was being performed repeatedly and optimized the data structure we used to support it. We started with a simple implementation (binary search trees) in the hopes that it would suffice, and then used profiling to reveal the trouble when it didn't. When an improved dictionary structure still did not suffice, we looked deeper into what kinds of queries we were performing, so that we could identify an even better data structure. Finally, we didn't give up until we had achieved the level of performance we needed. In algorithms, as in life, persistence usually pays off.

2.9 Exercises

2-1. Newt Gingrich is given the job of partitioning $2n$ players into two teams of n players each. Each player has a numerical rating that measures how good he/she is at the game. Newt seeks to divide the players as *unfairly* as possible, so as to create the biggest possible talent imbalance between team A and team B. Show how Newt can do the job in $O(n \lg n)$ time.

2-2. Take as input a sequence of $2n$ real numbers. Design an $O(n \log n)$ algorithm that partitions the numbers into n pairs, with the property that the partition minimizes the maximum sum of a pair. For example, say we are given the numbers (1,3,5,9). The possible partitions are ((1,3),(5,9)), ((1,5),(3,9)), and ((1,9),(3,5)). The pair sums for these partitions are (4,14), (6,12), and (10,8). Thus the third partition has 10 as its maximum sum, which is the minimum over the three partitions.

2-3. Assume that we are given as input n pairs of items, where the first item is a number and the second item is one of three colors (red, blue, or yellow). Further, assume that the items are sorted by number. Give an $O(n)$ algorithm to sort the items by color (all reds before all blues before all yellows) such that the numbers for identical colors stay sorted.

For example: (1,blue), (3,red), (4,blue), (6,yellow), (9,red) should become (3,red), (9,red), (1,blue), (4,blue), (6,yellow).

2-4. (*) The *mode* of a set of numbers is the number that occurs most frequently in the set. The set $(4, 6, 2, 4, 3, 1)$ has a mode of 4.

   (a) Give an efficient and correct algorithm to compute the mode of a set of n numbers.

   (b) Suppose we know that there is an (unknown) element that occurs $n/2 + 1$ times in the set. Give a worst-case linear-time algorithm to find the mode. For partial credit, your algorithm may run in expected linear time.

2-5. Given two sets S_1 and S_2 (each of size n), and a number x, describe an $O(n \log n)$ algorithm for finding whether there exists a pair of elements, one from S_1 and one from S_2, that add up to x. (For partial credit, give a $\Theta(n^2)$ algorithm for this problem.)

2-6. For each of the following problems, give an algorithm that finds the desired numbers within the given amount of time. To keep your answers brief, feel free to use algorithms from the book as subroutines. For the example, $S = \{6, 13, 19, 3, 8\}$, $19 - 3$ maximizes the difference, while $8 - 6$ minimizes the difference.

(a) Let S be an *unsorted* array of n integers. Give an algorithm that finds the pair $x, y \in S$ that *maximizes* $|x - y|$. Your algorithm must run in $O(n)$ worst-case time.

(b) Let S be a *sorted* array of n integers. Give an algorithm that finds the pair $x, y \in S$ that *maximizes* $|x - y|$. Your algorithm must run in $O(1)$ worst-case time.

(c) Let S be an *unsorted* array of n integers. Give an algorithm that finds the pair $x, y \in S$ that *minimizes* $|x - y|$, for $x \neq y$. Your algorithm must run in $O(n \lg n)$ worst-case time.

(d) Let S be a *sorted* array of n integers. Give an algorithm that finds the pair $x, y \in S$ that *minimizes* $|x - y|$, for $x \neq y$. Your algorithm must run in $O(n)$ worst-case time.

2-7. (*) Describe how to modify any balanced tree data structure such that search, insert, delete, minimum, and maximum still take $O(\lg n)$ time each, but successor and predecessor now take $O(1)$ time each. Which operations have to be modified to support this?

2-8. (*) In one of my research papers [Ski88], I give a comparison-based sorting algorithm that runs in $O(n \log(\sqrt{n}))$. Given the existence of an $\Omega(n \log n)$ lower bound for sorting, how can this be possible?

2-9. (*) Suppose you have access to a balanced dictionary data structure, which supports each of the operations search, insert, delete, minimum, maximum, successor, and predecessor in $O(\log n)$ time. Explain how to modify the insert and delete operations so they still take $O(\log n)$ but now minimum and maximum take $O(1)$ time. (Hint: think in terms of using the abstract dictionary operations, instead of mucking about with pointers and the like.)

2-10. (*) Mr. B. C. Dull claims to have developed a new data structure for priority queues that supports the operations *Insert*, *Maximum*, and *Extract-Max*, all in $O(1)$ worst-case time. Prove that he is mistaken. (Hint: the argument does not involve a lot of gory details–just think about what this would imply about the $\Omega(n \log n)$ lower bound for sorting.)

2-11. Use the partitioning step of Quicksort to give an algorithm that finds the *median* element of an array of n integers in expected $O(n)$ time.

2-12. (*) You are given the task of reading in n numbers and then printing them out in sorted order. Suppose you have access to a balanced dictionary data structure, which supports each of the operations search, insert, delete, minimum, maximum, successor, and predecessor in $O(\log n)$ time.

- Explain how you can use this dictionary to sort in $O(n \log n)$ time using only the following abstract operations: minimum, successor, insert, search.

- Explain how you can use this dictionary to sort in $O(n \log n)$ time using only the following abstract operations: minimum, insert, delete, search.

- Explain how you can use this dictionary to sort in $O(n \log n)$ time using only the following abstract operations: insert and in-order traversal.

2-13. The running time for Quicksort depends upon both the data being sorted and the partition rule used to select the pivot. Although Quicksort is $O(n \log n)$ on average, certain partition rules cause Quicksort to take $\Theta(n^2)$ time if the array is already sorted.

(a) Suppose we always pick the pivot element to be the key from the *last* position of the subarray. On a sorted array, does Quicksort now take $\Theta(n)$, $\Theta(n \log n)$, or $\Theta(n^2)$?

(b) Suppose we always pick the pivot element to be the key from the *middle* position of the subarray. On a sorted array, does Quicksort now take $\Theta(n)$, $\Theta(n \log n)$, or $\Theta(n^2)$?

(c) Suppose we always pick the pivot element to be the key of the *median* element of the *first three keys* of the subarray. (The median of three keys is the middle value, so the median of 5, 3, 8 is five.) On a sorted array, does Quicksort now take $\Theta(n)$, $\Theta(n \log n)$, or $\Theta(n^2)$?

(d) Suppose we always pick the pivot element to be the key of the *median* element of the first, last, and middle elements of the subarray. On a sorted array, does Quicksort now take $\Theta(n)$, $\Theta(n \log n)$, or $\Theta(n^2)$?

2-14. (*) Given a set S of n integers and an integer T, give an $O(n^{k-1} \lg n)$ algorithm to test whether k of the integers in S sum up to T.

2-15. (**) Design a data structure that allows one to search, insert, and delete an integer X in $O(1)$ time (i.e. constant time, independent of the total number of integers stored). Assume that $1 \leq X \leq n$ and that there are $m + n$ units of space available for the symbol table, where m is the maximum number of integers that can be in the table at any one time. (Hint: use two arrays $A[1..n]$ and $B[1..m]$.) You are not allowed to initialize either A or B, as that would take $O(m)$ or $O(n)$ operations. This means the arrays are full of random garbage to begin with, so you must be very careful.

2-16. (*) Let P be a simple, but not necessarily convex, polygon and q an arbitrary point not necessarily in P. Design an efficient algorithm to find a line segment originating from q that intersects the maximum number of edges of P. In other words, if standing at point q, in what direction should you aim a gun so the bullet will go through the largest number of walls. A bullet through a vertex of P gets credit for only one wall. An $O(n \lg n)$ algorithm is possible.

2-17. (**) The *onion* of a set of n points is the series of convex polygons that result from finding the convex hull, striping it from the point set, and repeating until no more points are left. Give an $O(n^2)$ algorithm for determining the onion of a point set.

Implementation Challenges

2-1. Implement versions of several different dictionary data structures, such as linked lists, binary trees, balanced binary search trees, and hash tables. Conduct experiments to assess the relative performance of these data structures in a simple application that reads a large text file and reports exactly one instance of each word that appears within it. This application can be efficiently implemented by maintaining a dictionary of all distinct words that have appeared thus far in the text and inserting/reporting each word that is not found. Write a brief report with your conclusions.

2-2. Implement versions of several different sorting algorithms, such as selection sort, insertion sort, heapsort, mergesort, and quicksort. Conduct experiments to assess the relative performance of these algorithms in a simple application that reads a large text file and reports exactly one instance of each word that appears within it. This application can be efficiently implemented by sorting all the words that occur in the text and then passing through the sorted sequence to identify one instance of each distinct word. Write a brief report with your conclusions.

3

Breaking Problems Down

One of the most powerful techniques for solving problems is to break them down into smaller, more easily solved pieces. Smaller problems are less overwhelming, and they permit us to focus on details that are lost when we are studying the entire problem. For example, whenever we can break the problem into smaller instances of the same type of problem, a recursive algorithm starts to become apparent.

Two important algorithm design paradigms are based on breaking problems down into smaller problems. *Dynamic programming* typically removes one element from the problem, solves the smaller problem, and then uses the solution to this smaller problem to add back the element in the proper way. *Divide and conquer* typically splits the problem in half, solves each half, then stitches the halves back together to form a full solution.

Both of these techniques are important to know about. Dynamic programming in particular is a misunderstood and underappreciated technique. To demonstrate its utility in practice, we present no fewer than three war stories where dynamic programming played the decisive role.

The "take-home" lessons for this chapter include:

- Many objects have an inherent left-to-right ordering among their elements, such as characters in a string, elements of a permutation, points around a polygon, or leaves in a search tree. For any optimization problem on such left-to-right objects, dynamic programming will likely lead to an efficient algorithm to find the best solution.

- Without an inherent left-to-right ordering on the objects, dynamic programming is usually doomed to require exponential space and time.

- Once you understand dynamic programming, it can be easier to work out such algorithms from scratch than to try to look them up.

- The global optimum (found, for example, using dynamic programming) is often noticeably better than the solution found by typical heuristics. How important this improvement is depends upon your application, but it can never hurt.

- Binary search and its variants are the quintessential divide-and-conquer algorithms.

3.1 Dynamic Programming

After you understand it, dynamic programming is probably the easiest algorithm design technique to apply in practice. In fact, I find that dynamic programming algorithms are usually easier to reinvent than to try to look up in a book. Until you understand it, however, dynamic programming seems like magic. You have to figure out the trick before you can use it.

In algorithms for problems such as sorting, correctness tends to be easier to verify than efficiency. This is not the case for *optimization problems*, where we seek to find a solution that maximizes or minimizes some function. In designing algorithms for an optimization problem, we must *prove* that our algorithm always gives the best possible solution.

Greedy algorithms, which make the best local decision at each step, occasionally happen to produce a global optimum for certain problems. These are typically efficient. However, you need a proof to show that you always end up with the best answer. Exhaustive search algorithms, which try all possibilities and select the best, by definition must always produce the optimum result, but usually at a prohibitive cost in terms of time complexity.

Dynamic programming combines the best of both worlds. The technique systematically considers all possible decisions and always selects the one that proves to be the best. By storing the *consequences* of all possible decisions to date and using this information in a systematic way, the total amount of work is minimized. Dynamic programming is best learned by carefully studying a number of examples until things start to click.

3.1.1 Fibonacci numbers

The tradeoff between space and time exploited in dynamic programming is best illustrated in evaluating recurrence relations, such as the Fibonacci numbers. The Fibonacci numbers were originally defined by the Italian mathematician Fibonacci in the thirteenth century to model the growth of rabbit populations. Rabbits breed, well, like rabbits. Fibonacci surmised that the number of pairs of rabbits born in a given year is equal to the number of pairs of rabbits born in each of the two previous years, if you start with one pair of rabbits in the first year. To count the number of rabbits born in the nth year, he defined the following recurrence relation:

$$F_n = F_{n-1} + F_{n-2}$$

with basis cases $F_0 = 0$ and $F_1 = 1$. Thus $F_2 = 1, F_3 = 2$, and the series continues $\{3, 5, 8, 13, 21, 34, 55, 89, 144, \ldots\}$. As it turns out, Fibonacci's formula didn't do a very good job of counting rabbits, but it does have a host of other applications and interesting properties.

Since they are defined by a recursive formula, it is easy to write a recursive program to compute the nth Fibonacci number. Most students have to do this in one of their first programming courses. Indeed, I have particularly fond memories of pulling my hair out writing such a program in 8080 assembly language. In pseudocode, the recursive algorithm looks like this:

Fibonacci[n]
       if ($n = 0$) then return(0)
       else if ($n = 1$) then return(1)
       else return(Fibonacci[$n - 1$]+Fibonacci[$n - 2$])

How much time does this algorithm take to compute Fibonacci[n]? Since $F_{n+1}/F_n \approx \phi = (1 + \sqrt{5})/2 \approx 1.61803$, this means that $F_n > 1.6^n$. Since our recursion tree, illustrated in Figure 3-1, has only 0 and 1 as leaves, summing up to such a large number means we must have at least 1.6^n leaves or procedure calls! This humble little program takes exponential time to run!

In fact, we can do much better. We can calculate F_n in linear time by storing all values. We trade space for time in the algorithm below:

Fibonacci[n]
       $F_0 = 0$
       $F_1 = 1$
       For $i = 1$ to n, $F_i = F_{i-1} + F_{i-2}$

Because we evaluate the Fibonacci numbers from smallest to biggest and store all the results, we know that we have F_{i-1} and F_{i-2} ready whenever we need to compute F_i. Thus each of the n values is computed as the simple sum of two integers in total $O(n)$ time, which is quite an improvement over exponential time.

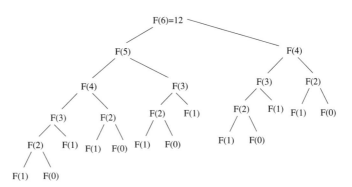

Figure 3-1. The computation tree for computing Fibonacci numbers recursively

3.1.2   The Partition Problem

Suppose that three workers are given the task of scanning through a shelf of books in search of a given piece of information. To get the job done fairly and efficiently, the books are to be partitioned among the three workers. To avoid the need to rearrange the books or separate them into piles, it would be simplest to divide the shelf into three regions and assign each region to one worker.

But what is the fairest way to divide the shelf up? If each book is the same length, say 100 pages, the job is pretty easy. Just partition the books into equal-sized regions,

$$100 \ 100 \ 100 \mid 100 \ 100 \ 100 \mid 100 \ 100 \ 100$$

so that everyone has 300 pages to deal with.

But what if the books are not the same length? Suppose we used the same partition when the book sizes looked like this:

$$100 \ 200 \ 300 \mid 400 \ 500 \ 600 \mid 700 \ 800 \ 900$$

I, for one, would volunteer to take the first section, with only 600 pages to scan, instead of the last one, with 2,400 pages. The fairest possible partition for this shelf would be

$$100 \ 200 \ 300 \ 400 \ 500 \mid 600 \ 700 \mid 800 \ 900$$

where the largest job is only 1,700 pages and the smallest job 1,300.

In general, we have the following problem:

Input: A given arrangement S of nonnegative numbers $\{s_1, \ldots, s_n\}$ and an integer k.
Output: Partition S into k or fewer ranges, so as to minimize the maximum sum over all the ranges.

This so-called *linear partition* problem arises often in parallel processing, since we seek to balance the work done across processors so as to minimize the total elapsed run time. Indeed, the war story of Section 5.8 revolves around a botched solution to this problem.

Stop for a few minutes and try to find an algorithm to solve the linear partition problem.

The beginning algorist might suggest a heuristic as the most natural approach to solve the partition problem. Perhaps they would compute the average size of a partition, $\sum_{i=1}^{n} s_i/k$, and then try to insert the dividers so as to come close to this average. However, such heuristic methods are doomed to fail on certain inputs, because they do not systematically evaluate all possibilities.

Instead, consider a recursive, exhaustive search approach to solving this problem. Notice that the kth partition starts right after we placed the $(k-1)$st divider. Where can we place this last divider? Between the ith and $(i+1)$st elements for some i, where $1 \leq i \leq n$. What is the cost of this? The total cost will be the larger of two quantities, (1) the cost of the last partition $\sum_{j=i+1}^{n} s_j$ and (2) the cost of the largest partition cost formed to the left of i. What is the size of this left partition? To minimize our total, we would want to use the $k-2$ remaining dividers to partition the elements $\{s_1, \ldots, s_i\}$ as equally as possible. *This is a smaller instance of the same problem, and hence can be solved recursively!*

Therefore, let us define $M[n,k]$ to be the minimum possible cost over all partitionings of $\{s_1, \ldots, s_n\}$ into k ranges, where the cost of a partition is the largest sum of elements in one of its parts. Thus defined, this function can be evaluated:

$$M[n,k] = \min_{i=1}^{n} \max(M[i, k-1], \sum_{j=i+1}^{n} s_j)$$

with the natural basis cases of

$$M[1,k] = s_1, \text{ for all } k > 0 \text{ and,}$$

$$M[n,1] = \sum_{i=1}^{n} s_i$$

By definition, this recurrence must return the size of the optimal partition. But how long does it take? If we evaluate the recurrence without storing partial results, we will be doomed to spend exponential time, repeatedly recalculating the same quantities. However, by storing the computed values and looking them up as needed, we can save an enormous amount of time.

How long does it take to compute this when we store the partial results? Well, how many results are computed? A total of $k \cdot n$ cells exist in the table. How much time does it take to compute the result $M[n', k']$? Well, calculating this quantity involves finding the minimum of n' quantities, each of which is the maximum of the table lookup and a sum of at most n' elements. If filling each of kn boxes takes at most n^2 time per box, the total recurrence can be computed in $O(kn^3)$ time.

To complete the implementation, we must specify the boundary conditions of the recurrence relation and an order to evaluate it in. These boundary conditions always settle the smallest possible values for each of the arguments of the recurrence. For this problem, the smallest reasonable value of the first argument is $n = 1$, meaning that the first partition consists of a single element. We can't create a first partition smaller than s_1 regardless of how many dividers are used. The smallest reasonable value of the second argument is $k = 1$, implying that we do not partition S at all.

The evaluation order computes the smaller values before the bigger values, so that each evaluation has what it needs waiting for it. Full details are provided in the pseudocode below:

Partition$[S, k]$
   (* compute prefix sums: $p[k] = \sum_{i=1}^{k} s_i$ *)
   $p[0] = 0$
   for $i = 1$ to n do $p[i] = p[i - 1] + s_i$

   (* initialize boundary conditions *)
   for $i = 1$ to n do $M[i, 1] = p[i]$
   for $j = 1$ to k do $M[1, j] = s_1$

   (* evaluate main recurrence *)
   for $i = 2$ to n do
      for $j = 2$ to k do
         $M[i, j] = \infty$
         for $x = 1$ to $i - 1$ do
            $s = \max(M[x, j - 1], p[i] - p[x])$
            if $(M[i, j] > s)$ then
               $M[i, j] = s$
               $D[i, j] = x$

The implementation above in fact runs faster than advertised. Our original analysis assumed that it took $O(n^2)$ time to update each cell of the matrix. This is because we selected the best of up to n possible points to place the divider, each of which requires the sum of up to n possible terms. In fact, it is easy to avoid the need to compute these sums by storing the set of n prefix sums $p[i] = \sum_{k=1}^{i} s_k$, since $\sum_{k=i}^{j} s_k = p[j] - p[k]$. This enables us to evaluate the recurrence in linear time per cell, yielding an $O(kn^2)$ algorithm.

By studying the recurrence relation and the dynamic programming matrices of Figures 3-2 and 3-3, you should be able to convince yourself that the final value of $M(n, k)$ will be the cost of the largest range in the optimal partition. However, what good is that? For most applications, what we need is the actual partition that does the job. Without it, all we are left with is a coupon for a great price on an out-of-stock item.

The second matrix, D, is used to reconstruct the optimal partition. Whenever we update the value of $M[i, j]$, we record which divider position was required to achieve that value. To reconstruct the path used to get to the optimal solution, we work backward from $D[n, k]$ and add a divider at each specified position. This backwards walking is best achieved by a recursive subroutine:

M	k		
n	1	2	3
1	1	1	1
2	3	2	2
3	6	3	3
4	10	6	4
5	15	9	6
6	21	11	9
7	28	15	11
8	36	21	15
9	45	24	17

D	k		
n	1	2	3
1	–	–	–
2	–	1	1
3	–	2	2
4	–	3	3
5	–	3	4
6	–	4	5
7	–	5	6
8	–	5	6
9	–	6	7

Figure 3–2. Dynamic programming matrices M and D in partitioning $\{1,2,3,4,5,6,7,8,9\}$

```
ReconstructPartition(S, D, n, k)
    If (k = 1) then print the first partition {s_1, ..., s_n}
    else
        ReconstructPartition(S, D, D[n, k], k − 1)
        Print the kth partition {s_D[n,k]+1, ..., s_m}
```

It is important to catch the distinction between storing the value of a cell and what decision/move it took to get there. The latter is not used in the computation but is presumably the real thing that you are interested in. For most of the examples in this chapter, we will not worry about reconstructing the answer. However, study this example closely to ensure that you know how to obtain the winning configuration when you need it.

M	k		
n	1	2	3
1	1	1	1
1	2	1	1
1	3	2	1
1	4	2	2
1	5	3	2
1	6	3	2
1	7	4	3
1	8	4	3
1	9	5	3

D	k		
n	1	2	3
1	–	–	–
1	–	1	1
1	–	1	2
1	–	2	2
1	–	2	3
1	–	3	4
1	–	3	4
1	–	4	5
1	–	4	6

Figure 3–3. Dynamic programming matrices M and D in partitioning $\{1,1,1,1,1,1,1,1,1\}$

3.1.3   Approximate String Matching

An important task in text processing is string matching – finding all the occurrences of a word in the text. Unfortunately, many words in documents are mispelled (sic). How can we search for the string closest to a given pattern in order to account for spelling errors?

To be more precise, let P be a pattern string and T a text string over the same alphabet. The *edit distance* between P and T is the smallest number of changes sufficient to transform a substring of T into P, where the changes may be:

1. *Substitution* – two corresponding characters may differ: KAT → CAT.
2. *Insertion* – we may add a character to T that is in P: CT → CAT.
3. *Deletion* – we may delete from T a character that is not in P: CAAT → CAT.

For example, $P=abcdefghijkl$ can be matched to $T=bcdeffghixkl$ using exactly three changes, one of each of the above types.

Approximate string matching arises in many applications, as discussed in Section 8.7.4. It seems like a difficult problem, because we have to decide where to delete and insert characters in pattern and text. But let us think about the problem in reverse. What information would we like to have in order to make the final decision; i.e. what should happen with the last character in each string? The last characters may be either be matched, if they are identical, or otherwise substituted one for the other. The only other options are inserting or deleting a character to take care of the last character of either the pattern or the text.

More precisely, let $D[i, j]$ be the minimum number of differences between $P_1, P_2, \ldots, P_i$ and the segment of T ending at j. $D[i, j]$ is the *minimum* of the three possible ways to extend smaller strings:

1. If $(P_i = T_j)$, then $D[i-1, j-1]$, else $D[i-1, j-1] + 1$. This means we either match or substitute the ith and jth characters, depending upon whether they do or do not match.
2. $D[i-1, j] + 1$. This means that there is an extra character in the pattern to account for, so we do not advance the text pointer and pay the cost of an insertion.
3. $D[i, j-1] + 1$. This means that there is an extra character in the text to remove, so we do not advance the pattern pointer and pay the cost of a deletion.

The alert reader will notice that we have not specified the boundary conditions of this recurrence relation. It is critical to get the initialization right if our program is to return the correct edit distance. The value of $D[0, i]$ will correspond to the cost of matching the first i characters of the text with none of the pattern. What this value should be depends upon what you want to compute. If you seek to match the entire pattern against the entire text, this means that we must delete the first i characters of the text, so $D[0, i] = i$ to pay the cost of the deletions. But what if we want to find where the pattern occurs in a long text? It should not cost more if the matched pattern starts far into the text than if it is near the front. Therefore, the starting cost should be equal for all positions. In this case, $D[0, i] = 0$, since we pay no cost for deleting the first i characters of the text. In both cases, $D[i, 0] = i$, since we cannot excuse deleting the first i characters of the pattern without penalty.

Once you accept the recurrence, it is straightforward to turn it into a dynamic programming algorithm that creates an $n \times m$ matrix D, where $n = |P|$ and $m = |T|$. Here it is, initialized for full pattern matching:

EditDistance(P, T)
    (*initialization*)
    For $i = 0$ to n do $D[i, 0] = i$
    For $i = 0$ to m do $D[0, i] = i$

    (*recurrence*)
    For $i = 1$ to n do
        For $j = 1$ to m do
            $D[i, j] = \min(D[i - 1, j - 1] + matchcost(s_i, t_j),$
                    $D[i - 1, j] + 1, D[i, j - 1] + 1)$

How much time does this take? To fill in cell $D[i, j]$, we need only compare two characters and look at three other cells. Since it requires only constant time to update each cell, the total time is $O(mn)$.

The value to return as the answer to our pattern matching problem depends on what we are interested in. If we only needed the cost of comparing all of the pattern against all of the text, such as in comparing the spelling of two words, all we would need is the cost of $D[n, m]$, as shown in Figure 3-4. But what if

		b	c	d	e	f	f	g	h	i	x	k	l
	0	1	2	3	4	5	6	7	8	9	10	11	12
a	**1**	1	2	3	4	5	6	7	8	9	10	11	12
b	2	**1**	2	3	4	5	6	7	8	9	10	11	12
c	3	2	**1**	2	3	4	5	6	7	8	9	10	11
d	4	3	2	**1**	2	3	4	5	6	7	8	9	10
e	5	4	3	2	**1**	2	3	4	5	6	7	8	9
f	6	5	4	3	2	**1**	**2**	3	4	5	6	7	8
g	7	6	5	4	3	2	2	**2**	3	4	5	6	7
h	8	7	6	5	4	3	3	3	**2**	3	4	5	6
i	9	8	7	6	5	4	4	4	3	**2**	3	4	5
j	10	9	8	7	6	5	5	5	4	3	**3**	4	5
k	11	10	9	8	7	6	6	6	5	4	4	**3**	4
l	12	11	10	9	8	7	7	7	6	5	5	4	**3**

Figure 3–4. Example dynamic programming matrix for edit distance computation, with the optimal alignment path highlighted in bold

we need to identify the best matching substring in the text? Assuming that the initialization was performed correctly for such substring matching, we seek the cheapest matching of the full pattern ending anywhere in the text. This means the cost equals $\min_{1 \leq i \leq m} D[n, i]$, i.e. the smallest cost on the last row of D.

Of course, this only gives the cost of the optimal matching, while we are often interested in reconstructing the actual alignment – which characters got matched, substituted, and deleted. These can be reconstructed from the pattern/text and table without an auxiliary storage, once we have identified the cell with the lowest cost. From this cell, we want to walk upwards and backwards through the matrix. Given the costs of its three neighbors and the corresponding characters, we can reconstruct which choice was made to get to the goal cell. The direction of each backwards step (to the left, up, or diagonal to the upper left) identifies whether it was a deletion, insertion, or match/substitution. Ties corresponding to legal transitions can be broken arbitrarily, since either way costs the same. We keep walking backwards until we hit the end of the matrix, specifying the starting point. This backwards-walking phase takes $O(n + m)$ time, since we traverse only the cells involved in the alignment.

The alert reader will notice that it is unnecessary to keep all $O(mn)$ cells to compute the cost of an alignment. If we evaluate the recurrence by filling in the columns of the matrix from left to right, we will never need more than two columns of cells to store what is necessary for the computation. Thus $O(m)$ space is sufficient to evaluate the recurrence without changing the time complexity at all. Unfortunately, without the full matrix we cannot reconstruct the alignment.

Saving space in dynamic programming is very important. Since memory on any computer is limited, $O(nm)$ space proves more of a bottleneck than $O(nm)$ time. Fortunately, there is a clever divide-and-conquer algorithm that computes the actual alignment in $O(nm)$ time and $O(m)$ space. This algorithm is discussed in Section 8.7.4.

3.1.4   Longest Increasing Sequence

Hopefully, a pattern is emerging. Every dynamic programming solution has three components:

1. Formulate the answer as a recurrence relation or recursive algorithm.
2. Show that the number of different values of your recurrence is bounded by a (hopefully small) polynomial.
3. Specify an order of evaluation for the recurrence so you always have the partial results you need available when you need them.

To see how this is done, let's see how we would develop an algorithm to find the longest monotonically increasing sequence in a sequence of n numbers. This problem arises in pattern matching on permutations, as described in Section 8.7.8. We distinguish an increasing sequence from a *run*, in that the selected elements need not be neighbors of each other. The selected elements must be in sorted order from left to right. For example, consider the sequence

$$S = (9,5,2,8,7,3,1,6,4)$$

The longest increasing subsequence of s has length 3 and is either $(2,3,4)$ or $(2,3,6)$. The longest increasing run is of length 2, either $(2,8)$ or $(1,6)$.

Finding the longest increasing run in a numerical sequence is straightforward, indeed you should be able to devise a linear-time algorithm fairly easily. However, finding the longest increasing subsequence is considerably trickier. How can we identify which scattered elements to skip? To apply dynamic programming, we need to construct a recurrence computing the length of the longest sequence. To find the right recurrence, ask what information about the first $n-1$ elements of S, coupled with the last element s_n, would enable you to find the answer for the entire sequence?

- The length of the longest increasing sequence in $s_1, s_2, \ldots, s_{n-1}$ seems a useful thing to know. In fact, this will be the longest increasing sequence in S, unless s_n extends it or another increasing sequence of the same length. Unfortunately, knowing just this length is not enough. Suppose I told you that the longest increasing sequence in $s_1, s_2, \ldots, s_{n-1}$ was of length 5 and that $s_n = 9$. Will the length of the final longest increasing subsequence of S be 5 or 6?
- Therefore, we also need the length of the longest sequence that s_n will extend. To be certain we know this, we really need the length of the longest sequence that *any* possible number can extend.

This provides the idea around which to build a recurrence. Define l_i to be the length of the longest sequence ending with s_i. Verify that the following table is correct:

sequence s_i	9	5	2	8	7	3	1	6	4
length l_i	1	1	1	2	2	2	1	3	3
predecessor p_i	–	–	–	2	2	3	–	6	6

The longest increasing sequence containing the nth number will be formed by appending it to the longest increasing sequence to the left of n that ends on a number smaller than s_n. The following recurrence computes l_i:

$$l_i = \max_{0 < j < i} l_j + 1, \text{ where } (s_j < s_i),$$

$$l_0 = 0$$

These values define the length of the longest increasing sequence ending at each number. The length of the longest increasing subsequence of the entire permutation is given by $\max_{1 \le i \le n} l_i$, since the winning sequence will have to end somewhere.

What is the time complexity of this algorithm? Each one of the n values of l_i is computed by comparing s_i against up to $i - 1 \leq n$ values to the left of it, so this analysis gives a total of $O(n^2)$ time. In fact, by using dictionary data structures in a clever way, we can evaluate this recurrence in $O(n \lg n)$ time. However, the simple recurrence would be easy to program and therefore is a good place to start.

What auxiliary information will we need to store in order to reconstruct the actual sequence instead of its length? For each element s_i, we will store the index p_i of the element that appears immediately before s_i in the longest increasing sequence ending at s_i. Since all of these pointers go towards the left, it is a simple matter to start from the last value of the longest sequence and follow the pointers so as to reconstruct the other items in the sequence.

3.1.5 Minimum Weight Triangulation

A *triangulation* of a polygon $P = \{v_1, \ldots, v_n, v_1\}$ is a set of nonintersecting diagonals that partitions the polygon into triangles. We say that the *weight* of a triangulation is the sum of the lengths of its diagonals. As shown in Figure 3-5, any given polygon may have many different triangulations. For any given polygon, we seek to find its minimum weight triangulation. Triangulation is a fundamental component of most geometric algorithms, as discussed in Section 8.6.3.

To apply dynamic programming, we need a way to carve up the polygon into smaller pieces. A first idea might be to try all $\binom{n}{2}$ possible chords, each of which partitions the polygon into two smaller polygons. Using dynamic programming, this will work to give a polynomial-time algorithm. However, there is a slicker approach.

Observe that every edge of the input polygon must be involved in exactly one triangle. Turning this edge into a triangle means identifying the third vertex, as shown in Figure 3-6. Once we find the correct connecting vertex, the polygon will be partitioned into two smaller pieces, both of which need to be triangulated optimally. Let $T[i, j]$ be the cost of triangulating from vertex v_i to vertex v_j, ignoring the length of the chord d_{ij} from v_i to v_j. The latter clause avoids double counting these internal chords in the following recurrence:

$$T[i, j] = \min_{k=i}^{j} \left(T[i, k] + T[k, j] + d_{ik} + d_{kj} \right)$$

 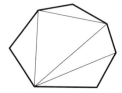

Figure 3–5. Two different triangulations of a given convex seven-gon

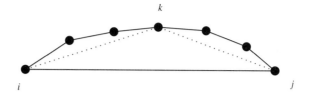

Figure 3–6. Selecting the vertex k to pair with an edge (i, j) of the polygon

The basis condition applies when i and j are immediate neighbors, as

$$T[i, i + 1] = 0.$$

Since the number of vertices in each subrange of the right side of the recurrence is smaller than that on the left side, evaluation can proceed in terms of the gap size from i to j:

Minimum-Weight-Triangulation(P)
    for $i = 1$ to n do $T[i, j] = 0$
    for $gap = 1$ to $n - 1$
        for $i = 1$ to $n - gap$ do
            $j = i + gap$
            $T[i, j] = \min_{k=i}^{j}(T[i, k] + T[k + 1, j] + d_{ik} + d_{kj})$
    return $T[1, n]$

There are $\binom{n}{2}$ values of T, each of which takes $O(j - i)$ time if we evaluate the sections in order of increasing size. Since $j - i = O(n)$, complete evaluation takes $O(n^3)$ time and $O(n^2)$ space.

What if there are points in the interior of the polygon? Then dynamic programming does not apply in the same way, because each of the triangulation edges does not necessarily cut the boundary into two distinct pieces as before. Instead of only $\binom{n}{2}$ possible subregions, the number of subregions now grows exponentially. In fact, no efficient algorithm for this problem is known. More surprisingly, there is also no known proof of hardness. Minimum weight triangulation is one of the few well-studied algorithmic problems that remain in this limbo state.

3.2 Limitations of Dynamic Programming

Dynamic programming can be applied to any problem that observes the *principle of optimality*. Roughly stated, this means that partial solutions can be optimally extended with regard to the *state* after the partial solution, instead of the partial solution itself. For example, to decide whether to extend an approximate string matching by a substitution, insertion, or deletion, we do not need to know exactly which sequence of operations was performed to date. In fact, there may be several different edit sequences that achieve a cost of C on the first p characters of pattern P and t characters of string T. Future decisions will be made based on the *consequences* of previous decisions, not the actual decisions themselves.

Problems do not satisfy the principle of optimality if the actual operations matter, as opposed to just the cost of the operations. Consider a form of edit distance where we are not allowed to use combinations of operations in certain particular orders. Properly formulated, however, most combinatorial problems respect the principle of optimality.

The biggest limitation on using dynamic programming is the number of partial solutions we must keep track of. For all of the examples we have seen, the partial solutions can be completely described by specifying the stopping *places* in the input. This is because the combinatorial objects being worked on (strings, numerical sequences, and polygons) all have an implicit order defined upon their elements. This order cannot be scrambled without completely changing the problem. Once the order is fixed, there are relatively few possible stopping places or states, so we get efficient algorithms. If the objects are not firmly ordered, however, we have an exponential number of possible partial solutions and are doomed to need an infeasible amount of memory.

To illustrate this, consider the following dynamic programming algorithm for the traveling salesman problem, discussed in greater detail in [RND77]. Recall that solving a TSP means finding the order that visits each site exactly once, while minimizing the total distance traveled or cost paid. Let $C(i, j)$ to be the edge cost to travel directly from i to j. Define $T(i; j_1, j_2, \ldots, j_k)$ to be the cost of the optimal tour from i to 1 that goes through each of the cities $j_1, j_2, \ldots, j_k$ exactly once, in any order. The cost of the optimal TSP tour is thus defined to be $T(1; 2, \ldots, n)$ and can be computed recursively by identifying the first edge in this sequence:

$$T(i; j_1, j_2, \ldots, j_i) = \min_{1 \leq m \leq k} C(i, j_m) + T(j_m; j_1, j_2, \ldots, j_k)$$

using the basis cases

$$T(i; j) = C(i, j) + C(j, 1)$$

This recurrence, although somewhat complicated to understand, is in fact correct. However, each partial solution is described by a vertex subset $j_1, j_2, \ldots, j_k$. Since there are 2^n subsets of n vertices, we require $\Omega(2^n)$ time and space to evaluate this recurrence. Whenever the input objects do not have an inherent left-right order, we are typically doomed to having an exponential-sized state space. Occasionally this is manageable – indeed, $O(2^n)$ is a big improvement over enumerating all $O(n!)$ possible TSP tours. Still, dynamic programming is most effective on well-ordered objects.

3.3 War Story: Evolution of the Lobster

I'll confess that I'm always surprised to see graduate students stop by my office early in the morning. Something about working through the night makes that time inconvenient for them. But there they were, two future Ph.D.s working in the field of high-performance computer graphics. They studied new techniques for rendering pretty computer images. The picture they painted for me that morning was anything but pretty, however.

"You see, we want to build a program to morph one image into another," they explained.

"What do you mean by morph?" I asked.

"For special effects in movies, we want to construct the intermediate stages in transforming one image into another. Suppose we want to turn you into Humphrey Bogart. For this to look realistic, we must construct a bunch of in-between frames that start out looking like you and end up looking like him."

"If you can realistically turn me into Bogart, you have something," I agreed.

"But our problem is that it isn't very realistic." They showed me a dismal morph between two images. "The trouble is that we must find the right correspondence between features in the two images. It looks real bad when we get the correspondence wrong and try to morph a lip into an ear."

"I'll bet. So you want me to give you an algorithm for matching up lips?"

"No, even simpler. We just morph each row of the initial image into the identical row of the final image. You can assume that we give you two lines of pixels, and you have to find the best possible match between the dark pixels in a row from object A to the dark pixels in the corresponding row of object B. Like this," they said, showing me images of successful matchings like those in Figure 3-7.

"I see," I said. "You want to match big dark regions to big dark regions and small dark regions to small dark regions."

"Yes, but only if the matching doesn't shift them too much to the left or the right. We might prefer to merge or break up regions rather than shift them too far away, since that might mean matching a chin to an eyebrow. What is the best way to do it?"

"One last question. Will you ever want to match two intervals to each other in such a way that they cross?"

"No, I guess not. Crossing intervals can't match. It would be like switching your left and right eyes."

I scratched my chin and tried to look puzzled, but I'm just not as good an actor as Bogart. I'd had a hunch about what needed to be done the instant they started

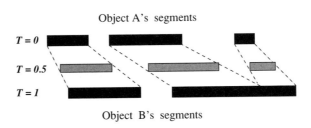

Object A's segments

$T = 0$

$T = 0.5$

$T = 1$

Object B's segments

Figure 3–7. A successful alignment of two lines of pixels

talking about lines of pixels. They want to transform one array of pixels into another array, with the minimum amount of changes. That sounded like editing one string of pixels into another string, which is a classic application of dynamic programming. See Sections 3.1.3 and 8.7.4 for discussions of approximate string matching.

The fact that the intervals couldn't cross just sealed things up. It meant that whenever a stretch of dark pixels from A was mapped to a stretch from B, the problem would be split into two smaller subproblems, i.e. the pixels to the left of the match and the pixels to the right of the match. The cost of the global match would ultimately be the cost of this match plus those of matching all the pixels to the left and of matching all the pixels to the right. Constructing the optimal match on the left side is a smaller problem and hence simpler. Further, there can be only $O(n^2)$ possible left subproblems, since each is completely described by the pair of one of n top pixels and one of n bottom pixels.

"Your algorithm will be based on dynamic programming," I pronounced. "However, there are several possible ways to do things, depending upon whether you want to edit pixels or runs. I would probably convert each row into a list of black pixel runs, with the runs sorted by right endpoint and each run labeled with its starting position and length. You will maintain the cost of the cheapest match between the leftmost i runs and the leftmost j runs for all i and j. The possible edit operations are:

- *Full run match:* We may match run i on top to run j on the bottom for a cost that is a function of the difference in the lengths of the two runs and their positions.

- *Merging runs:* We may match a string of consecutive runs on top to a run on the bottom. The cost will be a function of the number of runs, their relative positions, and their lengths.

- *Splitting runs:* We may match a big run on top to a string of consecutive runs on the bottom. This is just the converse of the merge. Again, the cost will be a function of the number of runs, their relative positions, and their lengths.

"For each pair of runs (i, j) and all the cases that apply, we compute the cost of the edit operation and add to the (already computed and stored) edit cost to the left of the start of the edit. The cheapest of these cases is what we will take for the cost of $c[i, j]$."

The pair of graduate students scribbled this down, then frowned. "So we are going to have a cost measure for matching two runs that is a function of their lengths and positions. How do we decide what the relative costs should be?"

"That is your business. The dynamic programming serves to optimize the matchings *once* you know the cost functions. It is up to your aesthetic sense to decide what penalties there should be for line length changes or offsets. My recommendation is that you implement the dynamic programming and try different values for the constants effecting the relative penalties on each of several different images. Then pick the setting that seems to do what you want."

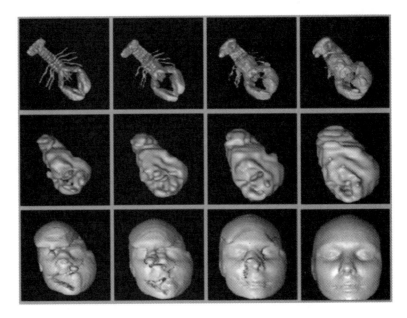

Figure 3–8. Morphing a lobster into a head via dynamic programming

They looked at each other and smiled, then ran back into the lab to implement it. Using dynamic programming to do their alignments, they completed their morphing system, which is described in [HWK94]. They produced the images shown in Figure 3-8, morphing a lobster into a man. Unfortunately, they never got around to turning me into Humphrey Bogart.

3.4 War Story: What's Past Is Prolog

"But our heuristic works very, very well in practice." He was simultaneously boasting and crying out for help.

Unification is the basic computational mechanism in logic programming languages like Prolog. A Prolog program consists of a set of rules, where each rule has a head and an associated action whenever the rule head matches or unifies with the current computation.

An execution of a Prolog program starts by specifying a goal, say $p(a, X, Y)$, where a is a constant and X and Y are variables. The system then systematically matches the head of the goal with the head of each of the rules that can be *unified* with the goal. Unification means binding the variables with the constants, if it is possible to match them. For the nonsense program below, $p(X, Y, a)$ unifies with either of the first two rules, since X and Y can be bound to match the extra characters. The goal $p(X, X, a)$ would only match the first rule, since the variable bound to the first and second positions must be the same.

$$p(a,a,a) := h(a);$$
$$p(b,a,a) := h(a) * h(b);$$
$$p(c,b,b) := h(b) + h(c);$$
$$p(d,b,b) := h(d) + h(b);$$

"In order to speed up unification, we want to preprocess the set of rule heads so that we can quickly determine which rules match a given goal. We must organize the rules in a trie data structure for fast unification."

Tries are extremely useful data structures in working with strings, as discussed in Section 8.1.3. Every leaf of the trie represents one string. Each node on the path from root to leaf is labeled with exactly one character of the string, with the ith node of the path corresponding to the ith character of the string.

"I agree. A trie is a natural way to represent your rule heads. Building a trie on a set of strings of characters is straightforward – just insert the strings one after another starting from the root. So what is your problem?" I asked.

"The efficiency of our unification algorithm depends very much on minimizing the number of edges in the trie. Since we know all the rules in advance, we have the freedom to reorder the character positions in the rules. Instead of the root node always representing the first argument in the rule, we can choose to have it represent the third argument. We would like to use this freedom to build a minimum-size trie for the set of rules."

He showed me the example in Figure 3-9. A trie constructed according to the original string position order $(1,2,3)$ uses a total of 12 edges. However, by permuting the character order to $(2,3,1)$, we can obtain a trie with only 8 edges.

"Why does the speed of unification depend on minimizing the number of edges in the trie?"

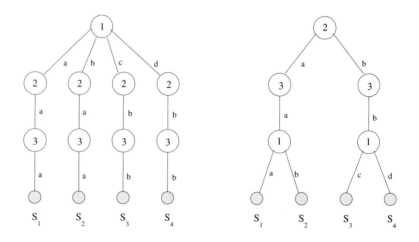

Figure 3–9. Two different tries for the given set of rule heads

"An open goal is one with all distinct variables in its arguments, like $p(X, Y, Z)$. Open goals match everything. In unifying an open goal against a set of clause heads, each symbol in all the clause heads will be bound to some variable. By doing a depth-first traversal of this minimum-edge trie, we minimize the number of operations in unifying the goal with all of the clause heads," he explained quickly.

"Interesting..." I started to reply before he cut me off again.

"One other constraint. For most applications we must keep the leaves of the trie ordered, so that the leaves of the underlying tree go left-to-right in the same order as the rules appear on the page."

"But why do you have to keep the leaves of the trie in the given order?" I asked.

"The order of rules in Prolog programs is very, very important. If you change the order of the rules, the program returns different results."

Then came my mission.

"We have a greedy heuristic for building good but not optimal tries, based on picking as the root the character position that minimizes the degree of the root. In other words, it picks the character position that has the smallest number of distinct characters in it. This heuristic works very, very well in practice. But we need you to prove that finding the best trie is NP-complete so our paper is, well, complete."

I agreed to think about proving the hardness of the problem and chased him from my office. The problem did seem to involve some nontrivial combinatorial optimization to build the minimal tree, but I couldn't see how to factor the left-to-right order of the rules into a hardness proof. In fact, I couldn't think of any NP-complete problem that had such a left-right ordering constraint. After all, if a given set of rules contained a character position in common to all the rules, that character position must be probed first in any minimum-size tree. Since the rules were ordered, each node in the subtree must represent the root of a run of consecutive rules, so there were only $\binom{n}{2}$ possible nodes to choose from for this tree....

Bingo! That settled it.

The next day I went back to the professor and told him. "Look, I cannot prove that your problem is NP-complete. But how would you feel about an efficient dynamic programming algorithm to find the best trie!" It was a pleasure watching his frown change to a smile as the realization took hold. An efficient algorithm to compute what he needed was infinitely better than a proof saying you couldn't do it!

My recurrence looked something like this. Suppose that we are given n ordered rule heads $s_1, \ldots, s_n$, each with m arguments. Probing at the pth position, $1 \le p \le m$, partitioned the rule heads into runs $R_1, \ldots, R_r$, where each rule in a given run $R_x = s_i, \ldots, s_j$ had the same character value of $s_i[p]$. The rules in each run must be consecutive, so there are only $\binom{n}{2}$ possible runs to worry about. The cost of probing at position p is the cost of finishing the trees formed by each of the created runs, plus one edge per tree to link it to probe p:

$$C[i, j] = \min_{p=1}^{m} \sum_{k=1}^{r} (C[i_k, j_k] + 1)$$

A graduate student immediately set to work implementing dynamic programming to compare with their heuristic. On many programs the optimal and greedy algorithms constructed the exact same trie. However, for some examples, dynamic programming gave a 20% performance improvement over greedy, i.e. 20% better than very, very well in practice. The run time spent in doing the dynamic programming was sometimes a bit larger than with greedy, but in compiler optimization you are always happy to trade off a little extra compilation time for better execution time performance of your program. Is a 20% improvement worth it? That depends upon the situation. For example, how useful would you find a 20% increase in your salary?

The fact that the rules had to remain ordered was the crucial property that we exploited in the dynamic programming solution. Indeed, without it, I was able to prove that the problem *was* NP-complete, something we put in the paper [DRR$^+$95] to make it complete.

3.5 War Story: Text Compression for Bar Codes

Ynjiun waved his laser wand over the torn and crumpled fragments of a bar code label. The system hesitated for a few seconds, then responded with a pleasant *blip* sound. He smiled at me in triumph. "Virtually indestructible."

I was visiting the research laboratories of Symbol Technologies, of Bohemia NY, the world's leading manufacturer of bar code scanning equipment. Next time you are in the checkout line at a grocery store, check to see what type of scanning equipment they are using. Likely it will say Symbol on the housing.

Although we take bar codes for granted, there is a surprising amount of technology behind them. Bar codes exist primarily because conventional optical character recognition (OCR) systems are not sufficiently reliable for inventory operations. The bar code symbology familiar to us on each box of cereal or pack of gum encodes a ten-digit number with sufficient error correction such that it is virtually impossible to scan the wrong number, even if the can is upside-down or dented. Occasionally, the cashier won't be able to get a label to scan at all, but once you hear that *blip* you know it was read correctly.

The ten-digit capacity of conventional bar code labels means that there is only room to store a single ID number in a label. Thus any application of supermarket bar codes must have a database mapping 11141-47011 to a particular size and brand of soy sauce. The holy grail of the bar code world has long been the development of higher-capacity bar code symbologies that can store entire documents, yet still be read reliably. Largely through the efforts of Theo Pavlidis and Ynjiun Wang at Stony Brook [PSW92], Symbol Technologies was ready to introduce the first such product.

Figure 3-10. A two-dimensional bar-code label of the Gettysburg Address using PDF-417

"PDF-417 is our new, two-dimensional bar code symbology," Ynjiun explained. A sample label is shown in Figure 3-10.

"How much data can you fit in a typical one-inch square label?" I asked him.

"It depends upon the level of error correction we use, but about 1,000 bytes. That's enough for a small text file or image," he said.

"Interesting. You will probably want to use some data compression technique to maximize the amount of text you can store in a label." See Section 8.7.5 for a discussion of standard data compression algorithms.

"We do incorporate a data compaction method," he explained. "We figure there are several different types of files our customers will want to use our labels for. Some files will be all in uppercase letters, while others will use mixed-case letters and numbers. We provide four different text modes in our code, each with a different subset of ASCII characters available. So long as we stay within a mode, we can describe each character using only five bits. When we have to switch modes, we issue a mode switch command first (taking an extra five bits) and then the new code."

"I see. So you designed the mode character sets to try to minimize the number of mode switch operations on typical text files." The modes are illustrated in Figure 3-11.

"Right. We put all the digits in one mode and all the punctuation characters in another. We also included both mode *shift* and mode *latch* commands. In a mode shift, we switch into a new mode just for the next character, say to produce a punctuation mark. This way, we don't pay a cost for returning back to text mode after a period. Of course, we can also latch permanently into a different mode if we will be using a run of several characters from there."

"Wow!" I said. "With all of this mode switching going on, there must be many different ways to encode any given text as a label. How do you find the smallest such encoding."

"We use a greedy-type algorithm. We look a few characters ahead and then decide which mode we would be best off in. It works fairly well."

I pressed him on this. "How do you know it works fairly well? There might be significantly better encodings that you are simply not finding."

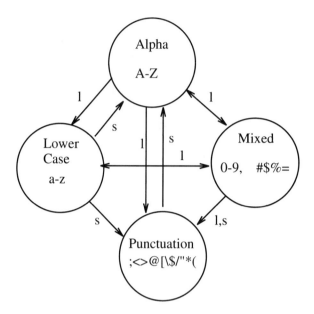

Figure 3–11. Mode switching in PDF-417

"I guess I don't know. But it's probably NP-complete to find the optimal coding." Ynjiun's voice trailed off. "Isn't it?"

I started to think. Every encoding started in a given mode and consisted of a sequence of intermixed character codes and mode shift/latch operations. At any given position in the text, we could output the next character code (if it was available in our current mode) or decide to shift. As we moved from left to right through the text, our current state would be completely reflected by our current character position and current mode state. For a given position/mode pair, we would have been interested in the cheapest way of getting there, over all possible encodings getting to this point. . . .

My eyes lit up so bright they cast shadows on the walls.

"The optimal encoding for any given text in PDF-417 can be found using dynamic programming. For each possible mode $1 \le m \le 4$ and each character position $1 \le i \le n$, we will maintain the cheapest encoding found of the string to the left of i ending in mode m. From each mode/position, we can either match, shift, or latch, so there are only a few possible operations to consider. Each of the $4n$ cells can be filled in constant time, so it takes time linear in the length of the string to find the optimal encoding."

Ynjiun was skeptical, but he encouraged us to implement an optimal encoder. A few complications arose due to weirdnesses of PDF-417 mode switching, but my student Yaw-Ling Lin rose to the challenge. Symbol compared our encoder to theirs on 13,000 labels and concluded that dynamic programming lead to an 8% tighter encoding on average. This was significant, because no one wants to waste 8% of their potential storage capacity, particularly in an environment where the capacity is only a few hundred bytes. For certain applications, this 8% margin permitted one bar code label to suffice where previously two had been required. Of course, an 8% *average* improvement meant that it did much better than that on certain labels. While our encoder took slightly longer to run than the greedy encoder, this was not significant, since the bottleneck would be the time needed to print the label, anyway.

Our observed impact of replacing a heuristic solution with the global optimum is probably typical of most applications. Unless you really botch your heuristic, you are probably going to get a decent solution. But replacing it with an optimal result usually gives a small but nontrivial improvement, which can have pleasing consequences for your application.

3.6 Divide and Conquer

Divide and conquer was a successful military strategy long before it became an algorithm design paradigm. Generals observed that it was easier to defeat one army of 50,000 men, followed by another army of 50,000 men than it was to beat a single 100,000 man army. Thus the wise general would attack so as to divide the enemy army into two forces and then mop up one after the other.

To use divide and conquer as an algorithm design technique, we must divide the problem into two smaller subproblems, solve each of them recursively, and then meld the two partial solutions into one solution to the full problem. Whenever the merging takes less time than solving the two subproblems, we get an efficient algorithm. Mergesort, discussed in Section 2.5.3, is the classic example of a divide-and-conquer algorithm. It takes only linear time to merge two sorted lists of $n/2$ elements each of which was obtained in $O(n \lg n)$ time.

Divide and conquer is a design technique with many important algorithms to its credit, including mergesort, the fast Fourier transform, and Strassen's matrix multiplication algorithm. However, with the exception of binary search, I find it to be a difficult technique to apply in practice. Therefore, the examples below will illustrate binary search and its variants, which yield simple algorithms to solve a surprisingly wide range of problems.

3.6.1 Fast Exponentiation

Suppose that we need to compute the value of a^n for some reasonably large n. Such problems occur in primality testing for cryptography, as discussed in Section 8.2.8.

The simplest algorithm performs $n - 1$ multiplications, by computing $a \times a \times \ldots \times a$. However, we can do better by observing that $n = \lfloor n/2 \rfloor + \lceil n/2 \rceil$. If n is even, then $a^n = (a^{n/2})^2$. If n is odd, then $a^n = a(a^{\lfloor n/2 \rfloor})^2$. In either case, we have halved the size of our exponent at the cost of at most two multiplications, so $O(\lg n)$ multiplications suffice to compute the final value.

function power(a, n)
　　if ($n = 0$) return(1)
　　x = power($a, \lfloor n/2 \rfloor$)
　　if (n is even) then return(x^2)
　　　　else return($a \times x^2$)

This simple algorithm illustrates an important principle of divide and conquer. It always pays to divide a job as evenly as possible. This principle applies to real life as well. When n is not a power of two, the problem cannot always be divided perfectly evenly, but a difference of one element between the two sides cannot cause any serious imbalance.

3.6.2   Binary Search

Binary search is a fast algorithm for searching in a sorted array S of keys. To search for key q, we compare q to the middle key $S[n/2]$. If q appears before $S[n/2]$, it must reside in the top half of our set; if not, it must reside in the bottom half of our set. By recursively repeating this process on the correct half, we find the key in a total of $\lceil \lg n \rceil$ comparisons, a big win over the $n/2$ we expect with sequential search.

This much you probably know. What is important is to have a sense for just how fast binary search is. *Twenty questions* is a popular children's game, where one player selects a word, and the other repeatedly asks true/false questions in an attempt to identify the word. If the word remains unidentified after 20 questions, the first party wins; otherwise, the second player takes the honors. In fact, the second player always has a winning strategy, based on binary search. Given a printed dictionary, the player opens it in the middle, selects a word (say "move"), and asks whether the unknown word is before "move" in alphabetical order. Since standard dictionaries contain 50,000 to 200,000 words, we can be certain that the process will always terminate within twenty questions.

Other interesting algorithms follow from simple variants of binary search. For example, suppose we have an array A consisting of a run of 0's, followed by an unbounded run of 1's, and would like to identify the exact point of transition between them. Binary search on the array would provide the transition point in $\lceil \lg n \rceil$ tests, if we had a bound n on the number of elements in the array. In the absence of such a bound, we can test repeatedly at larger intervals ($A[1]$, $A[2]$, $A[4]$, $A[8]$, $A[16]$, ...) until we find a first nonzero value. Now we have a window containing the target and can proceed with binary search. This *one-sided binary search* finds the transition point p using at most $2\lceil \lg p \rceil$ comparisons, regardless of how large the array actually is. One-sided binary search is most useful whenever we are looking for a key that probably lies close to our current position.

3.6.3   Square and Other Roots

The square root of n is the number r such that $r^2 = n$. Square root computations are performed inside every pocket calculator, but it is instructive to develop an efficient algorithm to compute square roots.

First, observe that the square root of $n \geq 1$ must be at least 1 and at most n. Let $l = 1$ and $r = n$. Consider the midpoint of this interval, $m = (l + r)/2$. How does m^2 compare to n? If $n \geq m^2$, then the square root must be greater than m, so the algorithm repeats with $l = m$. If $n < m^2$, then the square root must be less than m, so the algorithm repeats with $r = m$. Either way, we have halved the interval with only one comparison. Therefore, after only $\lg n$ rounds we will have identified the square root to within ± 1.

This bisection method, as it is called in numerical analysis, can also be applied to the more general problem of finding the roots of an equation. We say that x is a *root* of the function f if $f(x) = 0$. Suppose that we start with values l and r such that $f(l) > 0$ and $f(r) < 0$. If f is a continuous function, there must be a root between l and r. Depending upon the sign of $f(m)$, where $m = (l + r)/2$, we can cut this window containing the root in half with each test and stop when our estimate becomes sufficiently accurate.

Root-finding algorithms that converge faster than binary search are known for both of these problems. Instead of always testing the midpoint of the interval, these algorithms interpolate to find a test point closer to the actual root. Still, binary search is simple, robust, and works as well as possible without additional information on the nature of the function to be computed.

3.7 Exercises

3-1. Consider the problem of storing n books on shelves in a library. The order of the books is fixed by the cataloging system and so cannot be rearranged. Therefore, we can speak of a book b_i, where $1 \leq i \leq n$, that has a thickness t_i and height h_i. The length of each bookshelf at this library is L.

Suppose all the books have the same height h (i.e. $h = h_i = h_j$ for all i, j) and the shelves are all separated by a distance of greater than h, so any book fits on any shelf. The greedy algorithm would fill the first shelf with as many books as we can until we get the smallest i such that b_i does not fit, and then repeat with subsequent shelves. Show that the greedy algorithm always finds the optimal shelf placement, and analyze its time complexity.

3-2. (*) This is a generalization of the previous problem. Now consider the case where the height of the books is not constant, but we have the freedom to adjust the height of each shelf to that of the tallest book on the shelf. Thus the cost of a particular layout is the sum of the heights of the largest book on each shelf.

- Give an example to show that the greedy algorithm of stuffing each shelf as full as possible does not always give the minimum overall height.

- Give an algorithm for this problem, and analyze its time complexity. Hint: use dynamic programming.

3-3. (*) Consider a city whose streets are defined by an $X \times Y$ grid. We are interested in walking from the upper left-hand corner of the grid to the lower right-hand corner. Unfortunately, the city has bad neighborhoods, which are defined as intersections we do not want to walk in. We are given an $X \times Y$ matrix BAD, where $BAD[i,j]$

= "*yes*" if and only if the intersection between streets i and j is somewhere we want to avoid.

(a) Give an example of the contents of BAD such that there is no path across the grid avoiding bad neighborhoods.

(b) Give an $O(XY)$ algorithm to find a path across the grid that avoids bad neighborhoods.

(c) Give an $O(XY)$ algorithm to find the *shortest* path across the grid that avoids bad neighborhoods. You may assume that all blocks are of equal length. For partial credit, give an $O(X^2Y^2)$ algorithm.

3-4. (*) Consider the same situation as the previous problem. We have a city whose streets are defined by an $X \times Y$ grid. We are interested in walking from the upper left-hand corner of the grid to the lower right-hand corner. We are given an $X \times Y$ matrix BAD, where $BAD[i,j] =$ "*yes*" if and only if the intersection between streets i and j is somewhere we want to avoid.

If there were no bad neighborhoods to contend with, the shortest path across the grid would have length $(X - 1) + (Y - 1)$ blocks, and indeed there would be many such paths across the grid. Each path would consist of only rightward and downward moves.

Give an algorithm that takes the array BAD and returns the *number* of safe paths of length $X + Y - 2$. For full credit, your algorithm must run in $O(XY)$.

3-5. (*) Given an array of n real numbers, consider the problem of finding the maximum sum in any contiguous subvector of the input. For example, in the array

$$\{31, -41, 59, 26, -53, 58, 97, -93, -23, 84\}$$

the maximum is achieved by summing the third through seventh elements, where $59 + 26 + (-53) + 58 + 97 = 187$. When all numbers are positive, the entire array is the answer, while when all numbers are negative, the empty array maximizes the total at 0.

- Give a simple, clear, and correct $\Theta(n^2)$-time algorithm to find the maximum contiguous subvector.

- Now give a $\Theta(n)$-time dynamic programming algorithm for this problem. To get partial credit, you may instead give a *correct* $O(n \log n)$ divide-and-conquer algorithm.

3-6. In the United States, coins are minted with denominations of 1, 5, 10, 25, and 50 cents. Now consider a country whose coins are minted with denominations of $\{d_1, \ldots, d_k\}$ units. They seek an algorithm that will enable them to make change of n units using the minimum number of coins.

(a) The greedy algorithm for making change repeatedly uses the biggest coin smaller than the amount to be changed until it is zero. Show that the greedy algorithm does not always give the minimum number of coins in a country whose denominations are $\{1, 6, 10\}$.

(b) Give an efficient algorithm that correctly determines the minimum number of coins needed to make change of n units using denominations $\{d_1, \ldots, d_k\}$. Analyze its running time.

3-7. (*) In the United States, coins are minted with denominations of 1, 5, 10, 25, and 50 cents. Now consider a country whose coins are minted with denominations of $\{d_1, \ldots, d_k\}$ units. They want to count how many distinct ways $C(n)$ there are to make change of n units. For example, in a country whose denominations are $\{1, 6, 10\}$, $C(5) = 1$, $C(6)$ to $C(9) = 2$, $C(10) = 3$, and $C(12) = 4$.

(a) How many ways are there to make change of 20 units from $\{1, 6, 10\}$?

(b) Give an efficient algorithm to compute $C(n)$, and analyze its complexity. (Hint: think in terms of computing $C(n, d)$, the number of ways to make change of n units with highest denomination d. Be careful to avoid overcounting.)

3-8. (**) Consider the problem of examining a string $x = x_1 x_2 \ldots x_n$ of characters from an alphabet on k symbols, and a multiplication table over this alphabet, and deciding whether or not it is possible to parenthesize x in such a way that the value of the resulting expression is a, where a belongs to the alphabet. The multiplication table is neither commutative or associative, so the order of multiplication matters.

	a	b	c
a	a	c	c
b	a	a	b
c	c	c	c

For example, consider the following multiplication table and the string $bbbba$. Parenthesizing it $(b(bb))(ba)$ gives a, but $((((bb)b)b)a)$ gives c.

Give an algorithm, with time polynomial in n and k, to decide whether such a parenthesization exists for a given string, multiplication table, and goal element.

3-9. (*) Consider the following data compression technique. We have a table of m text strings, each of length at most k. We want to encode a data string D of length n using as few text strings as possible. For example, if our table contains $(a, ba, abab, b)$ and the data string is $bababbaababa$, the best way to encode it is $(b, abab, ba, abab, a)$ - a total of five code words. Give an $O(nmk)$ algorithm to find the length of the best encoding. You may assume that the string has an encoding in terms of the table.

3-10. A company database consists of 10,000 sorted names, 40% of whom are known as good customers and who together account for 60% of the accesses to the data base. There are two data structure options to consider for representing the database:

- Put all the names in a single array and use binary search.
- Put the good customers in one array and the rest of them in a second array. Only if we do not find the query name on a binary search of the first array do we do a binary search of the second array.

Demonstrate which option gives better expected performance. Does this change if linear search on an unsorted array is used instead of binary search for both options?

3-11. Suppose you are given an array A of n sorted numbers that has been *circularly shifted* k positions to the right. For example, $\{35, 42, 5, 15, 27, 29\}$ is a sorted array that has been circularly shifted $k = 2$ positions, while $\{27, 29, 35, 42, 5, 15\}$ has been shifted $k = 4$ positions.

- Suppose you know what k is. Give an $O(1)$ algorithm to find the largest number in A.

- Suppose you *do not* know what k is. Give an $O(\lg n)$ algorithm to find the largest number in A. For partial credit, you may give an $O(n)$ algorithm.

3-12. (*) Suppose that you are given a sorted sequence of *distinct* integers $\{a_1, a_2, \ldots, a_n\}$. Give an $O(\lg n)$ algorithm to determine whether there exists an index i such at $a_i = i$. For example, in $\{-10, -3, 3, 5, 7\}, a_3 = 3$. In $\{2, 3, 4, 5, 6, 7\}$, there is no such i.

Implementation Challenges

3-1. (*) Many types of products sold appeal more to members of one ethnic group than another. Perhaps Greeks eat more pasta per capita than Russians do, while Japanese find baseball more appealing than do Italians. A market researcher might be interested in having a program scan the names on a mailing list to select the ones most likely to be, say, Greek to target for a given mailing.

Develop a program that makes reasonable mappings between pairs of first/last names and ethnicities. One approach would be to compute the edit distance between query names and a family of names of known ethnicity. Feel free to experiment with other approaches. Test data is provided on the enclosed CD-ROM.

3-2. (*) In the game of Battleship, the first player hides a collection of, say, three 1×5 ships on a 10×10 grid. The second player guesses a series of grid positions and is informed whether they hit or miss a battleship. The second player continues to query until each of the 3×5 battleship positions has been probed. While the second player must succeed after making 100 different probes, we seek a strategy to use as few probes as possible to achieve the goal.

Develop a program that tries to efficiently sink all the battleships. One reasonable algorithmic approach would be based on divide-and-conquer or binary search.

3-3. (*) A Caesar shift (see Section 8.7.6) is a very simple class of ciphers for secret messages. Unfortunately, they can be broken using statistical properties of English. Develop a program capable of decrypting Caesar shifts of sufficiently long texts.

4

Graph Algorithms

A graph $G = (V, E)$ consists of a set of *vertices* V together with a set E of vertex pairs or *edges*. Graphs are important because they can be used to represent essentially *any* relationship. For example, graphs can model a network of roads, with cities as vertices and roads between cities as edges, as shown in Figure 4-1. Electronic circuits can also be modeled as graphs, with junctions as vertices and components as edges.

The key to understanding many algorithmic problems is to think of them in terms of graphs. Graph theory provides a language for talking about the properties of graphs, and it is amazing how often messy applied problems have a simple description and solution in terms of classical graph properties.

Designing truly novel graph algorithms is a very difficult task. The key to using graph algorithms effectively in applications lies in correctly modeling your problem as a standard graph property, so you can take advantage of existing algorithms. Becoming familiar with many different graph algorithmic *problems* is

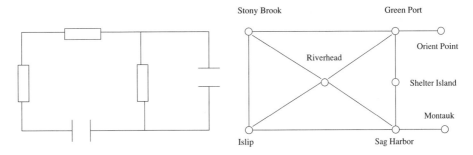

Figure 4-1. Modeling road networks and electronic circuits as graphs

more important than understanding the details of particular graph algorithms, particularly since Part II of this book can point you to an implementation as soon as you know the name of your problem.

In this chapter, we will present basic data structures and traversal operations for graphs, which will enable you to cobble together solutions to rudimentary graph problems. We will also describe more sophisticated algorithms for problems like shortest paths and minimum spanning trees in some detail. But we stress the primary importance of correctly modeling your problem. Time spent browsing through the catalog now will leave you better informed of your options when a real job arises.

The take-home lessons of this chapter include:

- Graphs can be used to model a wide variety of structures and relationships.

- Properly formulated, most applications of graphs can be reduced to standard graph properties and using well-known algorithms. These include minimum spanning trees, shortest paths, and several problems presented in the catalog.

- Breadth-first and depth-first search provide mechanisms to visit each edge and vertex of the graph. They prove the basis of most simple, efficient graph algorithms.

4.1 The Friendship Graph

To demonstrate the importance of proper modeling, let us consider a graph where the vertices are people, and there is an edge between two people if and only if they are friends. This graph is well-defined on any set of people—be they the people in your neighborhood, students in your class, or even the population of the entire world. There are many interesting aspects of people that are best understood as properties of this friendship graph.

We use this opportunity to define important graph theory terminology. "Talking the talk" proves to be an important part of "walking the walk".

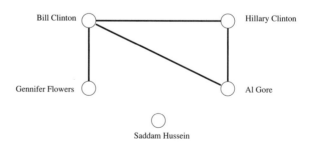

Figure 4-2. A portion of the friendship graph

- **If I am your friend, does that mean you are my friend?**

  A graph is *undirected* if edge (x, y) always implies (y, x). Otherwise, the graph is said to be *directed*. The "heard-of" graph is directed, since many famous people who I have heard of have never heard of me! The "had-sex-with" graph is presumably undirected, since the critical operation always requires a partner. I'd like to think that the "friendship" graph is always an undirected graph.

- **Am I my own friend?**

  An edge of the form (x, x) is said to be a *loop*. If x was y's friend several times over, we can model this relationship using *multiedges*, multiple edges between the same pair of vertices. A graph is said to be *simple* if it contains no loops and no multiple edges. Simple graphs really are often simpler to work with in practice. Therefore, we might be better off if no one was allowed to be their own friend.

- **How close a friend are you?**

  A graph is said to be *weighted* if each edge has an associated numerical attribute. We could model the strength of a friendship by associating each edge with an appropriate number, say from 0 (enemies) to 10 (blood brothers). The edges of a road network graph might be weighted with their length, drive-time, or speed limit, depending upon the application. A graph is said to be *unweighted* if all edges are assumed to be of equal weight.

- **Am I linked by some chain of friends to a star?**

  A *path* is a sequence of edges connecting two vertices. Since Mel Brooks is my father's sister's husband's cousin, there is a path in the friendship graph between me and him, shown in Figure 4-3. This is true even though the two of us have never met.

- **How close is my link to that star?**

  If I were trying to impress you with how tight I am with Mel Brooks, I would be much better off saying that my Uncle Lenny grew up with him than to go into the details of how connected I am to Uncle Lenny. Through Uncle Lenny, I have a path of length 2 to Cousin Mel, while the path is of length 4 by blood and marriage. I could make the path even longer by linking in people who know both me and my father, or are friends of Aunt Eve and Uncle Lenny. This multiplicity of paths hints at why finding the *shortest path* between two nodes is important and instructive, even in nontransportation applications.

Steve   Dad   Aunt Eve   Uncle Lenny   Cousin Mel

Figure 4–3. Mel Brooks is my father's sister's husband's cousin

- **Is there a path of friends between every two people in the world?**

  The "six degrees of separation" theory argues that there is always a short path linking every two people in the world. We say that a graph is *connected* if there is a path between any two vertices. A directed graph is *strongly connected* if there is always a directed path between any two vertices.

  If a graph is not connected, we call each connected piece a *connected component*. If we envision tribes in remote parts of the world that have yet not been encountered, each such tribe would form a connected component in the friendship graph. A remote hermit, or extremely unpleasant fellow (see Figure 4-2) would represent a connected component of one vertex, or an *isolated* vertex.

- **Who has the most friends? The fewest friends?**

  The *degree* of a vertex is the number of edges adjacent to it. The most popular person defines the vertex of highest degree in the friendship graph. Remote hermits will be associated with degree-zero vertices. In *dense* graphs, most vertices have high degree, as opposed to *sparse* graphs with relatively few edges. In *regular graphs*, each vertex has exactly the same degree. A regular friendship graph would truly be the ultimate in social-ism.

- **What is the largest clique?**

  A social clique is a group of mutual friends who all hang around together. A graph-theoretic *clique* is a complete subgraph, where each vertex pair has an edge between them. Cliques are the densest possible subgraphs. Within the friendship graph, we would expect to see large cliques corresponding to workplaces, neighborhoods, religious organizations, and schools.

- **How long will it take for my gossip to get back to me?**

  A *cycle* is a path where the last vertex is adjacent to the first. A cycle in which no vertex repeats (such as 1-2-3-1 verus 1-2-3-2-1) is said to be *simple*. The shortest cycle in the graph defines the graph's *girth*, while a simple cycle that passes through every vertex once is said to be a *Hamiltonian cycle*. An undirected graph with no cycles is said to be a *tree* if it is connected; otherwise it is a forest. A directed graph with no directed cycles is said to be a *DAG*, or *directed acyclic graph*.

4.2 Data Structures for Graphs

Selecting the right data structure to represent graphs can have an enormous impact on the performance of an algorithm. Your two basic choices are adjacency matrices and adjacency lists, illustrated in Figure 4-4.

An *adjacency matrix* is an $n \times n$ matrix M where (typically) $M[i, j] = 1$ if there is an edge from vertex i to vertex j and $M[i, j] = 0$ if there is not. Adjacency matrices are the simplest way to represent graphs. However, they doom you to

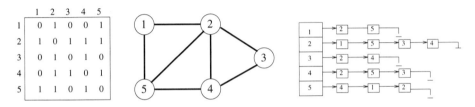

Figure 4–4. The adjacency matrix and adjacency list of a given graph

using $O(n^2)$ space no matter how many edges are in the graph. For large graphs, this will kill you. Remember that $1,000^2 = 1,000,000$, and work up from there. Although there is some potential for saving space by packing multiple bits per word or simulating a triangular matrix for undirected graphs, these cost some of the simplicity that makes adjacency matrices so appealing.

Beyond simplicity, there are certain algorithmic operations that prove faster on adjacency matrices than adjacency lists. In particular, it takes $\Theta(1)$ time to test whether edge (i, j) is in a graph represented by an adjacency matrix. All we must do is read the appropriate bit.

An *adjacency list* consists of an n-element array of pointers, where the ith element points to a linked list of the edges incident on vertex i. To test whether edge (i, j) is in the graph, we search the ith list for j. This takes $O(d_i)$, where d_i is the degree of the ith vertex. For a complete or almost complete graph, $d_i = \Theta(n)$, so testing the existence of an edge can be very expensive relative to adjacency matrices. However, d_i can be much less than n when the graph is sparse. Most of the graphs that one encounters in real life tend to be sparse. Recall the friendship graph as an example. Further, a surprising number of the most efficient graph algorithms can be and have been designed to avoid such edge-existence queries. The key is processing the edges in a systematic order like breadth-first or depth-first search.

For most applications, adjacency lists are the right way to go. The main drawback is the complexity of dealing with linked list structures. Things can be made arbitrarily hairy by adding extra pointers for special purposes. For example, the two versions of each edge in an undirected graph, (i, j) and (j, i), can be linked together by a pointer to facilitate deletions. Also, depending upon the operations you will perform on each list, you may or may not want it to be doubly linked, so that you can move backwards as easily as you move forwards.

It is a good idea to use a well-designed graph data type as a model for building your own, or even better as the foundation for your application. We recommend LEDA (see Section 9.1.1) as the best-designed general-purpose graph data structure currently available. It may be more powerful (and hence somewhat slower/larger) than what you need, but it does so many things right that you are likely to lose most of the potential do-it-yourself benefits through clumsiness.

In summary, we have the following tradeoffs between adjacency lists and matrices:

Comparison	Winner
Faster to test if (x, y) is in graph?	adjacency matrices
Faster to find the degree of a vertex?	adjacency lists
Less memory on small graphs?	adjacency lists $(m + n)$ vs. (n^2)
Less memory on big graphs?	adjacency matrices (a small win)
Edge insertion or deletion?	adjacency matrices $O(1)$ vs. $O(d)$
Faster to traverse the graph?	adjacency lists $\Theta(m + n)$ vs. $\Theta(n^2)$
Better for most problems?	adjacency lists

4.3 War Story: Getting the Graph

"It takes five minutes just to *read* the data. We will *never* have time to make it do something interesting."

The young graduate student was bright and eager, but green to the power of data structures. She would soon come to appreciate the power.

As described in a previous war story (see Section 2.6), we were experimenting with algorithms for extracting triangular strips for the fast rendering of triangulated surfaces. The task of finding a small number of strips that cover each triangle in a mesh could be modeled as a graph problem, where the graph has a vertex for every *triangle* of the mesh, and there is an edge between every pair of vertices representing adjacent triangles. This *dual graph* representation of the planar subdivision representing the triangulation (see Figure 4-5) captures all the information about the triangulation needed to partition it into triangle strips.

The first step in crafting a program that constructs a good set of strips was to build the dual graph of the triangulation. This I sent the student off to do. A few days later, she came back and announced that it took over five CPU minutes just to construct this dual graph of an object with a few thousand triangles.

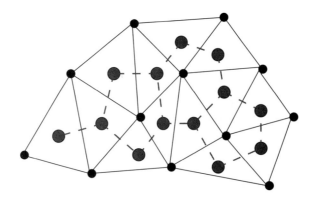

Figure 4–5. The dual graph (dashed lines) of a triangulation

"Nonsense!" I proclaimed. "You must be doing something very wasteful in building the graph. What format is the data in?"

"Well, it starts out with a list of the 3D-coordinates of the vertices used in the model and then follows with a list of triangles. Each triangle is described by a list of three indices into the vertex coordinates. Here is a small example:"

```
VERTICES 4
0.000000 240.000000 0.000000
204.000000 240.000000 0.000000
204.000000 0.000000 0.000000
0.000000 0.000000 0.000000
TRIANGLES 2
0  1  3
1  2  3
```

"I see. So the first triangle must use all but the third point, since all the indices start from zero. The two triangles must share an edge formed by points 1 and 3."

"Yeah, that's right," she confirmed.

"OK. Now tell me how you built your dual graph from this file."

"Well, I can pretty much ignore the vertex information, once I know how many vertices there will be. The geometric position of the points doesn't affect the structure of the graph. My dual graph is going to have as many vertices as the number of triangles. I set up an adjacency list data structure with that many vertices. As I read in each triangle, I compare it to each of the others to check whether it has two numbers in common. Whenever it does, I add an edge from the new triangle to this one."

I started to sputter. "But *that's* your problem right there! You are comparing each triangle against every other triangle, so that constructing the dual graph will be quadratic in the number of triangles. Reading in the input graph should take linear time!"

"I'm not comparing every triangle against every other triangle. On average, it only tests against half or a third of the triangles."

"Swell. But that still leaves us with an $O(n^2)$ algorithm. That is much too slow."

She stood her ground. "Well, don't just complain. Help me fix it!"

Fair enough. I started to think. We needed some quick method to screen away most of the triangles that would not be adjacent to the new triangle (i, j, k). What we really needed was just a list of all the triangles that go through each of the points i, j, and k. Since each triangle goes through three points, the average point is incident on three triangles, so this would mean comparing each new triangle against fewer than ten others, instead of most of them.

"We are going to need a data structure consisting of an array with one element for every vertex in the original data set. This element is going to be a list of all the triangles that pass through that vertex. When we read in a new triangle, we will look up the three relevant lists in the array and compare each of these against the new triangle. Actually, only two of the three lists are needed, since any adjacent triangle will share two points in common. For anything sharing two vertices, we will add an adjacency to our graph. Finally, we will add our new triangle to each of the three affected lists, so they will be updated for the next triangle read."

She thought about this for a while and smiled. "Got it, Chief. I'll let you know what happens."

The next day she reported that the graph could be built in seconds, even for much larger models. From here, she went on to build a successful program for extracting triangle strips, as reported in Section 2.6.

The take-home lesson here is that even elementary problems like initializing data structures can prove to be bottlenecks in algorithm development. Indeed, most programs working with large amounts of data have to run in linear or almost linear time. With such tight performance demands, there is no room to be sloppy. Once you focus on the need for linear-time performance, an appropriate algorithm or heuristic can usually be found to do the job.

4.4 Traversing a Graph

Perhaps the most fundamental graph problem is to traverse every edge and vertex in a graph in a systematic way. Indeed, most of the basic algorithms you will need for bookkeeping operations on graphs will be applications of graph traversal. These include:

- Printing or validating the contents of each edge and/or vertex.
- Copying a graph, or converting between alternate representations.
- Counting the number of edges and/or vertices.
- Identifying the connected components of the graph.
- Finding paths between two vertices, or cycles if they exist.

Since any maze can be represented by a graph, where each junction is a vertex and each hallway an edge, any traversal algorithm must be powerful enough to get us out of an arbitrary maze. For *efficiency*, we must make sure we don't get lost in the maze and visit the same place repeatedly. By being careful, we can arrange to visit each edge exactly twice. For *correctness*, we must do the traversal in a systematic way to ensure that we don't miss anything. To guarantee that we get out of the maze, we must make sure our search takes us through every edge and vertex in the graph.

The key idea behind graph traversal is to mark each vertex when we first visit it and keep track of what we have not yet completely explored. Although bread crumbs or unraveled threads are used to mark visited places in fairy-tale mazes, we will rely on Boolean flags or enumerated types. Each vertex will always be in one of the following three states:

- *undiscovered* – the vertex in its initial, virgin state.
- *discovered* – the vertex after we have encountered it, but before we have checked out all its incident edges.
- *completely-explored* – the vertex after we have visited all its incident edges.

Obviously, a vertex cannot be *completely-explored* before we discover it, so over the course of the traversal the state of each vertex progresses from *undiscovered* to *discovered* to *completely-explored*.

We must also maintain a structure containing all the vertices that we have discovered but not yet completely explored. Initially, only a single start vertex is considered to have been discovered. To completely explore a vertex, we must evaluate each edge going out of it. If an edge goes to an undiscovered vertex, we mark it *discovered* and add it to the list of work to do. If an edge goes to a *completely-explored* vertex, we will ignore it, since further contemplation will tell us nothing new about the graph. We can also ignore any edge going to a *discovered* but not *completely-explored* vertex, since the destination must already reside on the list of vertices to completely explore.

Regardless of which order we use to fetch the next vertex to explore, each undirected edge will be considered exactly twice, once when each of its endpoints is explored. Directed edges will be consider only once, when exploring the source vertex. Every edge and vertex in the connected component must eventually be visited. Why? Suppose the traversal didn't visit everything, meaning that there exists a vertex u that remains unvisited whose neighbor v *was* visited. This neighbor v will eventually be explored, and we will certainly visit u when we do so. Thus we must find everything that is there to be found.

The order in which we explore the vertices depends upon the container data structure used to store the *discovered* but not *completely-explored* vertices. There are two important possibilities:

- *Queue* – by storing the vertices in a first in, first out (FIFO) queue, we explore the oldest unexplored vertices first. Thus our explorations radiate out slowly from the starting vertex, defining a so-called *breadth-first search*.
- *Stack* – by storing the vertices in a last in, first out (LIFO) stack, we explore the vertices by lurching along a path, visiting a new neighbor if one is available, and backing up only when we are surrounded by previously discovered vertices. Thus our explorations quickly wander away from our starting point, defining a so-called *depth-first search*.

4.4.1 Breadth-First Search

The basic breadth-first search algorithm is given below. At some point during the traversal, every node in the graph changes state from *undiscovered* to *discovered*. In a breadth-first search of an undirected graph, we assign a direction to each edge, from the discoverer u to the discovered v. We thus denote u to be the parent $p[v]$. Since each node has exactly one parent, except for the root, this defines a

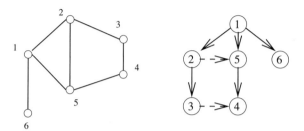

Figure 4–6. An undirected graph and its breadth-first search tree

tree on the vertices of the graph. This tree, illustrated in Figure 4-6, defines a shortest path from the root to every other node in the tree. This property makes breadth-first search very useful in shortest path problems.

```
BFS(G, s)
        for each vertex u ∈ V[G] − {s} do
                state[u] = "undiscovered"
                        p[u] = nil, i.e. no parent is in the BFS tree
                state[s] = "discovered"
                p[s] = nil
                Q = {s}
                while Q ≠ Ø do
                        u = dequeue[Q]
                        process vertex u as desired
                        for each v ∈ Adj[u] do
                                process edge (u, v) as desired
                                if state[v] = "undiscovered" then
                                        state[v] = "discovered"
                                        p[v] = u
                                        enqueue[Q, v]
                        state[u] = "completely-explored"
```

The graph edges that do not appear in the breadth-first search tree also have special properties. For undirected graphs, nontree edges can point only to vertices on the same level as the parent vertex or to vertices on the level directly below the parent. These properties follow easily from the fact that each path in the tree must be the shortest path in the graph. For a directed graph, a back-pointing edge $\overrightarrow{f}\,(u, v)$ can exist whenever v lies closer to the root than u does.

The breadth-first search algorithm above includes places to optionally process each vertex and edge, say to copy them, print them, or count them. Each vertex and directed edge is encountered exactly once, and each undirected edge is encountered exactly twice.

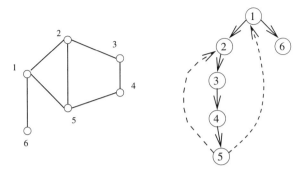

Figure 4–7. An undirected graph and its depth-first search tree

4.4.2 Depth-First Search

Depth-first search turns out to be, in general, even more useful than breadth-first search. The reason is that a depth-first search of a graph organizes the edges of the graph in a very precise way, which is quite different from breadth-first search. As with BFS, we assign a direction to each edge when we discover it, as shown in Figure 4-7.

Although there are four conceivable classes of edges resulting from such labelings, as shown in Figure 4-8, only two of them can occur with undirected graphs. In a DFS of an undirected graph, every edge is either in the tree or goes directly back to an ancestor. Why? Suppose we encountered a forward edge (x, y) directed toward a decendant vertex. In this case, we would have discovered (x, y) when exploring y, making it a back edge. Suppose we encounter a cross edge (x, y), linking two unrelated vertices. Again, we would have discovered this edge when we explored y, making it a tree edge. For directed graphs, depth-first search labelings can take on a wider range of possibilities.

Depth-first search has a neat recursive implementation, which eliminates the need to explicitly use a stack:

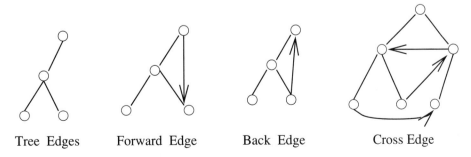

| Tree Edges | Forward Edge | Back Edge | Cross Edge |

Figure 4–8. Edge possibilities for search trees

```
DFS(G, u)
        state[u] = "discovered"
        process vertex u if desired
        for each v ∈ Adj[u] do
                process edge (u, v) if desired
                if state[v] = "undiscovered" then
                        p[v] = u
                        DFS(G, v)
        state[u] = "completely-explored"
```

As with BFS, this implementation of the depth-first search algorithm includes places to optionally process each vertex and edge, say to copy them, print them, or count them. Both algorithms will traverse all edges in the same connected component as the starting point. Since we need to start with a vertex in each component in order to traverse a disconnected graph, we must start from any vertex remaining undiscovered after a component search. With the proper initialization, this completes the traversal algorithm:

```
DFS-graph(G)
        for each vertex u ∈ V[G] do
                state[u] = "undiscovered"
        for each vertex u ∈ V[G] do
                if state[u] = "undiscovered" then
                        initialize new component, if desired
                        DFS[G, u]
```

4.5 Applications of Graph Traversal

Most elementary graph algorithms involve making one or two traversals of the graph, while we update our knowledge of the graph as we visit each edge and vertex. Properly implemented using adjacency lists, any such algorithm is destined to be very fast. Both BFS and DFS run in $O(n + m)$ on both directed and undirected graphs where, as usual, n is the number of vertices and m the number of edges in the graph. This is clearly optimal, since it is as fast as one can hope to read the graph. The trick is seeing when traversal approaches are destined to work. We present several examples below.

4.5.1 Connected Components

Either breadth-first or depth-first search can be used to identify the connected components of an undirected graph and label each vertex with the identifier of its components. In particular, we can modify the DFS-graph algorithm to increment a counter for the current component number and label each vertex accordingly as it is discovered in DFS.

For directed graphs, there are two distinct notions of connectivity, leading to algorithms for finding both weakly connected and strongly connected components. Both of these can be found in $O(n + m)$ time, as discussed in Section 8.4.1.

4.5.2 Tree and Cycle Detection

Trees are connected, undirected graphs that do not contain cycles. They are perhaps the simplest interesting class of graphs. Testing whether a graph is a tree is straightforward using depth-first search. During search, every edge will be labeled either a tree edge or a back edge, so the graph is a tree if and only if there are no back edges. Since $m = n - 1$ for any tree, this algorithm can be said to run in time linear in the number of vertices.

If the graph is not a tree, it must contain a cycle. Such a cycle can be identified as soon as the first back edge (u, v) is detected. If (u, v) is a back edge, then there must be a path in the tree from v to u. Coupled with edge (u, v), this defines a cycle.

4.5.3 Two-Coloring Graphs

In *vertex coloring*, we seek to assign a color to each vertex of a graph G such that no edge links two vertices of the same color. We can avoid all conflicts by assigning each vertex its own color. However, the goal is to use as few colors as possible. Vertex coloring problems arise often in scheduling applications, such as register allocation for compilers. See Section 8.5.7 for a full treatment of vertex coloring algorithms and applications.

A graph is *bipartite* if it can be colored without conflicts while using only two colors. Bipartite graphs are important because they arise often in practice and have more structure than arbitrary graphs. For example, consider the "had-sex-with" graph in a heterosexual world. Men have sex only with women, and vice versa. Thus gender defines a legal two-coloring. Irrespective of the accuracy of the model, it should be clear that bipartite graphs are simpler to work with than general graphs.

But how can we find an appropriate two-coloring of a graph, thus separating the men from the women? Suppose we assume that the starting vertex is male. All vertices adjacent to this man must be female, assuming the graph is indeed bipartite.

We can augment either breadth-first or depth-first search so that whenever we discover a new vertex, we color it the opposite of its parent. For each nondiscovery edge, we check whether it links two vertices of the same color. Such a conflict means that the graph cannot be two-colored. However, we will have constructed a proper two-coloring whenever we terminate without conflict. We can assign the first vertex in any connected component to be whatever color/sex we wish. Although we can separate the men from the women, we can't tell them apart just by using the graph.

4.5.4   Topological Sorting

A directed, acyclic graph, or *DAG*, is a directed graph with no directed cycles. Although undirected acyclic graphs are limited to trees, DAGs can be considerably more complicated. They just have to avoid directed cycles, as shown in Figure 4-9.

A *topological sort* of a directed acyclic graph is an ordering on the vertices such that all edges go from left to right. Only an acyclic graph can have a topological sort, because a directed cycle must eventually return home to the source of the cycle. However, every DAG has at least one topological sort, and we can use depth-first search to find such an ordering. Topological sorting proves very useful in scheduling jobs in their proper sequence, as discussed in catalog Section 8.4.2.

Depth-first search can be used to test whether a graph is a DAG, and if so to find a topological sort for it. A directed graph is a DAG if and only if no back edges are encountered during a depth-first search. Labeling each of the vertices in the reverse order that they are marked *completely-explored* finds a topological sort of a DAG. Why? Consider what happens to each directed edge $\{u, v\}$ as we encounter it during the exploration of vertex u:

- If v is currently *undiscovered*, then we then start a DFS of v before we can continue with u. Thus v is marked *completely-explored* before u is, and u appears before v in the topological order, as it must.

- If v is *discovered* but not *completely-explored*, then $\{u, v\}$ is a back edge, which is forbidden in a DAG.

- If v is *completely-explored*, then it will have been so labeled before u. Therefore, u appears before v in the topological order, as it must.

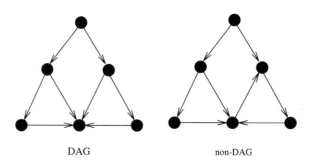

DAG   non-DAG

Figure 4-9. Directed acyclic and cyclic graphs

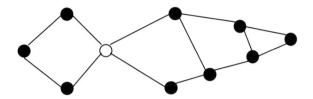

Figure 4–10. An articulation vertex is the weakest point in the graph

4.5.5   Articulation Vertices

Suppose you are a terrorist seeking to disrupt the telephone network. Which station in Figure 4-10 should you choose to blow up to cause the maximum amount of damage? An *articulation vertex* is a vertex of a connected graph whose deletion disconnects the graph. Any graph that contains an articulation vertex is inherently fragile, because deleting that single vertex causes a loss of connectivity.

In general, the *connectivity* of a graph is the smallest number of vertices whose deletion will disconnect the graph. For graphs with an articulation vertex, the connectivity is one. Connectivity is an important measure of robustness in network design, as discussed in catalog Section 8.4.8.

A simple application of either depth-first or breadth-first search suffices to find all the articulation vertices in a graph in $O(n(m + n))$. For each vertex v, delete it and then do a BFS traversal of the remaining graph to establish whether it is still connected. In fact, there is a clever $O(n + m)$ algorithm that tests all the vertices using only a single depth-first search. Additional information on edge and vertex connectivity testing appears in Section 8.4.8.

4.6   Modeling Graph Problems

Proper modeling is the key to making effective use of graph algorithms. We have seen a variety of definitions of graph properties, and algorithms for computing them. All told, about two dozen different graph problems are presented in the catalog, mostly in Sections 8.4 and 8.5. These problems provide a framework for modeling most applications.

The applications below demonstrate the power of proper modeling. Each of them arose in a real-world application as stated, and each can be modeled as a graph problem. Some of the modelings are quite clever, but they illustrate the versatility of graphs in representing relationships. As you read the problem, try to devise an appropriate graph representation before peeking to see how we did it.

- *"I'm looking for an algorithm to design natural routes for video-game characters to follow through an obstacle-filled room. How should I do it?"*

Presumably the route that is wanted is the path that looks most like the one that an intelligent being would choose. Since intelligent beings are either lazy or efficient, this should be modeled as some kind of shortest path problem.

But what is the graph? One approach would be to lay a grid of points in the room and have a vertex for each point that is a valid place for the character to stand, i.e. so it does not lie within an obstacle. There will be an edge between any pair of nearby vertices, weighted according to the distance between them. The shortest path between two vertices will be close to the shortest path between the points. Although direct geometric methods are known for shortest paths (see Section 8.4.4), it is easier to model this discretely as a graph.

- *"In DNA sequencing, we are given experimental data consisting of small fragments. For each fragment f, we have certain other fragments that are forced to lie to the left of f, certain fragments forced to be to the right of f, and the remaining fragments, which are free to go on either side. How can we find a consistent ordering of the fragments from left to right that satisfies all the constraints?"*

  Create a directed graph, where each fragment is assigned a unique vertex. Insert a directed edge (l, f) from any fragment l that is forced to be to the left of f, and a directed edge (f, r) to any fragment r forced to be to the right of f. We seek an ordering of the vertices such that all the edges go from left to right. This is exactly a *topological sort* of the resulting directed acyclic graph. The graph must be acyclic for this to work, because cycles make finding a consistent ordering impossible.

- *"In my graphics work I need to solve the following problem. Given an arbitrary set of rectangles in the plane, how can I distribute them into a minimum number of buckets such that the subset of rectangles in the same bucket do not intersect each other? In other words, there should not be any overlapping area between any two rectangles in the same bucket."*

  We can formulate a graph where each vertex is a rectangle, and there is an edge if two rectangles intersect. Each bucket corresponds to an *independent set* of rectangles, so there is no overlap between any two. A *vertex coloring* of a graph is a partition of the vertices into independent sets, so minimizing the number of colors is exactly what you want.

- *"In porting code from UNIX to DOS, I have to shorten the names of several hundred files down to at most 8 characters each. I can't just take the first eight characters from each name, because "filename1" and "filename2" will get assigned the exact same name. How can I shorten the names while ensuring that they do not collide?"*

  Construct a graph with vertices corresponding to each original file name f_i for $1 \leq i \leq n$, as well as a collection of acceptable shortenings for each name $f_{i1}, \ldots, f_{ik}$. Add an edge between each original and shortened name. Given such a formulation, we seek a set of n edges that have no vertices in common, because the file name of each is thus mapped to a distinct acceptable substitute. *Bipartite matching,* discussed in Section 8.4.6, is exactly this problem of finding an independent set of edges in a graph.

- "In organized tax fraud, criminals submit groups of phony tax returns in the hopes of getting undeserved refunds. These phony returns are all similar, but not identical. How can we detect clusters of similar forms so the IRS can nail the cheaters?"

   A natural graph model treats each form as a vertex and adds an edge between any two tax forms that are suspiciously similar. A cluster would correspond to a group of forms with many edges between them. In particular, a *clique* is a set of k vertices with all possible $\binom{k}{2}$ edges between them. Any sufficiently large clique identifies a cluster worth studying.

- "In the optical character-recognition system that we are building, we need a way to separate the lines of text. Although there is some white space between the lines, problems like noise and the tilt of the page makes it hard to find. How can we do line segmentation?

   Consider the following graph formulation. Treat each pixel in the image as a vertex in the graph, with an edge between two neighboring pixels. The weight of this edge should be proportional to how dark the pixels are. A segmentation between two lines is a path in this graph from the left to right side of the page. Of all possible paths, we seek a relatively straight path that avoids as much blackness as possible. This suggests that the *shortest path* in the pixel graph will likely find a good line segmentation.

4.7 Minimum Spanning Trees

A tree is a connected graph with no cycles. A spanning tree is a subgraph of G that has the same set of vertices of G and is a tree. A *minimum spanning tree* of a weighted graph G is the spanning tree of G whose edges sum to minimum weight.

Minimum spanning trees are useful in finding the least amount of wire necessary to connect a group of homes or cities, as illustrated in Figure 4-11. In such geometric problems, the point set $p_1, \ldots, p_n$ defines a complete graph, with

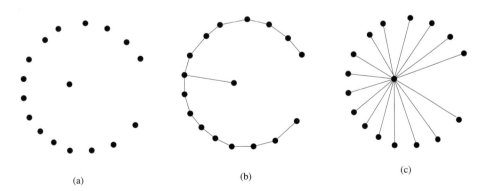

(a)   (b)   (c)

Figure 4-11. Two spanning trees of point set (a); the minimum spanning tree (b); and the shortest path from center tree (c)

edge (v_i, v_j) assigned a weight equal to the distance from p_i to p_j. Additional applications of minimum spanning trees are discussed in Section 8.4.3.

A minimum spanning tree minimizes the total length over all possible spanning trees. However, there can be more than one minimum spanning tree in any graph. Consider a graph G with m identically weighted edges. All spanning trees of G are minimum spanning trees, since each contains exactly $n - 1$ equal-weight edges. For general weighted graphs, however, the problem of finding a minimum spanning tree is more difficult. It can, however, be solved optimally using two different greedy algorithms. Both are presented below, to illustrate how we can demonstrate the optimality of certain greedy heuristics.

4.7.1   Prim's Algorithm

Every vertex will appear in the minimum spanning tree of any connected graph G. Prim's minimum spanning tree algorithm starts from one vertex and grows the rest of the tree one edge at a time.

In greedy algorithms, we make the decision of what to do next by selecting the best local option from all available choices without regard to the global structure. Since we seek the tree of minimum weight, the natural greedy algorithm for minimum spanning tree repeatedly selects the smallest weight edge that will enlarge the tree.

Prim-MST(G)
       Select an arbitrary vertex s to start the tree from.
       While (there are still nontree vertices)
              Select the edge of minimum weight between a tree and
                  nontree vertex
              Add the selected edge and vertex to the tree T_{prim}.

Prim's algorithm clearly creates a spanning tree, because no cycle can be introduced by adding edges between tree and nontree vertices. However, why should it be of minimum weight over all spanning trees? We have seen ample

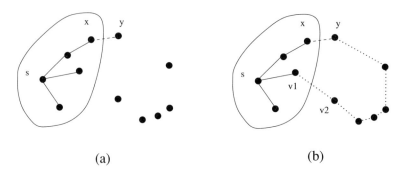

(a)   (b)

Figure 4–12. Where Prim's algorithm goes bad? No, because $d(v_1, v_2) \geq d(x, y)$

evidence of other natural greedy heuristics that do not yield a global optimium. Therefore, we must be particularly careful to demonstrate any such claim.

Suppose that there existed a graph G for which Prim's algorithm did not return a minimum spanning tree. Since we are building the tree incrementally, this means that there must have been some particular instant where we went wrong. Before we inserted edge (x, y), T_{prim} consisted of a set of edges that was a subtree of a minimum spanning tree T_{min}, but choosing edge (x, y) took us away from a minimum spanning tree. But how could it? There must be a path p from x to y in T_{min}, using an edge (v_1, v_2), where v_1 is in T_{prim} but v_2 is not. This edge (v_1, v_2) must have weight at least that of (x, y), or else Prim's algorithm would have selected it instead of (x, y) when it had the chance. Inserting (x, y) and deleting (v_1, v_2) from T_{min} leaves a spanning tree no larger than before, meaning that Prim's algorithm could not have made a fatal mistake in selecting edge (x, y). Therefore, by contradiction, Prim's algorithm has to construct a minimum spanning tree.

Prim's algorithm is correct, but how efficient is it? That depends on which data structures are used to implement it, but it should be clear that $O(nm)$ time suffices. In each of n iterations, we will scan through all the m edges and test whether the current edge joins a tree with a nontree vertex and whether this is the smallest edge seen thus far. By maintaining a Boolean flag along with each vertex to denote whether it is in the tree or not, this test can be performed in constant time. In fact, better data structures lead to a faster, $O(n^2)$, implementation by avoiding the need to sweep through more than n edges in any iteration.

4.7.2 Kruskal's Algorithm

Kruskal's algorithm is an alternative approach to finding minimum spanning trees that is more efficient on sparse graphs. Like Prim's, Kruskal's algorithm is greedy; unlike Prim's, it does not start with a particular vertex.

Kruskal's algorithm works by building up connected components of the vertices. Initially, each vertex forms its own separate component in the tree-to-be. The algorithm repeatedly considers the lightest remaining edge and tests whether the two endpoints lie within the same connected component. If so, the edge will be discarded, because adding it will create a cycle in the tree-to-be. If the endpoints are in different components, we insert the edge and merge the components. Since each connected component is always a tree, we need never explicitly test for cycles:

```
Kruskal-MST(G)
        Put the edges in a priority queue ordered by weight.
        count = 0
        while (count < n − 1) do
                get next edge (v, w)
                if (component (v) ≠ component(w))
                        add to T_kruskal
                        merge component(v) and component(w)
```

This algorithm adds $n - 1$ edges without creating a cycle, so clearly it creates a spanning tree of any connected graph. But why must this be a *minimum* spanning tree? Suppose it wasn't. As with the correctness proof of Prim's algorithm, there must be some graph for which it fails, and in particular there must a single edge (x, y) whose insertion first prevented the tree $T_{kruskal}$ from being a minimum spanning tree T_{min}. Inserting edge (x, y) in T_{min} will create a cycle with the path from x to y. Since x and y were in different components at the time of inserting (x, y), at least one edge on this path (v_1, v_2) would have been considered by Kruskal's algorithm after (x, y) was. But this means that $w(v_1, v_2) \geq w(x, y)$, so exchanging the two edges yields a tree of weight at most T_{min}. Therefore, we could not have made a mistake in selecting (x, y), and the correctness follows.

What is the time complexity of Kruskal's algorithm? Inserting and retrieving m edges from a priority queue such as a heap takes $O(m \lg m)$ time. The while loop makes at most m iterations, each testing the connectivity of two trees plus an edge. In the most simple-minded approach, this can be implemented by a breadth-first or depth-first search in a graph with at most n edges and n vertices, thus yielding an $O(mn)$ algorithm.

However, a faster implementation would result if we could implement the component test in faster than $O(n)$ time. In fact, the union-find data structure, discussed in Section 8.1.5 can support such queries in $O(\lg n)$ time. With this data structure, Kruskal's algorithm runs in $O(m \lg m)$ time, which is faster than Prim's for sparse graphs. Observe again the impact that the right data structure can have in implementing a straightforward algorithm.

4.8 Shortest Paths

The shortest path between two vertices s and t in an unweighted graph can be constructed using a breadth-first search from s. When we first encounter t in the search, we will have reached it from s using the minimum number of possible edges. This minimum-link path is recorded in the breadth-first search tree, and it provides the shortest path when all edges have equal weight. However, in an arbitrary weighted graph, the weight of a path between two vertices is the sum of the weights of the edges along the path. The shortest path might use a large number of edges, just as the shortest route (timewise) from home to office may involve shortcuts using backroads and many turns, as shown in Figure 4-13.

Shortest paths have a surprising variety of applications. See catalog Section 8.4.4 and the war story of Section 4.10 for further examples.

4.8.1   Dijkstra's Algorithm

We can use Dijkstra's algorithm to find the shortest path between any two vertices (s, t) in a weighted graph, where each edge has nonnegative edge weight. Although most applications of shortest path involve graphs with positive edge weights, such a condition is not needed for either Prim's or Kruskal's algorithm to work correctly. The problems that negative edges cause Dijkstra's algorithm will become apparent once you understand the algorithm.

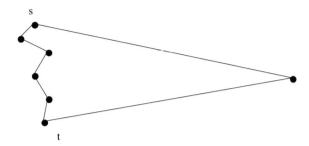

Figure 4–13. The shortest path from s to t can pass through many intermediate vertices.

The principle behind Dijkstra's algorithm is that given the shortest path between s and each of a given set of vertices $v_1, v_2, \ldots, v_k$, there must exist some other vertex x such that the shortest path from s to x must go from s to v_i to x, for some $1 \le i \le k$. Specifically, it is the vertex x that minimizes $dist(s, v_i) + w(v_i, x)$ over all $1 \le i \le k$, where $w(i, j)$ is the length of the edge from i to j and $dist(i, j)$ is the length of the shortest path between them.

This suggests a dynamic programming-like strategy. The shortest path from s to itself is trivial unless there are negative weight edges, so $dist(s, s) = 0$. Armed with the shortest path to s, if (s, y) is the lightest edge incident to s, then $d(s, y) = w(s, y)$. As soon as we decide that we have determined the shortest path to a node x, we search through all the outgoing edges of x to see whether there is a better path from s to some unknown vertex through x:

ShortestPath-Dijkstra(G, s, t)
    $known = \{s\}$
    for $i = 1$ to n, $dist[i] = \infty$
    for each edge (s, v), $dist[v] = w(s, v)$
    $last = s$
    while ($last \ne t$)
        select v_{next}, the unknown vertex minimizing $dist[v]$
        for each edge (v_{next}, x),
            $dist[x] = min[dist[x], dist[v_{next}] + w(v_{next}, x)]$
        $last = v_{next}$
        $known = known \cup \{v_{next}\}$

To be certain of finding the shortest path between s and t, we might have to first find the shortest path between s and all other vertices. This defines a shortest path spanning tree rooted in s. For undirected graphs, this will be the breadth-first search tree, but in general it provides the shortest path between s and all other vertices.

What is the running time of this algorithm? When implemented using adjacency lists and a Boolean array to mark what is known about each vertex, the complexity is $O(n^2)$. This is the same running time as a proper version of Prim's algorithm for minimum spanning trees; indeed, except for the extension condition, it *is* the same algorithm as Prim's.

4.8.2 All-Pairs Shortest Path

If we want to find the length of the shortest path between all $\binom{n}{2}$ pairs of vertices, we could run Dijkstra's algorithm n times, once from each possible starting vertex. This yields a cubic time algorithm for all-pairs shortest path, since $O(n^3) = n \times O(n^2)$.

Can we do better? Significantly improving the complexity is still an open question, but there is a superslick dynamic programming algorithm that also runs in $O(n^3)$.

There are several ways to characterize the shortest path between two nodes in a graph. The Floyd-Warshall algorithm starts by numbering the vertices of the graph from 1 to n. We use these numbers here not to label the vertices, but to order them. Define $D[i, j]^k$ to be the length of the shortest path from i to j using only vertices numbered from $1, 2, ..., k$ as possible intermediate vertices.

What does this mean? When $k = 0$, we are allowed no intermediate vertices, so that every path consists of at most one edge. Thus $D[i, j]^0 = w[i, j]$. In general, adding a new vertex $k + 1$ as a possible intermediary helps only if there is a short path that goes through it, so

$$D[i, j]^k = \min(D[i, j]^{k-1}, D[i, k]^{k-1} + D[k, j]^{k-1})$$

This recurrence performs only a constant amount of work per cell. The following dynamic programming algorithm implements the recurrence:

```
Floyd(G)
        Let D^0 = w(G), the weight matrix of G
        for k = 1 to n
            for i = 1 to n
                for j = 1 to n
                    D[i, j]^k = min(D[i, j]^{k-1}, D[i, k]^{k-1} + D[k, j]^{k-1})
```

The Floyd-Warshall all-pairs shortest path runs in $O(n^3)$ time, which asymptotically is no better than n calls to Dijkstra's algorithm. However, the loops are so tight and the program so short that it runs better in practice. It is also notable as one of the rare algorithms that work better on adjacency matrices than adjacency lists.

4.9 War Story: Nothing but Nets

I'd been tipped off about a small printed-circuit board testing company nearby that was in need of some algorithmic consulting. And so I found myself inside a typically nondescript building in a typically nondescript industrial park, talking with the president of Integri-Test, along with one of his lead technical people.

"We're the leaders in robotic printed-circuit board testing devices. Our customers have very high reliability requirements for their PC-boards. They must check that each and every board has no wire breaks *before* filling it with components. This means testing that each and every pair of points on the board that are supposed to be connected *are* connected."

"How do you do the testing?" I asked.

"We have a robot with two arms, each with electric probes. To test whether two points are properly connected, the arms simultaneously contact both of the points. If the two points are properly connected, then the probes will complete a circuit. For each net, we hold one arm fixed at one point and move the other to cover the rest of the points."

"Wait!" I cried. "What is a net?"

"On a circuit board there are certain sets of points that are all connected together with a metal layer. This is what we mean by a net. Sometimes a net is an isolated wire connecting two points. Sometimes a net can have 100 to 200 points, like all the connections to power or ground."

"I see. So you have a list of all the connections between pairs of points on the circuit board, and you want to trace out these wires."

He shook his head. "Not quite. The input for our testing program consists only of the net contact points, as shown in Figure 4-14(b). We don't know where the actual wires are, but we don't have to. All we have to do is verify that all the points in a net are connected together. We do this by putting the left robot arm on the leftmost point in the net, then having the right arm move around to all the other points in the net to test if they are connected to the left point. If they are all connected to the left point, it means that they must all be connected to each other."

I thought for a moment about what this meant. "OK. So your right arm has to visit all the other points in the net. How do you choose the order to visit them?"

The technical guy spoke up. "Well, we sort the points from left to right and then go in that order. Is that a good thing to do?"

"Have you ever heard of the traveling salesman problem?" I asked.

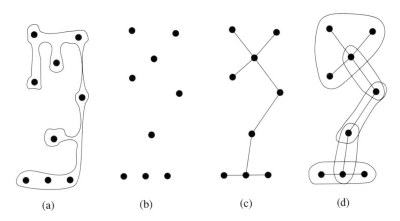

Figure 4–14. An sample net showing (a) the metal connection layer, (b) the contact points, (c) their minimum spanning tree, and (d) the points partitioned into clusters

He was an electrical engineer, not a computer scientist. "No, what's that?" he asked.

"Traveling salesman is the name of the exact problem that you are trying to solve. Given a set of points you have to visit, how do you order them so as to minimize the travel time. Algorithms for the traveling salesman problem have been extensively studied. For small nets, by doing an exhaustive search you will be able to find the optimal tour. For big nets, there are heuristics that will get you very close to the optimal tour." I would have pointed them to Section 8.5.4 if I had had this book handy.

The president scribbled down some notes and then frowned. "Fine. Maybe you can order the points in a net better for us. But that is not our real problem. When you watch our robot in action, the right arm sometimes has to run all the way to the right side of the board on a given net, while the left arm just sits there. It seems we would benefit by breaking a net into smaller pieces to balance things out."

I sat down and thought. The left and right arms were each going to have interlocking TSP problems to solve. The left arm would move between the leftmost points of each net, while the right arm was going to visit all the other points in each net as ordered by the left TSP tour. By breaking each net into smaller nets, so that each net occupies a small chunk of real estate, we would avoid making the right arm cross all the way across the board. Further, a lot of little nets meant there would be more points in the left TSP, so each left-arm movement was likely to be short, too.

"You are right. We should win if we can break big nets into small nets. We want the nets to be small, both in the number of points and in the area of the net. But we must be sure that if we validate the connectivity of each small net, we will have confirmed that the big net is connected. Whenever there is one point in common between two little nets, that is enough to show that the bigger net formed by the two little nets is connected, since current can flow between any pair of points."

Now we had to break each net into overlapping pieces, where each piece was small. This is a clustering problem. Minimum spanning trees are often used for clustering, as discussed in Section 8.4.3. In fact, that was the answer! We could find the minimum spanning tree of the net points and break it into little clusters whenever a spanning tree edge got too long. As shown in Figure 4-14(d), each cluster would share exactly one point in common with another cluster, with connectivity ensured because we are covering the edges of a spanning tree. The shape of the clusters would reflect the points in the net, exactly as we would want. If the points lay along a line across the board, the minimum spanning tree would be a path, and the clusters would be pairs of points. If the points all fell in a tight region, there would be one nice fat cluster that the right arm would just scoot around.

So I explained the idea of constructing the minimum spanning tree of a graph. The boss listened, scribbled more notes, and frowned again.

"I like your clustering idea. But these minimum spanning trees you talk about are defined on graphs. All you got are points. Where do the weights of the edges come from?"

"Oh, we can think of it as a complete graph, where every pair of points are connected. The weight of the edge defined by the two points is simply the distance. Or is it...?"

I went back to thinking. The edge cost between two points should reflect the travel time between them. While the distance was related to the travel time, it wasn't necessarily exactly the same thing.

"Hey. I have a question about your robot. Does it take the same amount of time to move the arm left-right as it does up-down?"

They thought a minute. "Yeah, it does. We use the same type of motors to control horizontal and vertical movements. Further, since the two motors for each arm are independent, we can simultaneously move each arm both horizontally and vertically."

"That so? The time to move both one foot left and one foot up is exactly the same as just moving one foot left? This means that the weight cost for each edge in the graph should not be the Euclidean distance between the two points, but the biggest difference between either the $x-$ or y-coordinate. This is something we call the L_∞ metric, but we can capture it by changing the edge weights in the graph. Anything else funny about your robots?" I asked.

"Well, it takes some time for the robot to come up to speed. We should probably also factor in acceleration and deceleration of the arms."

"Darn right. The more accurately you can model the time your arm takes to move between two points, the better our solution will be. But now we have a very clean formulation. Let's code it up and let's see how well it works!"

They were somewhat skeptical whether this approach would do any good, but they agreed to think about it. A few weeks later they called me back and reported that the new algorithm reduced testing time by about 30% over their previous approach, at a cost of a little more computational preprocessing. However, since their testing machine costs $200,000 a pop and a PC costs $2,000, this is an excellent tradeoff. It is particularly advantageous since the preprocessing need only be done once when testing multiple instances of the same board.

The key idea leading to the successful solution was knowing how to model the job in terms of classical algorithmic graph problems. I smelled TSP the instant they started talking about minimizing robot motion. Once I realized that they were implicitly forming a star-shaped spanning tree to ensure connectivity, it was natural to ask whether a minimum spanning tree would perform any better. This idea led to a natural way to think about clustering, and thus partitioning each net into smaller nets. Finally, by carefully constructing our distance metric to accurately model the costs of the robot itself, we get to incorporate quite complicated properties (such as acceleration and differences between horizontal and vertical speeds) without changing our fundamental graph model or algorithm design.

4.10 War Story: Dialing for Documents

I was part of a group visiting Periphonics, an industry leader in building telephone voice-response systems. These are more advanced versions of the *Press 1 for more options, Press 2 if you didn't press 1* telephone systems that have come to blight everyone's lives in recent years. We were being given the standard tour when someone from our group asked, "Why don't you guys use voice recognition for data entry. It would be a lot less annoying than typing things out on the keypad."

The tour guide reacted smoothly. "Our customers have the option of incorporating speech recognition into our products, but very few of them do. User-independent, connected-speech recognition is not accurate for most applications. Our customers prefer building systems around typing text on the telephone keyboards."

"Prefer typing, my pupik!", came a voice from the rear of our group. "I *hate* typing on a telephone. Whenever I call my brokerage house to get stock quotes, I end up talking to some machine, which asks me to type in the three letter code. To make it worse, I have to hit two buttons to type in one letter, in order to distinguish between the three letters printed on each key of the telephone. I hit the 2 key and it says Press 1 for 'A', Press 2 for 'B', Press 3 for 'C'. Pain in the neck if you ask me."

"Maybe you don't really have to hit two keys for each letter?" I chimed in. "Maybe the system could figure out the correct letter from context?"

"There isn't a whole lot of context when you type in three letters of stock market code."

"Sure, but there would be plenty of context if we were typing in English sentences. I'll bet that we could reconstruct English sentences correctly if they were typed in a telephone at one keystroke per letter."

The guy from Periphonics gave me a disinterested look, then continued the tour. But when I got back to the office, I decided to give it a try.

It was clear that not all letters were equally likely to be typed on a telephone. In fact, not all letters *can* be typed, since 'Q' and 'Z' are not labeled on the standard American telephone. Therefore, we adopted the convention that 'Q', 'Z', and space were all on the * key. We could take advantage of the uneven distribution of letter frequencies to help us decode the text. For example, if you hit the 3 key while typing English, you were more likely to have meant to type an 'E' than either a 'D' or 'F'. By taking into account the frequencies of a window of three characters, we could predict the typed text. Indeed, this is what happened when I tried it on the Gettysburg Address:

enurraore ane reten yeasr ain our ectherr arotght eosti on ugis aootinent a oey oation aoncdivee in licesty ane eedicatee un uhe rrorosition uiat all oen are arectee e ual

ony ye are enichde in a irect aitil yar uestini yhethes uiat oatioo or aoy oation ro aoncdivee ane ro eedicatee aan loni eneure ye are oet on a irect aattlediele oe uiat yar ye iate aone un eedicate a rostion oe uiat eiele ar a einal restini rlace eor uiore yin iere iate uhdis lives uiat uhe oation ogght live it is aluniethes eittini ane rrores uiat ye rioule en ugir

att in a laries reore ye aan oou eedicate ye aan oou aoorearate ye aan oou ialloy ugis iroune the arate oen litini ane eeae yin rustgilee iere iate aoorearatee it ear aante our roor rowes un ade or eeuraat the yople yill little oote oor loni renences yiat ye ray iere att it aan oetes eosiet yiat uhfy eie iere it is eor ur uhe litini rathes un ae eedicatee iere un uhe undiniside yopl yhici uhfy yin entght iere iate uiur ear ro onaky aetancde it is rathes eor ur un ae iere eedicatee un

uhe irect uarl rencinini adeore ur uiat eron uhere ioooree eeae ye uale inarearee eeuotion uo
tiat aaure eor yhici uhfy iere iate uhe lart eull oearure oe eeuotioo tiat ye iere iggily rerolue
uiat uhere eeae riall oou iate eide io

The trigram statistics did a decent job of translating it into Greek, but a terrible job of transcribing English. One reason was clear. This algorithm knew nothing about English words. If we coupled it with a dictionary, we might be on to something. The difficulty was that often two words in the dictionary would be represented by the exact same string of phone codes. For an extreme example, the code string "22737" collides with eleven distinct English words, including *cases, cares, cards, capes, caper,* and *bases.* As a first attempt, we reported the unambiguous characters of any words that collided in the dictionary, and used trigrams to fill in the rest of the characters. We were rewarded with:

eourscore and seven yearr ain our eatherr brought forth on this continent azoey nation conceivee in liberty and dedicatee uo uhe proposition that all men are createe equal

ony ye are engagee in azipeat civil yar uestioi whether that nation or aoy nation ro conceivee and ro dedicatee aan long endure ye are oet on azipeat battlefield oe that yar ye iate aone uo dedicate a rostion oe that field ar a final perthni place for those yin here iate their lives that uhe nation oight live it is altogether fittinizane proper that ye should en this

aut in a larges sense ye aan oou dedicate ye aan oou consecrate ye aan oou hallow this ground the arate men litioi and deae yin strugglee here iate consecratee it ear above our roor power uo ade or detract the world will little oote oor long remember what ye ray here aut it aan meter forget what uhfy die here it is for ur uhe litioi rather uo ae dedicatee here uo uhe toeioisgee york which uhfy yin fought here iate thus ear ro mocky advancee it is rather for ur uo ae here dedicatee uo uhe great task renagogoi adfore ur that from there honoree deae ye uale increasee devotion uo that aause for which uhfy here iate uhe last eull measure oe devotion that ye here highky resolve that there deae shall oou iate fide io vain that this nation under ioe shall iate azoey birth oe freedom and that ioternmenu oe uhe people ay uhe people for uhe people shall oou perish from uhe earth

If you were a student of American history, maybe you could recognize it, but you certainly couldn't read it. Somehow we had to distinguish between the different dictionary words that got hashed to the same code. We could factor in the relative popularity of each word, which would help, but this would still make too many mistakes.

At this point I started working with Harald Rau on the project, who proved a great collaborator for two reasons. First, he was a bright and peristent graduate student. Second, as a native German speaker he would believe every lie I told him about English grammar.

Harald built up a phone code reconstruction program on the lines of Figure 4-15. It worked on the input one sentence at a time, identifying dictionary words that matched each code string. The key problem was how to incorporate the grammatical constraints.

"We can get good word-use frequencies and grammatical information using this big text database called the Brown Corpus. It contains thousands of typical English sentences, each of which is parsed according to parts of speech. But how do we factor it all in?" Harald asked.

"Let's try to think about it as a graph problem," I suggested.

"*Graph problem?* What graph problem? Where is there even a graph?"

"Think of a sentence as a list of phone tokens, each representing a word in the sentence. For each phone token, we have a list of words from the dictionary that match it. How can we choose which one is right? Each possible sentence interpretation can be thought of as a path in a graph. The vertices of this graph will be the complete set of possible word choices. There will be an edge from a possible choice for the ith word to each possible choice for the $(i + 1)$st word. The cheapest path across this graph is the right interpretation of the sentence."

"But all the paths look the same. They have the same number of edges. Wait. Now I see! To make the paths different, we have to weight the edges."

"Exactly! The cost of an edge will reflect how likely it is that we will want to travel through the given pair of words. Maybe we can count how often that pair

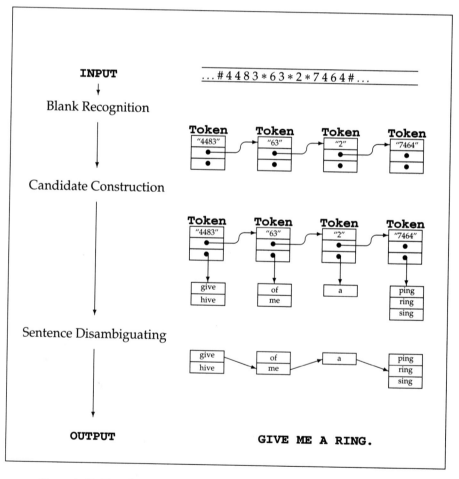

Figure 4–15. The phases of the telephone code reconstruction process

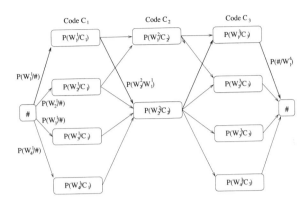

Figure 4–16. The minimum-cost path through the graph defines the best interpretation for a sentence

of words occurred together in previous texts. Or we can weight by what part of speech each word is. Maybe nouns don't like to be next to nouns as much as they like being next to verbs."

"It will be hard to keep track of word-pair statistics, since there are so many of them. But we certainly know the frequency of each word. How can we factor that into things?"

"We can pay a cost for walking through a particular vertex that depends upon the frequency of the word. Our best sentence will be given by the shortest path across the graph."

"But how do we figure out the relative weights of these factors?"

"Well, try what seems natural to you and then we can experiment with it."

Harald incorporated this shortest-path algorithm. With proper grammatical and statistical constraints, the system performed great, as reported in [RS96]. Look at the Gettysburg Address now, with all the reconstruction errors highlighted:

> FOURSCORE AND SEVEN YEARS AGO OUR FATHERS BROUGHT FORTH ON THIS CONTINENT A NEW NATION CONCEIVED IN LIBERTY AND DEDICATED TO THE PROPOSITION THAT ALL MEN ARE CREATED EQUAL. NOW WE ARE ENGAGED IN A GREAT CIVIL WAR TESTING WHETHER THAT NATION OR ANY NATION SO CONCEIVED AND SO DEDICATED CAN LONG ENDURE. WE ARE MET ON A GREAT BATTLEFIELD OF THAT **WAS**. WE HAVE COME TO DEDICATE A PORTION OF THAT FIELD AS A FINAL **SERVING** PLACE FOR THOSE WHO HERE **HAVE** THEIR LIVES THAT THE NATION MIGHT LIVE. IT IS ALTOGETHER FITTING AND PROPER THAT WE SHOULD DO THIS. BUT IN A LARGER SENSE WE CAN NOT DEDICATE WE CAN NOT CONSECRATE WE CAN NOT HALLOW THIS GROUND. THE BRAVE MEN LIVING AND DEAD WHO STRUGGLED HERE HAVE CONSECRATED IT FAR ABOVE OUR POOR POWER TO ADD OR DETRACT. THE WORLD WILL LITTLE NOTE NOR LONG REMEMBER WHAT WE SAY HERE BUT IT CAN NEVER FORGET WHAT THEY DID HERE. IT IS FOR US THE LIVING RATHER TO BE DEDICATED HERE TO THE UNFINISHED WORK WHICH THEY WHO FOUGHT HERE HAVE THUS FAR SO NOBLY ADVANCED. IT IS RATHER FOR US TO BE HERE DEDICATED TO THE GREAT TASK REMAINING BEFORE US THAT FROM THESE HONORED DEAD WE TAKE INCREASED DEVOTION TO THAT CAUSE FOR WHICH THEY HERE **HAVE** THE LAST FULL MEASURE OF DEVOTION THAT WE HERE HIGHLY

RESOLVE THAT THESE DEAD SHALL NOT HAVE DIED IN VAIN THAT THIS NATION
UNDER GOD SHALL HAVE A NEW BIRTH OF FREEDOM AND THAT GOVERNMENT
OF THE PEOPLE BY THE PEOPLE FOR THE PEOPLE SHALL NOT PERISH FROM THE
EARTH.

While we still made a few mistakes, the results are clearly good enough for a
variety of applications. Periphonics certainly thought so, for they later licensed
our program to incorporate into their products. Figure 4-17 shows that we
were able to reconstruct over 99% of the characters correctly on a megabyte of
President Clinton's speeches, so if Bill had phoned them in, we would certainly
still be able to understand it. The reconstruction time is fast enough, indeed
faster than you can type it in on the phone keypad.

The constraints associated with many different pattern recognition problems
can be formulated as shortest path problems in graphs. In fact, there is a particularly convenient dynamic programming solution for these problems known
as the Viterbi algorithm, which is used in speech and handwriting recognition
systems. Despite the fancy name, all the Viterbi algorithm is doing is solving a
shortest path problem. Hunting for a graph formulation for any given problem
is always a good way to proceed.

Text	characters	characters correct	non-blanks correct	words correct	time per character
Clinton Speeches	1,073,593	99.04%	98.86%	97.67%	0.97ms
Herland	278,670	98.24%	97.89%	97.02%	0.97ms
Moby Dick	1,123,581	96.85%	96.25%	94.75%	1.14ms
Bible	3,961,684	96.20%	95.39%	95.39%	1.33ms
Shakespeare	4,558,202	95.20%	94.21%	92.86%	0.99ms

Figure 4–17. Our telephone-code reconstruction system applied to various text samples

4.11 Exercises

4-1. Present correct and efficient algorithms to convert between the following graph
data structures, for an undirected graph G with n vertices and m edges. You must
give the time complexity of each algorithm.

(a) Convert from an adjacency matrix to adjacency lists.

(b) Convert from an adjacency list to an incidence matrix. An incidence matrix M
has a row for each vertex and a column for each edge, such that $M[i, j] = 1$ if
vertex i is part of edge j, otherwise $M[i, j] = 0$.

(c) Convert from an incidence matrix to adjacency lists.

4-2. Is the path between a pair of vertices in a minimum spanning tree necessarily
a shortest path between the two vertices in the full graph? Give a proof or a
counterexample.

4-3. Assume that all edges in the graph have distinct edge weights (i.e. no pair of edges have the same weight). Is the path between a pair of vertices in a minimum spanning tree necessarily a shortest path between the two vertices in the full graph? Give a proof or a counterexample.

4-4. Suppose G is a connected undirected graph. An edge e whose removal disconnects the graph is called a *bridge*. Must every bridge e be an edge in a depth-first search tree of G, or can e be a back edge? Give a proof or a counterexample.

4-5. (*) In breadth-first and depth-first search, an undiscovered node is marked *discovered* when it is first encountered, and marked *completely-explored* when it has been completely searched. At any given moment, several nodes might be simultaneously in the *discovered* state.

(a) Describe a graph on n vertices and a particular starting vertex v such that during a *breadth-first search* starting from v, $\Theta(n)$ nodes are simultaneously in the *discovered* state.

(b) Describe a graph on n vertices and a particular starting vertex v such that during a *depth-first search* starting from v, $\Theta(n)$ nodes are simultaneously in the *discovered* state.

(c) Describe a graph on n vertices and a particular starting vertex v such that at some point during a *depth-first search* starting from v, $\Theta(n)$ nodes remain *undiscovered*, while $\Theta(n)$ nodes have been *completely-explored*. (Note, there may also be *discovered* nodes.)

4-6. Given the pre-order and in-order traversals of a binary tree, is it possible to reconstruct the tree? If so, sketch an algorithm to do it. If not, give a counterexample. Repeat the problem if you are given the pre-order and post-order traversals.

4-7. Suppose an arithmetic expression is given as a tree. Each leaf is an integer and each internal node is one of the standard arithmetical operations $(+, -, *, /)$. For example, the expression $2 + 3 * 4 + (3 * 4)/5$ could be represented by the tree in Figure 4-18(a).

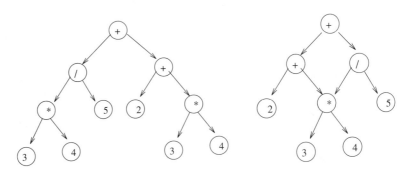

Figure 4–18. Expression $2 + 3 * 4 + (3 * 4)/5$ as a tree and a DAG.

Give an $O(n)$ algorithm for evaluating such an expression, where there are n nodes in the tree.

4-8. (*) Suppose an arithmetic expression is given as a DAG (directed acyclic graph) with common subexpressions removed. Each leaf is an integer and each internal node is one of the standard arithmetical operations $(+, -, *, /)$. For example, the expression $2 + 3 * 4 + (3 * 4)/5$ could be represented by the DAG in Figure 4-18(b). Give an $O(n + m)$ algorithm for evaluating such a DAG, where there are n nodes and m edges in the DAG. Hint: modify an algorithm for the tree case to achieve the desired efficiency.

4-9. (*) Given an undirected graph G with n vertices and m edges, and an integer k, give an $O(m + n)$ algorithm that finds the maximum induced subgraph H of G such that each vertex in H has degree $\geq k$, or prove that no such graph exists. An induced subgraph $F = (U, R)$ of a graph $G = (V, E)$ is a subset of U of the vertices V of G, and all edges R of G such that both vertices of each edge are in U.

4-10. (*) An articulation vertex of a graph G is a vertex whose deletion disconnects G. Let G be a graph with n vertices and m edges. Give a simple $O(n + m)$ algorithm for finding a vertex of G that is *not* an articulation vertex, i.e. whose deletion does not disconnect G.

4-11. (*) Following up on the previous problem, give an $O(n + m)$ algorithm that finds a deletion order for the n vertices such that no deletion disconnects the graph. (Hint: think DFS/BFS.)

4-12. (*) Let G be a weighted *directed* graph with n vertices and m edges, where all edges have positive weight. A directed cycle is a directed path that starts and ends at the same vertex and contains at least one edge. Give an $O(n^3)$ algorithm to find a directed cycle in G of minimum total weight. Partial credit will be given for an $O(n^2 m)$ algorithm.

4-13. (*) Suppose we are *given* the minimum spanning tree T of a given graph G (with n vertices and m edges) and a new edge $e = (u, v)$ of weight w that we will add to G. Give an efficient algorithm to find the minimum spanning tree of the graph $G + e$. Your algorithm should run in $O(n)$ time to receive full credit, although slower but correct algorithms will receive partial credit.

4-14. (*) (a) Let T be a minimum spanning tree of a weighted graph G. Construct a new graph G' by adding a weight of k to every edge of G. Do the edges of T form a minimum spanning tree of G'? Prove the statement or give a counterexample.
(b) Let $P = \{s, \ldots, t\}$ describe a shortest weighted path between vertices s and t of a weighted graph G. Construct a new graph G' by adding a weight of k to every edge of G. Does P describe a shortest path from s to t in G'? Prove the statement or give a counterexample.

4-15. (*) In certain graph problems, vertices have can have weights instead of or in addition to the weights of edges. Let C_v be the cost of vertex v, and $C_{(x,y)}$ the cost of the edge (x, y). This problem is concerned with finding the cheapest path between vertices a and b in a graph G. The cost of a path is the sum of the costs of the edges and vertices encountered on the path.

- Suppose that each edge in the graph has a weight of zero (while nonedges have a cost of ∞). Assume that $C_v = 1$ for all vertices $1 \leq v \leq n$ (i.e. all vertices have the same cost). Give an *efficient* algorithm to find the cheapest path from a to b and its time complexity. For partial credit, give a less efficient but correct algorithm.

- Now suppose that the vertex costs are not constant (but are all positive) and the edge costs remain as above. Give an *efficient* algorithm to find the cheapest path from a to b and its time complexity. For partial credit, give a less efficient but correct algorithm.

- Now suppose that both the edge and vertex costs are not constant (but are all positive). Give an *efficient* algorithm to find the cheapest path from a to b and its time complexity. For partial credit, give a less efficient but correct algorithm.

4-16. (*) Devise and analyze an algorithm that takes a weighted graph G and finds the smallest change in the cost of a nonMST edge that causes a change in the minimum spanning tree of G. Your algorithm must be correct and run in polynomial time.

4-17. An *arborescence* of a directed graph G is a rooted tree such that there is a directed path from the root to every other vertex in the graph. Give an efficient and correct algorithm to test whether G contains an arborescence, and its time complexity.

4-18. (**) The war story of Section 4.3 describes an algorithm for constructing the dual graph of the triangulation efficiently, although it does not guarantee linear time. Give a worst-case linear algorithm for the problem.

Implementation Challenges

4-1. Airline flight schedules define a natural graph, where the vertices are the airports and there is an edge between any two airports with direct flights between them whose weight is proportional to the distance between them. An extensive airplane data set due to Roberto Tammasia is available from http://www.cs.sunysb.edu/~algorith or the enclosed WWW/CD-ROM. Write a program that explicitly constructs the airport graph from this data set.

4-2. This problem is a follow-up to the exercise above. Changing planes repeatedly on connecting flights can be a hassle. Develop and implement an algorithm that finds the fewest flights needed to get from airport A to B, regardless of waiting time.

4-3. (*) This problem is a follow-up to the exercise above. Develop and implement an algorithm that finds the flight plan from airport A to B that minimizes the total distance traveled.

4-4. (*) This problem is a follow-up to the exercise above. Suppose that we must arrive at airport B at time T for an important scheduled meeting. Develop and implement an algorithm that finds the latest time one can leave airport A in time to make the meeting.

4-5. (*) This problem is a follow-up to the exercise above. In order to take advantage of a frequent flyer program, we might want to fly only on a particular airline. How can we modify the above solutions so as to accommodate such queries?

4-6. (**) This problem is a follow-up to the exercise above. A really devout frequent flyer might want to find the *longest* flight plan between airports A and B, so as to maximize the number of miles they get credit for. Develop and implement an algorithm to find the longest such route.

5

Combinatorial Search and Heuristic Methods

We have seen how clever algorithms can reduce the complexity of sorting from $O(n^2)$ to $O(n \log n)$, which is good. However, the algorithmic stakes can be even higher for combinatorially explosive problems, whose time grows exponentially in the size of the problem. Looking back at Figure 1-7 will make clear the limitations of exponential-time algorithms on even modest-sized problems.

By using exhaustive search techniques, we can solve small problems to optimality, although the time complexity may be enormous. For certain applications, it may well pay to spend extra time to be certain of the optimal solution. A good example occurs in testing a circuit or a program on all possible inputs. You can prove the correctness of the device by trying all possible inputs and verifying that they give the correct answer. Proving such correctness is a property to be proud of. However, claiming that it works correctly on all the inputs you tried is worth much, much less.

In this section, we present backtracking as a technique for listing all configurations representing possible solutions for a combinatorial algorithm problem. We then discuss techniques for pruning search that significantly improve efficiency by eliminating irrelevant configurations from consideration. We illustrate the power of clever pruning techniques to speed up real search applications. For problems that are too large to contemplate using brute-force combinatorial search, we introduce heuristic methods such as simulated annealing. Such heuristic methods are an important weapon in the practical algorist's arsenal.

The take-home lessons from this chapter are:

- Combinatorial search, augmented with tree pruning techniques, can be used to find the optimal solution of small optimization problems. How small depends upon the specific problem, but the size limit is likely to be somewhere between $15 \leq n \leq 50$ items.

- Clever pruning techniques can speed up combinatorial search to an amazing extent. Proper pruning will have a greater impact on search time than any other factor.

- Simulated annealing is a simple but effective technique to efficiently obtain good but not optimal solutions to combinatorial search problems.

5.1 Backtracking

Backtracking is a systematic way to go through all the possible configurations of a space. These configurations may be all possible arrangements of objects (permutations) or all possible ways of building a collection of them (subsets). Other applications may demand enumerating all spanning trees of a graph, all paths between two vertices, or all possible ways to partition the vertices into color classes.

What these problems have in common is that we must generate each one of the possible configurations exactly once. Avoiding both repetitions and missing configurations means that we must define a systematic generation order among the possible configurations. In combinatorial search, we represent our configurations by a vector $A = (a_1, a_2, ..., a_n)$, where each element a_i is selected from an ordered set of possible candidates S_i for position i. As shown below, this representation is general enough to encode most any type of combinatorial object naturally.

The search procedure works by growing solutions one element at a time. At each step in the search, we will have constructed a partial solution with elements fixed for the first k elements of the vector, where $k \leq n$. From this partial solution $(a_1, a_2, ..., a_k)$, we will construct the set of possible candidates S_{k+1} for the $(k + 1)$st position. We will then try to extend the partial solution by adding the next element from S_{k+1}. So long as the extension yields a longer partial solution, we continue to try to extend it.

However, at some point, S_{k+1} might be empty, meaning that there is no legal way to extend the current partial solution. If so, we must *backtrack*, and replace a_k, the last item in the solution value, with the next candidate in S_k. It is this backtracking step that gives the procedure its name:

Backtrack(A)

    Compute S_1, the set of candidate first elements of solution A.
    $k = 1$
    while $k > 0$ do
        while $S_k \neq \emptyset$ do (*advance*)
            a_k = the next element from S_k
            $S_k = S_k - a_k$
            if $A = (a_1, a_2, ..., a_k)$ is a solution, report it.
            $k = k + 1$
            compute S_k, the set of candidate kth elements of
                solution A.
        $k = k - 1$ (*backtrack*)

Backtracking constructs a tree of partial solutions, where each vertex is a partial solution. There is an edge from x to y if node y was created by advancing from x. This tree of partial solutions provides an alternative way to think about backtracking, for the process of constructing the solutions corresponds exactly to doing a depth-first traversal of the backtrack tree. Viewing backtracking as depth-first search yields a natural recursive implementation of the basic algorithm:

Backtrack-DFS(A, k)
     if $A = (a_1, a_2, ..., a_k)$ is a solution, report it.
     else
          $k = k + 1$
          compute S_k
          while $S_k \neq \emptyset$ do
               a_k = an element in S_k
               $S_k = S_k - a_k$
               Backtrack(a, k)

Although a breadth-first search could also be used to enumerate all solutions, depth-first search is greatly preferred because of the amount of storage required. In depth-first search, the current state of the search is completely represented by the path from the root to the current search node, which requires space proportional to the *height* of the tree. In breadth-first search, the queue stores all the nodes at the current level, which is proportional to the *width* of the search tree. For most interesting problems, the width of the tree will grow exponentially in its height.

To really understand how backtracking works, you must see how such objects as permutations and subsets can be constructed by defining the right state spaces. Examples of several state spaces are described below.

5.1.1 Constructing All Subsets

To design a suitable state space for representing a collection of combinatorial objects, it is important to know how many objects you will need to represent. How many subsets are there of an n-element set, say the integers $\{1, \ldots, n\}$? There are exactly two subsets for $n = 1$ namely $\{\}$ and $\{1\}$, four subsets for $n = 2$, and eight subsets for $n = 3$. Since each new element doubles the number of possibilities, there are 2^n subsets of n elements.

Each subset is described by stating which elements are in it. Therefore, to construct all 2^n subsets, we can set up an array/vector of n cells, where the value of a_i is either true or false, signifying whether the ith item is or is not in the given subset. To use the notation of the general backtrack algorithm, $S_k = (\text{true}, \text{false})$, and A is a solution whenever $k \geq n$.

Using this state space representation, the backtracking algorithm constructs the following sequence of partial solutions in finding the subsets of $\{1, 2, 3\}$. Final solutions, i.e. complete subsets, are marked with a $*$. False choices correspond to dashes in the partial solution, while true in position i is denoted by i itself:

$$(1) \rightarrow (1,2) \rightarrow (1,2,3)* \rightarrow (1,2,-)* \rightarrow (1,-) \rightarrow (1,-,3)* \rightarrow$$

$$(1,-,-)* \rightarrow (1,-) \rightarrow (1) \rightarrow (-) \rightarrow (-,2) \rightarrow (-,2,3)* \rightarrow (-,2,-)* \rightarrow$$

$$(-,-) \rightarrow (-,-,3)* \rightarrow (-,-,-)* \rightarrow (-,-) \rightarrow (-) \rightarrow ()$$

Trace through this example carefully to make sure you understand the backtracking procedure. The problem of generating subsets is more thoroughly discussed in Section 8.3.5.

5.1.2   Constructing All Permutations

To design a suitable state space for representing permutations, we start by counting them. There are n distinct choices for the value of the first element of a permutation of $\{1,\ldots,n\}$. Once we have fixed this value of a_1, there are $n-1$ candidates remaining for the second position, since we can have any value except a_1 (repetitions are forbidden). Repeating this argument yields a total of $n! = \prod_{i=1}^{n} i$ distinct permutations.

This counting argument suggests a suitable representation. To construct all $n!$ permutations, set up an array/vector A of n cells. The set of candidates for the ith position will be the set of elements that have not appeared in the $(i-1)$ elements of the partial solution, corresponding to the first $i-1$ elements of the permutation. To use the notation of the general backtrack algorithm, $S_k = \{1,\ldots,n\} - A$. The vector A contains a full solution whenever $k = n+1$. This representation generates the permutations of $\{1,2,3\}$ in the following order:

$$(1) \rightarrow (1,2) \rightarrow (1,2,3)* \rightarrow (1,2) \rightarrow (1) \rightarrow (1,3) \rightarrow (1,3,2)* \rightarrow$$

$$(1,3) \rightarrow (1) \rightarrow () \rightarrow (2) \rightarrow (2,1) \rightarrow (2,1,3)* \rightarrow (2,1) \rightarrow$$

$$(2) \rightarrow (2,3) \rightarrow (2,3,1)* \rightarrow (2,3) \rightarrow ()(2) \rightarrow () \rightarrow (3) \rightarrow$$

$$(3,1) \rightarrow (3,1,2)* \rightarrow (3,1) \rightarrow (3) \rightarrow (3,2)(3,2,1)* \rightarrow (3,2) \rightarrow (3) \rightarrow ()$$

The problem of generating permutations is more thoroughly discussed in Section 8.3.4.

5.1.3   Constructing All Paths in a Graph

Enumerating all the simple paths from s to t through a given graph is a somewhat more complicated problem than listing permutations or subsets. Unlike the earlier problems, there is no explicit formula that counts the number of solutions as a function of the number of edges or vertices, because the number of paths depends upon the structure of the graph.

Since the starting point of any path from s to t is always s, S_1 must be $\{s\}$. The set of possible candidates for the second position are the vertices v such that (s,v) is an edge of the graph, for the path wanders from vertex to vertex using edges to define the legal steps. In general, S_{k+1} consists of the set of vertices

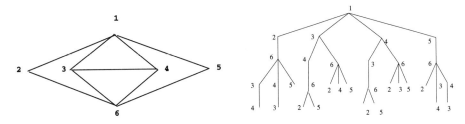

Figure 5-1. The search tree enumerating all simple paths from vertex 1 in the graph

adjacent to a_k that have not been used in the partial solution A. We can report a successful path whenever $a_k = t$. The solution vector A must have room for all n vertices, although most paths are likely to be shorter than this. Figure 5-1 shows the search tree giving all paths from a particular vertex in an example graph.

5.2 Search Pruning

Backtracking ensures correctness by enumerating all possibilities. For example, a correct algorithm to find the optimal traveling salesman tour could enumerate all $n!$ permutations of n vertices of the graph and select the best one. For each permutation, we could check whether each of the n edges implied in the tour really exists in the graph G, and if so, sum the weights of these edges together.

For most graphs, however, it would be pointless to construct all the permutations first and then analyze them later. Suppose we started our search from vertex v_1, and it happened that edge (v_1, v_2) was not in G. Enumerating all the $(n - 2)!$ permutations beginning with v_1, v_2 would be a complete waste of effort. Much better would be to prune the search after v_1, v_2 and continue next with v_1, v_3. By carefully restricting the set of next elements to reflect only the moves that are legal from the current partial configuration, we reduce the search complexity significantly.

Pruning is the technique of cutting off search the instant we have established that this partial solution cannot be extended into the solution that we want. For example, in our traveling salesman search program, we seek the cheapest tour that visits all vertices before returning to its starting position. Suppose that in the course of our search we find a tour t whose cost is C_t. As the search continues, perhaps we will find a partial solution $a_1, \ldots, a_k$, where $k < n$ and the sum of the edges on this partial tour is $C_A > C_t$. Can there be any reason to continue exploring this node any further? No, assuming all edges have positive cost, because any tour with the prefix $a_1, \ldots, a_k$ will have cost greater than tour t, and hence is doomed to be nonoptimal. Cutting away such failed partial tours as soon as possible can have an enormous impact on running time.

Exploiting symmetry is a third avenue for reducing combinatorial search. It is clearly wasteful to evaluate the same candidate solution more than once, because we will get the exact same answer each time we consider it. Pruning away partial solutions identical to those previously considered requires recognizing underlying symmetries in the search space. For example, consider the state of our search for an optimal TSP tour after we have tried all partial positions beginning with v_1. Can it pay to continue the search with partial solutions beginning with v_2? No. Any tour starting and ending at v_2 can be viewed as starting and ending at v_1 or any other vertex, for these tours are cycles. There are thus only $(n - 1)!$ distinct tours on n vertices, not $n!$. By restricting the first element of the tour to always be v_1, we save a factor of n in time without missing any interesting solutions. Detecting such symmetries can be subtle, but once identified they can usually be easily exploited by a search program.

5.3 Bandwidth Minimization

To better demonstrate the power of pruning and symmetry detection, let's apply these ideas to producing a search program that solves the *bandwidth minimization* problem, discussed in detail in catalog Section 8.2.2. I annually run competitions for the fastest bandwidth-minimization program for students in my algorithms courses; the timings below are drawn from these experiences.

The *bandwidth problem* takes as input a graph G, with n vertices and m edges. The goal is to find a permutation of the vertices on the line that minimizes the maximum length of any edge. Figure 5-2 gives two distinct layouts of a complete binary tree on 15 vertices. The clean, neat layout on the top has a longest edge of length 4, but the seemingly cramped layout on the bottom realizes the optimal bandwidth of 3.

The bandwidth problem has a variety of applications, including circuit layout, linear algebra, and optimizing memory usage in hypertext documents. The problem is NP-complete, which implies that no polynomial time worst-case algorithm is known for the problem. It remains NP-complete even for very restricted classes of trees.

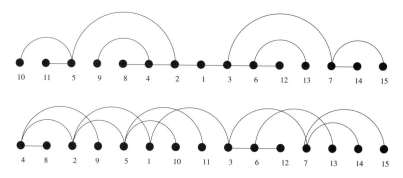

Figure 5-2. A pretty bandwidth-4 layout of a binary tree atop an ugly bandwidth-3 layout

Since the bandwidth problem seeks a particular permutation, a backtracking program that iterates through all the $n!$ possible permutations and computes the length of the longest edge for each gives a straightforward $O(n! \cdot m)$ algorithm. Depending upon how well it is programmed, and how fast a machine it is running on, such an algorithm can expect to solve instances of approximately 8 to 12 vertices within one CPU minute.

To speed up this search, we can try to exploit symmetry. For any permutation p, its reverse permutation will realize the exact same bandwidth, since the length of each edge is the same. Hence we can immediately eliminate half of our search space. The reverse copies are easily removed by placing the leftmost and rightmost elements of the permutation as the first two elements of the vector and then pruning if $a_1 > a_2$. Because we are dealing with an exponential search, removing a single factor of two can only be of limited usefulness. Such symmetry elimination might add one to the size of the problem we can do within one CPU minute.

For more serious speedups, we need to prune partial solutions. Say we have found a permutation p that yields a longest edge of b_p. By definition, $b_{opt} \leq b_p$. Suppose that among the elements in a partial layout $a_1, \ldots, a_k$, where $k < n$, there is an edge that is at least b_p in length. Can this partial solution expand to provide a better bandwidth solution? Of course not! By pruning the search the instant we have created a long edge, typical instances of 15 to 20 vertices can be solved in one CPU minute, thus providing a substantial improvement.

Efforts to further improve the search algorithm must strive for even greater pruning. By using a heuristic method to find the best solution we can before starting to search, we save time by realizing early length cutoffs. In fact, most of the effort in a combinatorial search is typically spent *after* the optimal solution is found, in the course of proving that no better answer exists. By observing that the optimal bandwidth solution must always be at least half the degree of any vertex (think about the incident edges), we have a lower bound on the size of the optimal solution. We can terminate search soon as we find a solution matching the lower bound.

One limitation of this pruning strategy is that only partial solutions of length $> b$ can be pruned, where b is the bandwidth of the best solution to date, since we must place $b + 1$ vertices before we can generate any edges of length at least b. To achieve earlier cutoffs, we can alternately fill in the leftmost and rightmost slots of the configuration, instead of always proceeding from the left. This way, whenever there is an edge between a vertex on the left side and a vertex on the right side, this edge is likely long enough to achieve a cutoff. Pruning can easily occur while positioning the second vertex in the solution vector.

Using these enhancements, top-notch programs are capable of solving typical problems on up to 30 vertices consistently within one CPU minute, operating literally millions of times faster than unpruned, untuned efforts. The speed difference between the final and initial versions of the program dwarf the difference between a supercomputer and a microcomputer. Clever search algorithms can easily have a bigger impact on performance than expensive hardware.

5.4 War Story: Covering Chessboards

Every researcher dreams of solving a classical problem, one that has remained open and unsolved for over a hundred years. There is something romantic about communicating across the generations, being part of the evolution of science, helping to climb another rung up the ladder of human progress. There is also a pleasant sense of smugness that comes from figuring out how to do something that nobody else could do before you.

There are several possible reasons why a problem might stay open for such a long period of time. Perhaps the problem is so difficult and profound that it requires a uniquely powerful intellect to solve. A second reason is technological— the ideas or techniques required to solve the problem may not have existed when the problem was first posed. A final possibility is that no one may have cared enough about the problem in the interim to seriously bother with it. Once, I was involved in solving a problem that had been open for over a hundred years. Decide for yourself which reason best explains why.

Chess is a game that has fascinated mankind for thousands of years. In addition, it has inspired a number of combinatorial problems of independent interest. The combinatorial explosion was first recognized in the legend that the inventor of chess demanded as payment one grain of rice for the first square of the board, and twice the amount of the ith square for the $(i + 1)$st square, for a total of $\sum_{i=1}^{64} 2^i = 2^{65} - 1 = 36,893,488,147,419,103,231$ grains. In beheading him, the wise king first established pruning as a technique for dealing with the combinatorial explosion.

In 1849, Kling posed the question of whether all 64 squares on the board can be simultaneously threatened by an arrangement of the eight main pieces on the chess board—the king, queen, two knights, two rooks, and two oppositely colored bishops. Configurations that simultaneously threaten 63 squares, such as those in Figure 5-3, have been known for a long time, but whether this was the best possible remained an open problem. This problem seemed ripe for solution by exhaustive combinatorial searching, although whether it was solvable would depend upon the size of the search space.

Consider the 8 main pieces in chess (king, queen, two rooks, two bishops, two knights). How many ways can they be positioned on a chessboard? The trivial bound is $64!/(64 - 8)! = 178,462,987,637,760 \approx 10^{15}$ positions. Anything much

Figure 5-3. Configurations covering 63 but not 64 squares

Figure 5-4. The ten unique positions for the queen, with respect to symmetry

larger than about 10^9 positions would be unreasonable to search on a modest computer in a modest amount of time.

Getting the job done would require significant pruning. The first idea is to remove symmetries. Considering the orthogonal and diagonal symmetries, there are only ten distinct positions for the queen, shown in Figure 5-4.

Once the queen is placed, there are 2,080 distinct ways to position a pair of rooks or knights, 64 places to locate the king, and 32 spots for each of the white and black bishops. Thus to perform an exhaustive search, we must test $2,835,349,504,000 \approx 10^{13}$ distinct positions, still much too large to try.

We could use backtracking to construct all of the positions, but we had to find a way to prune the search space significantly if we could hope to finish in our lifetime. Pruning the search meant that we needed a quick way to prove, for a partially filled-in position, that there was no possible way to complete it so as to cover all 64 squares. Suppose we had already placed seven pieces on the board, and together they covered all but 10 squares of the board. Say the remaining piece was the king. Is there any possible position to place the king so that all squares are threatened? The answer must be no, because according to the rules of chess the king can threaten at most eight squares. There can be no reason to bother testing any of the subsequent positions. By pruning away these positions, we might win big.

Optimizing this pruning strategy required carefully ordering the evaluation of the pieces. Each piece could threaten a certain maximum number of squares: the queen 27, the king 8, the rook 14, and the bishop 13. To maximize the chances of a cutoff, we would want to insert the pieces in decreasing order of mobility. Whenever the number of unthreatened squares exceeds the sum of the maximum coverage of the unplaced pieces, we can prune. This sum is minimized by using the decreasing order of mobility.

When we implemented backtrack search with this pruning strategy, we found that it eliminated over 95% of the search space. After optimizing our move generation, our program could search over 1,000 positions per second. But this was still too slow, for $10^{12}/10^3 = 10^9$ seconds meant 1,000 days! Although we might further tweak the program to speed it up by an order of magnitude or so, what we really needed was to find a way to prune more nodes.

Effective pruning meant eliminating large numbers of positions at a single stroke. Our previous attempt was too weak. What if instead of placing up to

Figure 5–5. Weakly covering 64 squares

eight pieces on the board simultaneously, we placed *more* than eight pieces. Obviously, the more pieces we placed simultaneously, the less likely it would be that they didn't threaten all 64 squares. But *if* they didn't cover, all subsets of eight distinct pieces from the set couldn't possibly threaten all squares. The potential existed to eliminate a vast number of positions by pruning a single node.

Thus the nodes of our search tree corresponded to chessboards that could have any number of pieces, and more than one piece on a square. For a given board, we would distinguish two kinds of attack on a square: *strong* and *weak*. The notion of strong attack corresponds to the usual notion of attack in chess. A square is *weakly attacked* if the square is strongly attacked by some subset of the board, that is, weak attack ignores any possible blocking effects of intervening pieces. All 64 squares can be weakly attacked with eight pieces, as shown in Figure 5-5.

Our algorithm consists of two passes. The first pass lists all boards such that every square is weakly attacked. The second pass filters the list by considering blocking and reports any boards with *n* or fewer safe squares. The advantage of separating weak and strong attack computations is that weak attack is faster to compute (no blocking to worry about), and yet the strong attack set is always a subset of the weak attack set. Whenever there was a non-weakly-threatened square, the position could be pruned.

Figure 5–6. Seven pieces suffice when superimposing queen and knight.

This program was efficient enough to complete the search on a machine as slow as a 1988-era IBM PC-RT in under one day. More details of our searching procedure and results appear in our paper [RHS89]. It did not find a single position covering all 64 squares with the bishops on opposite colored squares. However, our program showed that it is possible to cover the board with *seven* pieces if a queen and a knight can occupy the same square, as shown in Figure 5-6.

The take-home lesson of this war story should be clear. Clever pruning can make short work of surprisingly hard combinatorial search problems.

5.5 Heuristic Methods

The techniques we have discussed thus far seek to find the optimal answer to a combinatorial problem as quickly as possible. Traditional algorithmic methods fail whenever the problem is provably hard (as discussed in Chapter 6), or the problem is not clean enough to lead to a nice formulation.

Heuristic methods provide a way to approach difficult *combinatorial optimization* problems. Combinatorial search gives us a method to construct possible solutions and find the best one, given a function that measures how good each candidate solution is. However, there may be no algorithm to find the best solution short of searching all configurations. Heuristic methods such as simulated annealing, genetic algorithms, and neural networks provide general ways to search for good but not optimal solutions.

In this section we discuss such heuristic methods. Each of these three techniques relies on a simple model of a real-world physical process. We devote the bulk of our attention to simulated annealing, which is the easiest method to apply in practice, as well as the most reliable.

5.5.1 Simulated Annealing

The inspiration for simulated annealing comes from the physical process of cooling molten materials down to the solid state. When molten steel is cooled too quickly, cracks and bubbles form, marring its surface and structural integrity. To end up with the best final product, the steel must be cooled slowly and evenly. *Annealing* is a metallurgical technique that uses a disciplined cooling schedule to efficiently bring the steel to a low-energy, optimal state.

In thermodynamic theory, the energy state of a system is described by the energy state of each of the particles constituting it. The energy state of each particle jumps about randomly, with such transitions governed by the temperature of the system. In particular, the probability $P(e_i, e_j, T)$ of transition from energy e_i to e_j at temperature T is given by

$$P(e_i, e_j, T) = e^{(e_i - e_j)/(k_B T)}$$

where k_B is a constant, called Boltzmann's constant.

What does this formula mean? Consider the value of the exponent under different conditions. The probability of moving from a high-energy state to a lower-energy state is very high. However, there is also a nonzero probability of accepting a transition into a high-energy state, with small energy jumps much more likely than big ones. The higher the temperature, the more likely such energy jumps will occur.

What relevance does this have for combinatorial optimization? A physical system, as it cools, seeks to go to a minimum-energy state. For any discrete set of particles, minimizing the total energy is a combinatorial optimization problem. Through random transitions generated according to the above probability distribution, we can simulate the physics to solve arbitrary combinatorial optimization problems.

```
Simulated-Annealing()
        Create initial solution S
        Initialize temperature t
        Repeat
                for i = 1 to iteration-length do
                        Generate a random transition from S to S_i
                        If (C(S) ≥ C(S_i)) then S = S_i
                        else if (e^{(C(S)−C(S_i))/(k·t)} > random[0, 1)) then S = S_i
                Reduce temperature t
        until (no change in C(S))
        Return S
```

There are three components to any simulated annealing algorithm for combinatorial search:

- *Concise problem representation* – The problem representation includes both a representation of the solution space and an appropriate and easily computable cost function $C(s)$ measuring the quality of a given solution.

- *Transition mechanism between solutions* – To move from one state to the next, we need a collection of simple transition mechanisms that slightly modify the current solution. Typical transition mechanisms include swapping the position of a pair of items or inserting/deleting a single item. Ideally, the effect that these incremental changes have on measuring the quality of the solution can be computed incrementally, so cost function evaluation takes time proportional to the size of the change (typically constant) instead of linear in the size of the solution.

- *Cooling schedule* – These parameters govern how likely we are to accept a bad transition as a function of time. At the beginning of the search, we are eager to use randomness to explore the search space widely, so the probability of accepting a negative transition is high. As the search progresses, we seek to limit transitions to local improvements and optimizations. The cooling schedule can be regulated by the following parameters:

- *Initial system temperature* – Typically $t_1 = 1$.
- *Temperature decrement function* – Typically $t_k = \alpha \cdot t_{k-1}$, where $0.8 \leq \alpha \leq 0.99$. This implies an exponential decay in the temperature, as opposed to a linear decay.
- *Number of iterations between temperature change* – Typically, 100 to 1,000 iterations might be permitted before lowering the temperature.
- *Acceptance criteria* – A typical criterion is to accept any transition from s_i to s_{i+1} when $C(s_{i+1}) > C(s_i)$ and to accept a negative transition whenever

$$e^{-\frac{(C(s_{i+1})-C(s_i))}{c \cdot t_i}} \geq r,$$

where r is a random number $0 \leq r < 1$. The constant c normalizes this cost function, so that almost all transitions are accepted at the starting temperature.

- *Stop criteria* – Typically, when the value of the current solution has not changed or improved within the last iteration or so, the search is terminated and the current solution reported.

Creating the proper cooling schedule is somewhat of a trial and error process. It might pay to start from an existing implementation of simulated annealing, pointers to which are provided in Section 8.2.5.

We provide several examples below to demonstrate how these components can lead to elegant simulated annealing algorithms for real combinatorial search problems.

5.5.1.1 Traveling Salesman Problem

The solution space for traveling salesman consists of the set of all $(n-1)!$ possible circular permutations of the vertices. A candidate solution can thus be represented using an array S of $n-1$ vertices, where S_i defines the $(i+1)$st vertex on the tour starting from v_1. The cost function evaluating a candidate solution is equally straightforward, for we can sum up the costs of the edges defined by S.

The most obvious transition mechanism would be to swap the current tour positions of a random pair of vertices S_i and S_j. This changes up to eight edges on the tour, deleting the edges currently adjacent to both S_i and S_j, and adding their replacements. Better would be to swap two edges on the tour with two others that replace it, as shown in Figure 5-7. Since only four edges change in the tour, the transitions can be performed and evaluated faster. Faster transitions mean that we can evaluate more positions in the given amount of time.

In practice, problem-specific heuristics for TSP outperform simulated annealing, but the simulated annealing solution works admirably, considering it uses very little knowledge about the problem.

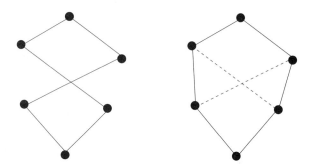

Figure 5-7. Improving a TSP tour by swapping a pair of edges

5.5.1.2   Maximum Cut

For a weighted graph G, the maximum cut problem seeks to partition the vertices into sets V_1 and V_2 so as to maximize the weight (or number) of edges with one vertex in each set. When the graph specifies an electronic circuit, the maximum cut in the graph defines the largest amount of data communication that can take place in the circuit simultaneously. As discussed in catalog Section 8.5.6, maximum cut is an NP-complete version of graph partitioning.

How can we formulate maximum cut for simulated annealing? The solution space consists of all 2^{n-1} possible vertex partitions; we save a factor of two over all vertex subsets because we can assume that vertex v_1 is fixed to be on the left side of the partition. The subset of vertices accompanying it can be represented using a bit vector. The cost of a solution will be the sum of the weights cut in the current configuration. A natural transition mechanism is to select one vertex at random and move it across the partition by simply flipping the corresponding bit in the bit vector. The change in the cost function will be the weight of its old neighbors minus the weight of its new neighbors, so it can be computed in time proportional to the degree of the vertex.

This kind of simple, natural modeling is the right type of heuristic to seek in practice.

5.5.1.3   Independent Set

An independent set of a graph G is a subset of vertices S such that there is no edge with both endpoints in S. The maximum independent set of a graph is the largest such empty induced subgraph. The need to find large independent sets arises in dispersion problems associated with facility location and coding theory, as discussed in catalog Section 8.5.2.

The natural state space for a simulated annealing solution would be all 2^n subsets of the vertices, represented as a bit vector. As with maximum cut above, a simple transition mechanism would be to add or delete one vertex from S.

One natural cost function for subset S might be 0 if S contains an edge, and $|S|$ if it is indeed an independent set. This function ensures that we work towards an independent set at all times. However, this condition is strict enough that we are liable to move only in a narrow portion of the possible search space. More flexibility in the search space and quicker cost function computations can result from allowing nonempty graphs at the early stages of cooling. Better in practice would be a cost function like $C(S) = |S| - \lambda \cdot m_S/T$, where λ is a constant, T is the temperature, and m_S is the number of edges in the subgraph induced by S. The dependence of $C(S)$ on T ensures that the search will drive the edges out faster as the system cools.

5.5.1.4   Circuit Board Placement

In designing printed circuit boards, we are faced with the problem of positioning modules (typically integrated circuits) on the board. Desired criteria in a layout include (1) minimizing the area or aspect ratio of the board, so that it properly fits within the allotted space, and (2) minimizing the total or longest wire length in connecting the components. Circuit board placement is an example of the kind of messy, multicriterion optimization problems for which simulated annealing is ideally suited.

Formally, we are given a collection of a rectangular modules $r_1, \ldots, r_n$, each with associated dimensions $h_i \times l_i$. Further, for each pair of modules r_i, r_j, we are given the number of wires w_{ij} that must connect the two modules. We seek a placement of the rectangles that minimizes area and wire-length, subject to the constraint that no two rectangles overlap each other.

The state space for this problem must describe the positions of each rectangle. To provide a discrete representation, the rectangles can be restricted to lie on vertices of an integer grid. Reasonable transition mechanisms including moving one rectangle to a different location, or swapping the position of two rectangles. A natural cost function would be

$$C(S) = \lambda_{area}(S_{height} \cdot S_{width}) + \sum_{i=1}^{n} \sum_{j=1}^{n} (\lambda_{wire} \cdot w_{ij} \cdot d_{ij} + \lambda_{overlap}(r_i \cap r_j))$$

where λ_{area}, λ_{wire}, and $\lambda_{overlap}$ are constants governing the impact of these components on the cost function. Presumably, $\lambda_{overlap}$ should be an inverse function of temperature, so after gross placement it adjusts the rectangle positions so they are distinct.

Simulated annealing performs well on such module placement problems. Indeed, a similar application appeared in the original paper on simulated annealing [KGV83]. More details on these and other applications appear in [AK89].

5.5.2   Neural Networks

Neural networks are a computational paradigm inspired by the architecture of the human brain. The intuition is that since brains are good at solving problems, machines built in the same way should be, too.

The basic computational component of the brain is a neuron, a simple unit that produces a nonlinear, weighted sum of its inputs, which are connections from other neurons. *Neural networks* are weighted digraphs with neurons as vertices and weights on edges denoting the connection strength of the pair.

Brains are very good at learning and recognizing certain patterns. Learning in brains seems to work by adding connections between different pairs of neurons and changing the strengths of the connections. Modifying connection strength in response to training examples provides a natural way to "teach" a neural network.

Although there have been attempts to apply neural networks to solving combinatorial optimization problems, the successes have been rather limited. Simulated annealing is a much more straightforward and efficient approach to optimization.

Neural networks have been more successful in classification and forecasting applications, such as optical character recognition, gene prediction, and stock-market time-series prediction. A set of features for the given patterns is selected, and each training example is represented in terms of its features. The network is trained on a series of positive and negative examples, with the strengths of the connections adjusted to recognize these examples. Output cells for each class of item are provided and the strength of these cells on a given input used to determine the classification. Once the network is trained, feature vectors corresponding to unknown items can be entered and a classification made.

Because neural networks are black boxes, with the strength of edges adjusted only by the training examples, there is usually no way to figure out exactly why they are making the decisions that they are. A particularly amusing instance where this led to trouble is reported in Section 8.6.13. Still, they can be useful in certain pattern-recognition applications.

5.5.3   Genetic Algorithms

Genetic algorithms draw their inspiration from evolution and natural selection. Through the process of natural selection, organisms adapt to optimize their chances for survival in a given environment. Random mutations occur to the genetic description of an organism, which is then passed on to its children. Should a mutation prove helpful, these children are more likely to survive to reproduce. Should it be harmful, these children are less likely to reproduce, so the bad trait will die with them.

Genetic algorithms maintain a "population" of solution candidates for the given problem. Elements are drawn at random from this population and allowed to "reproduce", by combining some aspects of the two parent solutions. The probability that an element is chosen to reproduce is based on its "fitness", essentially a function of the cost of the solution it represents. Eventually, unfit elements die from the population, to be replaced by successful-solution offspring.

The idea behind genetic algorithms is extremely appealing. However, they just don't seem to work as well on practical combinatorial optimization problems as simulated annealing does. There are two primary reasons for this. First, it is quite unnatural to model most applications in terms of genetic operators like mutation and crossover on bit strings. The pseudobiology adds another level of complexity between you and your problem. Second, genetic algorithms take a very long time on nontrivial problems. The crossover and mutation operations make no real use of problem-specific structure, so a large fraction of transitions lead to inferior solutions, and convergence is slow. Indeed, the analogy with evolution, where significant improvements require millions of years, can be quite appropriate.

We will not discuss genetic algorithms further, in order to discourage you from considering them for your applications. However, pointers to implementations of genetic algorithms are provided in Section 8.2.5 if you really insist on playing with them.

5.6 War Story: Annealing Arrays

The war story of Section 2.8 reported how we used advanced data structures to simulate a new method for sequencing DNA. Our method, interactive sequencing by hybridization (SBH), involved building arrays of specific oligonucleotides on demand. Although the simulation results were very promising to us, most biologists we encountered were suspicious. They needed to see our technique proven in the lab before they would take it seriously.

But we got lucky. A biochemist at Oxford University got interested in our technique, and moreover he had in his laboratory the equipment we needed to test it out. The Southern Array Maker [Sou96], manufactured by Beckman Instruments, could prepare discrete oligonucleotide sequences in 64 parallel rows across a polypropylene substrate. The device constructs arrays by appending single characters to each cell along specific rows and columns of arrays. Figure 5-8 shows how to construct an array of all $2^4 = 16$ purine (A or G) 4-mers by building the prefixes along rows and the suffixes along columns. This technology provided an ideal environment for testing the feasibility of interactive SBH in a laboratory, because with proper programming it gave an inexpensive way to fabricate a wide variety of oligonucleotide arrays on demand.

But we had to provide the proper programming. Fabricating complicated arrays requires solving a difficult combinatorial problem. We were given as input a set S of n strings (representing oligonucleotides) to fabricate an $m \times m$ array (where $m = 64$ on the Southern apparatus). We had to produce a schedule of row and column commands to realize the set of strings S. We proved that

	AA	AG	GA	GG
AA	AAAA	AAAG	AAGA	AAGG
AG	AGAA	AGAG	AGGA	AGGG
GA	GAAG	GAAG	GAGA	GAGG
GG	GGAA	GGAG	GGGA	GGGG

Figure 5-8. A prefix-suffix array of all purine 4-mers.

the problem of designing dense arrays was NP-complete, but that didn't really matter. My student Ricky Bradley and I had to solve it anyway.

"If it's hard, it's hard. We are going to have to use a heuristic," I told him. "So how do we model this problem?"

"Well, for each string we can identify the possible prefix and suffix pairs that will realize it. For example, the string 'ACC' can be realized in four different ways: prefix '' and suffix 'ACC', prefix 'A' and suffix 'CC', prefix 'AC' and suffix 'C', or prefix 'ACC' and suffix ''. We seek the smallest set of prefixes and suffixes that together realize all the given strings," Ricky said.

"Good. This gives us a natural representation for simulated annealing. The state space will consist of all possible subsets of prefixes and suffixes. The natural transitions between states might include inserting or deleting strings from our subsets, or swapping a pair in or out."

"What's a good cost function?" he asked.

"Well, we need as small an array as possible that covers all the strings. How about something like the maximum of number of rows (prefixes) or columns (suffixes) used in our array, plus the number of strings from S that are not yet covered. Try it and let's see what happens."

Ricky went off and implemented a simulated annealing program along these lines. Printing out the state of the solution each time a transition was accepted, it was fun to watch. Starting from a random solution, the program quickly kicked out unnecessary prefixes and suffixes, and the array began shrinking rapidly in size. But after several hundred iterations, progress started to slow. A transition would knock out an unnecessary suffix, wait a while, then add a different suffix back again. After a few thousand iterations, no real improvement was happening.

"The program doesn't seem to recognize when it is making progress. The evaluation function only gives credit for minimizing the larger of the two dimensions. Why not add a term to give some credit to the other dimension."

Ricky changed the evaluation function, and we tried again. This time, the program did not hesitate to improve the shorter dimension. Indeed, our arrays started to be skinny rectangles instead of squares.

"OK. Let's add another term to the evaluation function to give it points for being roughly square."

Ricky tried again. Now the arrays were the right shape, and progress was in the right direction. But the progress was slow.

"Too many of the prefix/suffix insertion moves don't really seem to affect many strings. Maybe we should skew the random selections so that the important prefix/suffixes get picked more often."

Ricky tried again. Now it converged faster, but sometimes it still got stuck. We changed the cooling schedule. It did better, but was it doing well? Without a lower bound knowing how close we were to optimal, it couldn't really tell how good our solution was. We tweaked and tweaked until our program stopped improving.

Our final solution refined the initial array by applying the following random moves:

- *swap* – swap a prefix/suffix on the array with one that isn't.
- *add* – add a random prefix/suffix to the array.
- *delete* – delete a random prefix/suffix from the array.
- *useful add* – add the prefix/suffix with the highest usefulness to the array.
- *useful delete* – delete the prefix/suffix with the lowest usefulness from the array.
- *string add* – randomly select a string not on the array, and add the most useful prefix and/or suffix that covers this string (additional preference is given to a prefix/suffix whose corresponding suffix/prefix is already on the array).

A standard annealing schedule was used, with an exponentially decreasing temperature (dependent upon the problem size) and a temperature-dependent Boltzmann criterion for accepting states that have higher costs. Our final cost function was defined as

$$cost = 2 \times max + min + \frac{(max - min)^2}{4} + 4(str_{total} - str_{in})$$

where *max* is the size of the maximum chip dimension, *min* is the size of the minimum chip dimension, $str_{total} = |S|$, and str_{in} is the number of strings in S currently on the chip.

Careful analysis of successful moves over the course of the annealing process suggested a second phase of annealing to speed convergence. Once the temperature reaches a predetermined cutoff point, the temperature schedule was changed to force the temperature to decrease more rapidly, and the probability distribution for choosing moves was altered to include only swap, add, and delete, with preference given to swap moves. This modification sped up late convergence, which had been slower than it was in the early stages of the annealing process.

How well did we do? As reported in our paper [BS97], Figure 5-9 shows the convergence of a custom array consisting of the 5,716 unique 7-mers of the HIV-virus. Figure 5-9 shows snapshots of the state of the chip at four points during the annealing process (0, 500, 1,000, and the final chip at 5,750 iterations). Black pixels represent the first occurrence of an HIV 7-mer, while white pixels represent either duplicated HIV 7-mers or strings not in the HIV input set. The

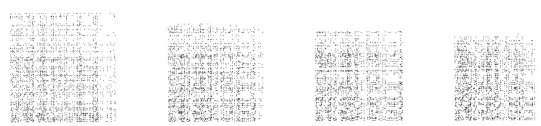

Figure 5–9. Compression of the HIV array by simulated annealing—after 0, 500, 1,000, and 5,750 iterations

final chip size here is 130×132, quite an improvement over the initial size of 192×192. It took about fifteen minutes' worth of computation on a desktop workstation to complete the optimization, which was perfectly acceptable for the application.

But how well did we do? Since simulated annealing is only a heuristic, we really don't know how close to optimal our solution is. I think we did pretty well, but I can't really be sure. In conclusion, simulated annealing can be the right way to handle complex optimization problems. However, to get the best results, expect to spend more time tweaking and refining your program than you did in writing it in the first place. This is dirty work, but sometimes you have to do it.

5.7 Parallel Algorithms

Two heads are better than one, and more generally, n heads are better than $n - 1$. In our era of computing plenty, parallel processing seems like an exciting technique for solving combinatorial optimization problems. Today many facilities contain networks of workstations, most of them idle at night and underutilized during the day. Why not put them to work?

Parallelism seems like the easy way out of hard problems. Indeed, sometimes, for some problems, parallel algorithms are the most effective solution. High-resolution, real-time graphics applications must render thirty frames per second for realistic animation. Assigning each frame to a distinct processor, or dividing each image into regions assigned to different processors might be the only way to get the job done in time. Large systems of linear equations for scientific applications are routinely solved in parallel on high-speed machines.

However, there are several pitfalls associated with parallel algorithms that one should be aware of:

- *There is often a small upper bound on the potential win* – Suppose that you have access to twenty workstations that can be devoted exclusively to your job. Potentially, these could be used to speed up the fastest sequential program by a factor of twenty. That is nice, but much greater performance gains are potentially possible by finding a better sequential algorithm and implementing that. Your time spent parallelizing a code might well be better spent enhancing the sequential version. Performance-tuning tools such as profilers are better developed for sequential machines than for parallel models.

- *Speedup means nothing* – Suppose my parallel program runs 16 times faster on a 16-processor machine then it does on one processor. That's great, isn't it? If you always get linear speedup and have an arbitrary number of processors, you will eventually beat any sequential algorithm. However, a carefully designed sequential algorithm can often beat an easily parallelized code running on a typical parallel machine. The one-processor parallel version of your algorithm is likely to be a crummy sequential algorithm, so measuring speedup typically provides an unfair test of the benefits of parallelism.

The classic example of this occurs in the minimax game-tree search algorithms used in computer chess programs. Brute-force tree search is embarrassingly easy to parallelize; just put each subtree on a different processor. However, a lot of work gets wasted because the same positions get considered on different machines. Moving from brute-force search to the more clever alpha-beta pruning algorithm can easily save over 99.99% of the work, thus dwarfing any benefits of parallel brute-force search. Alpha-beta can be parallelized, but not easily, and speedups are typically limited to a factor of six or so regardless of how many processors you have.

- *Parallel algorithms are tough to debug* – Unless your problem can be decomposed into several independent jobs, the different processors will have to communicate with each other in order to end up with the correct final result. Unfortunately, the nondeterministic nature of this communication makes parallel programs notoriously difficult to debug. Perhaps the best example is *Deep Blue*, the world-champion chess computer. Before it beat Kasparov, it lost several games in embarrassing fashion due to bugs, mostly associated with its extensive parallelism.

I recommend considering parallel processing only after repeated attempts at solving the problem sequentially prove too slow. Even then, I would restrict attention to algorithms that parallelize the problem by partitioning the input into distinct tasks, where no communication is needed between the processors, except in collecting the final results. Such large-grain, naive parallelism can be simple enough to be readily implementable and debuggable, because it really reduces to producing a good sequential implementation. Still, there can be pitfalls in this approach, as discussed in the war story below.

5.8 War Story: Going Nowhere Fast

In Section 2.7, I related our efforts to build a fast program to test Waring's conjecture for pyramidal numbers. At that point, my code was fast enough that it could complete the job in a few weeks running in the background on a desktop workstation. This option did not appeal to my supercomputing colleague, however.

"Why don't we do it in parallel?" he suggested. "After all, you have an outer loop doing the same type of calculation on each integer from 1 to 1,000,000,000. I can split this range of numbers into different intervals and run each one of these on a different processor. Watch, it will be easy."

He set to work trying to do our computations on an Intel IPSC-860 hypercube using 32 nodes, with 16 megabytes of memory per node. However, instead of getting answers, over the next few weeks I was treated to a regular stream of e-mail about system reliability:

- "Our code is running fine, except one processor died last night. I will rerun."
- "This time the machine was rebooted by accident, so our long-standing job was killed."
- "We have another problem. The policy on using our machine is that nobody can command the entire machine for more than thirteen hours, under any condition."

Still, eventually, he rose to the challenge. Waiting until the machine was stable, he locked out 16 processors (half the computer), divided the integers from 1 to 1,000,000,000 into 16 equal-sized intervals, and ran each interval on its own processor. He spent the next day fending off angry users who couldn't get their work done because of our rogue job. The instant the first processor completed analyzing the numbers from 1 to 62,500,000, he announced to all the people yelling at him that the other processors would soon follow.

But they didn't. He failed to realize that the time to test each integer increased as the numbers got larger. After all, it would take longer to test whether 1,000,000,000 could be expressed as the sum of three pyramidal number than it would for 100. Thus at slower and slower intervals, each new processor would announce its completion. Because of the architecture of the hypercube, he couldn't return any of the processors until our entire job was completed. Eventually, half the machine and most of its users were held hostage by one, final interval.

When the job finally completed, the numbers were passed on to the Nobel Prize winner who had requested them. It turns out he had been curious about the problem because his father had made the conjecture back in 1928. There had never been a more important scientific reason to justify the computation in the first place. Indeed, no results from the computation ever appeared in print.

What conclusions can be drawn from this? Before devoting heroic efforts to solve a problem efficiently, make sure that it really needs to be solved, and solved quickly. If you are going to parallelize a problem, be sure to balance the load carefully among the processors. Proper load balancing, using either back-of-the-envelope calculations or the partition algorithm of Section 3.1.2, would have significantly reduced the time we needed the machine, and his exposure to the wrath of his colleagues.

5.9 Exercises

5-1. (*) A *derangement* is a permutation p of $\{1, \ldots, n\}$ such that no item is in its proper position, i.e. $p_i \neq i$ for all $1 \leq i \leq n$. Write an efficient backtracking program with pruning that constructs all the derangements of n items.

5-2. (*) *Multisets* are allowed to have repeated elements. A multiset of n items may thus have fewer than $n!$ distinct permutations. For example, $\{1, 1, 2, 2\}$ has only six different permutations: $\{1, 1, 2, 2\}, \{1, 2, 1, 2\}, \{1, 2, 2, 1\}, \{2, 1, 1, 2\}, \{2, 1, 2, 1\}$, and $\{2, 2, 1, 1\}$. Design and implement an efficient algorithm for constructing all permutations of a multiset.

5-3. (*) Design and implement an algorithm for testing whether two graphs are isomorphic to each other. The graph isomorphism problem is discussed in Section 8.5.9. With proper pruning, graphs on hundreds of vertices can be tested reliably.

5-4. (**) Design and implement an algorithm for solving the subgraph isomorphism problem. Given graphs G and H, does there exist a subgraph H' of H such that G is isomorphic to H'. How does your program perform on such special cases of subgraph isomorphism as Hamiltonian cycle, clique, independent set, and graph isomorphism.

5-5. (*) Design and implement an algorithm for solving the set cover problem, discussed in Section 8.7.1. Use it to solve special-case vertex cover problems as well as general set cover problems.

5-6. (**) In the turnpike reconstruction problem, you are given $n(n-1)/2$ distances in sorted order. The problem is to find the positions of the points on the line that give rise to these distances. For example, the distances $\{1,2,3,4,5,6\}$ can be determined by placing the second point 1 unit from the first, the third point 3 from the second, and the fourth point 2 from the third. Design and implement an efficient algorithm to report all solutions to the turnpike reconstruction problem. Exploit additive constraints when possible to minimize search. With proper pruning, problems with hundreds of points can be solved reliably.

Implementation Challenges

5-1. (*) Anagrams are rearrangements of the letters of a word or phrase into a different word or phrase. Sometimes the results are quite striking. For example, "MANY VOTED BUSH RETIRED" is an anagram of "TUESDAY NOVEMBER THIRD," which correctly predicted the outcome of the 1992 U.S. presidential election. Design and implement an algorithm for finding anagrams using combinatorial search and a dictionary.

5-2. (*) For any of the exercises above, design and implement a simulated annealing heuristic to get reasonable solutions. How well does your program perform in practice?

5-3. (**) Design and implement a parallel sorting algorithm that distributes data across several processors. An appropriate variation of mergesort is a likely candidate. Measure the speedup of this algorithm as the number of processors increases. Later, compare the execution time to that of a purely sequential mergesort implementation. What are your experiences?

CHAPTER 6

Intractable Problems and Approximations

In this chapter, we will concentrate on techniques for proving that *no* efficient algorithm exists for a given problem. The practical reader is probably squirming at the notion of proving anything and will be particularly alarmed at the idea of investing time to prove that something does not exist. Why will you be better off knowing that something you don't know how to do in fact can't be done at all?

The truth is that the theory of NP-completeness is an immensely useful tool for the algorithm designer, even though all it does is provide negative results. That noted algorithm designer Sherlock Holmes once said, "When you have eliminated the impossible, what remains, however improbable, must be the truth." The theory of NP-completeness enables the algorithm designer to focus her efforts more productively, by revealing that the search for an efficient algorithm for this particular problem is doomed to failure. When one *fails* to show that a problem is hard, that means there is likely an algorithm that solves it efficiently. Two of the war stories in this book describe happy results springing from bogus claims of hardness.

The theory of NP-completeness also enables us to identify exactly what properties make a particular problem hard, thus providing direction for us to model it in different ways or exploit more benevolent characteristics of the problem. Developing a sense for which problems are hard and which are not is a fundamental skill for algorithm designers, and it can come only from hands-on experience proving hardness.

We will not discuss the complexity-theoretic aspects of NP-completeness in depth, limiting our treatment to the fundamental concept of *reductions*, which show the equivalence of pairs of problems. For a discussion, we refer the reader to [GJ79], the truly essential reference on the theory of intractability.

The take-home lessons from this chapter are:

- Reductions are a way to show that two problems are essentially identical. A fast algorithm for one of the problems implies a fast algorithm for the other.
- In practice, a small set of NP-complete problems (3-SAT, vertex cover, integer partition, and Hamiltonian cycle) suffice to prove the hardness of most other hard problems.
- Approximation algorithms guarantee answers that are always close to the optimal solution and can provide an approach to dealing with NP-complete problems.

6.1 Problems and Reductions

Throughout this book we have encountered problems, such as the traveling salesman problem, for which we couldn't find any efficient algorithm. By the early 1970s, literally hundreds of problems were stuck in this swamp. The theory of NP-completeness provided the tools needed to show that all of these problems were really the same thing.

The key idea behind demonstrating the hardness of a problem is that of a *reduction*. Suppose that I gave you the following algorithm to solve the *Bandersnatch* problem:

Bandersnatch(G)
      Translate the input G to an instance of the Bo-billy problem Y.
      Call the subroutine Bo-billy on Y to solve this instance.
      Return the answer of Bo-billy(Y) as the answer to Bandersnatch(G).

It is important to see that this algorithm *correctly* solves the Bandersnatch problem provided that the translation to Bo-billy always preserves the correctness of the answer. In other words, the translation has the property that for any instance of G, *Bandersnatch(G)* = *Bo-billy(Y)*. A translation of instances from one type of problem to instances of another type such that the answers are preserved is called a *reduction*.

Now suppose this reduction translates G to Y in $O(P(n))$ time. There are two possible implications:

- *If* my Bo-billy subroutine ran in $O(P'(n))$, this means I could solve the Bandersnatch problem in $O(P(n) + P'(n))$ by spending the time to translate the problem and then the time to execute the Bo-Billy subroutine.
- *If* I know that $\Omega(P'(n))$ is a lower bound on computing Bandersnatch, meaning there definitely exists no faster way to solve it, then $\Omega(P'(n) - P(n))$ *must* be a lower bound to compute Bo-billy. Why? If I could solve Bo-billy any faster,

then I could solve Bandersnatch in faster time by using the above simulation, thus violating my lower bound. This implies that there can be no way to solve Bo-billy any faster than claimed.

This second argument is the approach that we will use to prove problems hard. Essentially, this reduction shows that Bo-billy is at least as hard as Bandersnatch, and therefore once we believe that Bandersnatch is hard, we have a tool for proving other problems hard.

Reductions, then, are operations that convert one problem into another. To describe them, we must be somewhat rigorous in our definition of a problem. A *problem* is a general question, with parameters for the input and conditions on what constitutes a satisfactory answer or solution. An *instance* is a problem with the input parameters specified. The difference can be made clear by an example. The traveling salesman problem is defined thus:

Input: A weighted graph G.
Output: Which tour $\{v_1, v_2, ..., v_n\}$ minimizes $\sum_{i=1}^{n-1} d[v_i, v_{i+1}] + d[v_n, v_1]$?

Thus any weighted graph defines an instance of TSP. Each particular instance has at least one minimum cost tour. The general traveling salesman problem asks for an algorithm to find the optimal tour for all possible instances.

Any problem with answers restricted to yes and no is called a *decision problem*. Most interesting optimization problems can be phrased as decision problems that capture the essence of the computation. For example, the traveling salesman decision problem can be defined thus:

Input: A weighted graph G and integer k.
Output: Does there exist a TSP tour with cost $\leq k$?

It should be clear that the decision version captures the heart of the traveling salesman problem, for if you had a program that gave fast solutions to the decision problem, you could do a binary search with different values of k to quickly hone in on the correct solution.

Therefore, from now on we will talk only about decision problems, because it is easier to reduce one problem to another when the only possible answers to both are true or false.

6.2 Simple Reductions

Since they can be used either to prove hardness or to give efficient algorithms, reductions are powerful tools for the algorithm designer to be familiar with. The best way to understand reductions is to look at some simple ones.

6.2.1   Hamiltonian Cycle

The Hamiltonian cycle problem is one of the most famous in graph theory. It seeks a tour that visits each vertex of a given graph exactly once. It has a long history and many applications, as discussed in Section 8.5.5. Formally, it is defined:

Input: An unweighted graph G.
Output: Does there exist a simple tour that visits each vertex of G without repetition?

Hamiltonian cycle has some obvious similarity to the traveling salesman problem. Both problems ask for a tour to visit each vertex exactly once. There are also differences between the two problems. TSP works on weighted graphs, while Hamiltonian cycle works on unweighted graphs. The following reduction from Hamiltonian cycle to traveling salesman shows that the similarities are greater than the differences:

> HamiltonianCycle($G = (V, E)$)
>   Construct a complete weighted graph $G' = (V', E')$ where $V' = V$
> $n = |V|$
>   for $i = 1$ to n do
>   for $j = 1$ to n do,   if $(i, j) \in E$ then $w(i, j) = 1$ else $w(i, j) = 2$
>   Return the answer to Traveling-Salesman(G', n)

The actual reduction is quite simple, with the translation from unweighted to weighted graph easily performed in linear time. Further, this translation is designed to ensure that the answers of the two problems will be identical. If the graph G has a Hamiltonian cycle $\{v_1, \ldots, v_n\}$, then this exact same tour will correspond to n edges in E', each with weight 1. Therefore, this gives a TSP tour of G' of weight exactly n. If G does not have a Hamiltonian cycle, then there can be no such TSP tour in G', because the only way to get a tour of cost n in G would be to use only edges of weight 1, which implies a Hamiltonian cycle in G.

This reduction is both efficient and truth preserving. A fast algorithm for TSP would imply a fast algorithm for Hamiltonian cycle, while a hardness proof for Hamiltonian cycle would imply that TSP is hard. Since the latter is the case, this reduction shows that TSP is hard, at least as hard as Hamiltonian cycle.

6.2.2   Independent Set and Vertex Cover

The vertex cover problem, discussed more thoroughly in Section 8.5.3, asks for a small set of vertices that contacts each edge in a graph. More formally:

Input: A graph $G = (V, E)$ and integer $k \le |V|$.
Output: Is there a subset S of at most k vertices such that every $e \in E$ has at least one vertex in S?

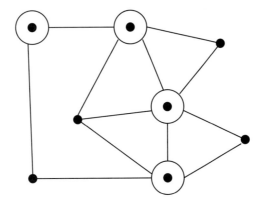

Figure 6-1. Circled vertices form a vertex cover, the dark vertices an independent set

It is trivial to find *a* vertex cover of a graph, for the cover can consist of all of the vertices. More tricky is to cover the edges using as small a set of vertices as possible. For the graph of Figure 6-1, four of the eight vertices are sufficient to cover.

A set of vertices S of graph G is *independent* if there are no edges (x, y) where $x \in S$ and $y \in S$, meaning there are no edges between any two vertices in the independent set. As discussed in Section 8.5.2, the independent set problem arises in facility location problems. The maximum independent set decision problem is formally defined:

Input: A graph G and integer $k \le |V|$.
Output: Does there exist an independent set of k vertices in G?

Both vertex cover and independent set are problems that revolve around finding special subsets of vertices, the first with representatives of every edge, the second with no edges. If S is the vertex cover of G, the remaining vertices $S - V$ must form an independent set, for if there were an edge with both vertices in $S - V$, then S could not have been a vertex cover. This gives us a reduction between the two problems:

> VertexCover(G, k)
> $G' = G$
> $k' = |V| - k$
>   Return the answer to IndependentSet(G', k')

Again, a simple reduction shows that the problems are identical. Notice how this translation occurs without any knowledge of the answer. We transform the input, not the solution. This reduction shows that the hardness of vertex cover imples that independent set must also be hard. It is easy to reverse the roles of the two problems in this reduction, thus proving that both of these problems are equally hard.

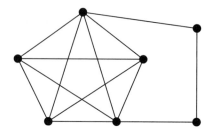

Figure 6–2. A small graph with a five-vertex clique

6.2.3   Clique and Independent Set

Consider the clique problem, further discussed in Section 8.5.1:

Input: A graph $G = (V, E)$ and integer $j \leq |V|$.
Output: Does the graph contain a clique of j vertices; i.e. is there a subset $S \subset V$, where $|S| \leq j$, such that every pair of vertices in S defines an edge of G?

For example, the graph in Figure 6-2 contains a clique of five vertices. In the independent set problem, we looked for a subset S with no edges between two vertices of S. However, for a clique, we insist that there always be an edge between two vertices. A reduction between these problems results by reversing the roles of edges and nonedges, an operation known as *complementing* the graph:

IndependentSet(G, k)
   Construct a graph $G = (V', E')$ where $V' = V$, and
     For all (i, j) not in E, add (i, j) to E'
   Return the answer to Clique(G', k)

These last two reductions provide a chain linking three different problems. The hardness of clique is implied by the hardness of independent set, which is implied by the hardness of vertex cover. By constructing reductions in a chain, we link together pairs of problems in implications of hardness. Our work is done as soon as all these chains begin with a single problem that is accepted as hard. Satisfiability is the problem that serves as the first link in this chain.

6.3   Satisfiability

To prove the hardness of different problems using reductions, we need to start with a single problem that is absolutely, certifiably, undeniably hard. The mother of all NP-complete problems is a logic problem named *satisfiability*:

Input: A set of Boolean variables V and a set of clauses C over V.

Output: Does there exist a satisfying truth assignment for C, i.e. a way to set the variables $v_1, \ldots, v_n$ either true or false so that each clause contains at least one true literal?

This can be made clearer with two examples. Suppose that $C = \{\{v_1, \bar{v}_2\}, \{\bar{v}_1, v_2\}\}$ over the Boolean variables $V = \{v_1, v_2\}$. We use $\bar{v}_i$ to denote the complement of the variable v_i, so we would get credit for satisfying a particular clause containing v_i if $v_i = $ true , or a clause containing $\bar{v}_i$ if $v_i = $ false . Therefore, satisfying a particular set of clauses involves making a series of n true or false decisions, trying to find the right truth assignment to satisfy all of them.

This example set of clauses $C = \{\{v_1, \bar{v}_2\}, \{\bar{v}_1, v_2\}\}$ can be satisfied by simply setting $v_1 = v_2 = $ true or $v_1 = v_2 = $ false . However, consider the set of clauses $C = \{\{v_1, v_2\}, \{v_1, \bar{v}_2\}, \{\bar{v}_1\}\}$. There can be no satisfying assignment because v_1 must be false in order to satisfy the third clause, which means that v_2 must be false to satisfy the second clause, which then leaves the first clause unsatisfiable. Although you try, and you try, and you try, and you try, you can't get no satisfaction.

6.3.1   The Theory of NP-Completeness

For a variety of social and technical reasons, it is well accepted that satisfiability is a hard problem, one for which no worst-case polynomial-time algorithm exists. Literally every top-notch algorithm expert in the world (and countless lesser lights) has directly or indirectly tried to come up with a fast algorithm to test whether a given set of clauses is satisfiable, but all have failed. Further, many strange and impossible-to-believe things in the field of computational complexity have been shown to be true if there exists a fast satisfiability algorithm. Satisfiability is a hard problem, and it is important to accept this. See Section 8.3.10 for more on the satisfiability problem and its applications.

The theory of NP-completeness rests on a foundation of rigorous but subtle definitions from automata and formal language theory. This terminology is typically confusing to or misused by beginners who lack a mastery of these foundations, and it is not really essential to the practical aspects of designing and applying reductions. For completeness, however, we briefly define the key terms below.

A problem is said to be polynomial (or in the class P) if it can be solved in time polynomial in its size. A problem is said to be nondeterministically polynomial (or in the class NP) if a conjectured answer can be verified in time polynomial in its size. The traveling salesman decision problem is not known to be in P, because there is no known polynomial-time algorithm for it. However, the problem *is* in NP, because if we are given a candidate tour, we can efficiently add up the cost of the associated edges and verify whether the total is at most the cost bound k. It is typically straightforward to verify whether the answer to a problem is correct, and it certainly can be no harder than actually finding the answer in the first place.

Through a complicated proof, it has been established that satisfiability is at least as hard as any problem in NP. This means that if a fast (i.e. polynomial-time) algorithm is discovered for solving satisfiability, this will yield a fast algorithm for *every* problem in NP. Since essentially every problem mentioned in this book is in NP, this would be an enormously powerful and surprising result. We say that a problem is *NP-hard* if, like satisfiability, it is at least as hard as any problem in NP. We say that a problem is *NP-complete* if it is NP-hard, and also in NP itself. Because NP is such a large class of problems, most NP-hard problems you encounter will in fact be complete, and the issue can always be settled by giving a (usually simple) verification strategy for the problem.

6.3.2   3-Satisfiability

Satisfiability's role as the first NP-complete problem implies that the problem is hard to solve in the worst case, but certain instances of the problem are not necessarily so tough. Suppose that each clause contains exactly one literal. To satisfy such a clause, we have to appropriately set that literal, so we can repeat this argument for every clause in the problem instance. Only when we have two clauses that directly contradict each other, such as $C = \{\{v_1\}, \{\overline{v}_1\}\}$, will the set not be satisfiable.

Since clause sets with only one literal per clause are easy to satisfy, we are interested in slightly larger classes. Exactly what is the clause size at which the problem turns from polynomial to hard? This transition occurs when each clause contains three literals, the so-called 3-satisfiability problem, or 3-SAT:

Input: A collection of clauses C where each clause contains exactly 3 literals, over a set of Boolean variables V.

Output: Is there a truth assignment to V such that each clause is satisfied?

Since this is a more restricted problem than satisfiablity, the hardness of 3-SAT implies that satisfiability is hard. The converse isn't true, as the hardness of general satisfiability might depend upon having long clauses. We can show the hardness of 3-SAT using a reduction that translates every instance of satisfiability into an instance of 3-SAT without changing the result of whether it is satisfiable.

This reduction transforms each clause independently based on its *length*, by adding new Boolean variables along the way. Suppose clause C_i contained k literals:

- If $k = 1$, meaning that $C_i = \{z_1\}$, we create two new variables v_1, v_2 and four new 3-literal clauses: $\{v_1, v_2, z_1\}, \{v_1, \overline{v}_2, z_1\}, \{\overline{v}_1, v_2, z_1\}, \{\overline{v}_1, \overline{v}_2, z_1\}$. Note that the only way that all four of these clauses can be simultaneously satisfied is if $z_1 = \text{true}$, which must be the case if the original C_i were to be satisfied.

- If $k = 2$, meaning that $C_i = \{z_1, z_2\}$, we create one new variable v_1 and two new clauses: $\{v_1, z_1, z_2\}, \{\overline{v}_1, z_1, z_2\}$. Again, the only way to satisfy both of these clauses is to have at least one of z_1 and z_2 be true.

- If $k = 3$, meaning that $C_i = \{z_1, z_2, z_3\}$, we copy C_i into the 3-SAT instance unchanged: $\{z_1, z_2, z_3\}$.
- If $k > 3$, meaning that $C_i = \{z_1, z_2, ..., z_n\}$, create $n - 3$ new variables and $n - 2$ new clauses in a chain, where for $3 \leq j \leq n - 3$, $C_{i,j} = \{v_{i,j-1}, z_j, \overline{v}_{i,j}\}$, $C_{i,1} = \{z_1, z_2, \overline{v}_{i,1}\}$, and $C_{i,n-2} = \{v_{i,n-2}, z_{n-1}, z_n\}$.

The most complicated case is that of the large clauses. If none of the original variables C_i are true , then there are not enough additional variables to be able to satisfy all of the new subclauses. You can satisfy $C_{i,1}$ by setting $v_{i,1} = $ false , but this forces $v_{i,2} = $ false , and so on until finally $C_{i,n-2}$ cannot be satisfied. But if any single literal $z_i = $ true , then we have $n - 3$ free variables and $n - 3$ remaining 3-clauses, so we can satisfy each of them.

This transform takes $O(m + n)$ time if there were n clauses and m total literals in the SAT instance. Since any SAT solution also satisfies the 3-SAT instance and any 3-SAT solution sets the variables giving a SAT solution, the transformed problem is equivalent to the original.

Note that a slight modification to this construction would serve to prove that 4-SAT, 5-SAT, or any $(k \geq 3)$-SAT is also NP-complete. However, this construction breaks down when we try to use it for 2-SAT, since there is no way to stuff anything into the chain of clauses. It turns out that resolution gives a polynomial-time algorithm for 2-SAT, as discussed in Section 8.3.10.

6.4 Difficult Reductions

Now that both satisfiability and 3-SAT are known to be hard, we can use either of them in reductions. What follows are a pair of more complicated reductions, designed to serve both as examples for how to proceed and to increase our repertoire of known hard problems from which we can start. Many reductions are quite intricate, because we are essentially programming one problem in the language of a significantly different problem.

One perpetual point of confusion is getting the direction of the reduction right. Recall that we must transform *every* instance of a known NP-complete problem into an instance of the problem we are interested in. If we perform the reduction the other way, all we get is a slow way to solve the problem of interest, by using a subroutine that takes exponential time. This always is confusing at first, for this direction of reduction seems bass-ackwards. Check to make sure you understand the direction of reduction now, and think back to this whenever you get confused.

6.4.1 Integer Programming

As discussed in Section 8.2.6, integer programming is a fundamental combinatorial optimization problem. It is best thought of as linear programming with the variables restricted to take only integer (instead of real) values.

Input: A set V of integer variables, a set of inequalities over V, a maximization function $f(V)$, and an integer B.

Output: Does there exist an assignment of integers to V such that all inequalities are true and $f(V) \geq B$?

Consider the following two examples. Suppose

$$v_1 \geq 1, \quad v_2 \geq 0$$

$$v_1 + v_2 \leq 3$$

$$f(v) : 2v_2, \quad B = 3$$

A solution to this would be $v_1 = 1$, $v_2 = 2$. Not all problems have realizable solutions, however. For the following problem:

$$v_1 \geq 1, \quad v_2 \geq 0$$

$$v_1 + v_2 \leq 3$$

$$f(v) : 2v_2, \quad B = 5$$

the maximum value of $f(v)$ is $2 \times 2 = 4$ (given the constraints), and so there can be no solution to the associated decision problem.

We show that integer programming is hard using a reduction from 3-SAT. For this particular reduction, general satisfiability would work just as well, although usually 3-SAT makes reductions easier.

In which direction must the reduction go? We want to prove integer programming that is hard, and we know that 3-SAT is hard. If I could solve 3-SAT using integer programming and integer programming were easy, this would mean that satisfiability would be easy. Now the direction should be clear; we have to translate 3-SAT into integer programming.

What should the translation look like? Every satisfiability instance contains Boolean (true/false) variables and clauses. Every integer programming instance contains integer variables (values restricted to $0, 1, 2, \ldots$) and constraints. A reasonable idea is to make the integer variables correspond to Boolean variables and have constraints serve the same role as the clauses do in the original problem.

Our translated integer programming problem will have twice as many variables as the SAT instance, one for each variable and one for its complement. For each variable v_i in the set problem, we will add the following constraints:

- To restrict each integer programming variable V_i to values of 0 or 1, we add constraints $1 \geq V_i \geq 0$ and $1 \geq \overline{V}_i \geq 0$. Thus they correspond to values of true and false .

- To ensure that exactly one of the two integer programming variables associated with a given SAT variable is true , add constraints so that $1 \geq V_i + \overline{V}_i \geq 1$.

For each clause $C_i = \{z_1, z_2, z_3\}$ in the 3-SAT instance, construct a constraint: $V_1 + V_2 + V_3 \geq 1$. To satisfy this constraint, at least one the literals per clause must be set to 1, thus corresponding to a true literal. Satisfying this constraint is therefore equivalent to satisfying the clause.

The maximization function and bound prove relatively unimportant, since we have already encoded the entire 3-SAT instance. By using $f(v) = V_1$ and $B = 0$, we ensure that they will not interfere with any variable assignment satisfying all the inequalities. Clearly, this reduction can be done in polynomial time. To establish that this reduction preserves the answer, we must verify two things:

- *Any SAT solution gives a solution to the IP problem* – In any SAT solution, a true literal corresponds to a 1 in the integer program, since the clause is satisfied. Therefore, the sum in each clause inequality is ≥ 1.

- *Any IP solution gives a SAT solution* – In any solution to this integer programming instance, all variables must be set to either 0 or 1. If $V_i = 1$, then set literal $z_i =$ true. If $V_i = 0$, then set literal $z_i =$ false. No Boolean variable and its complement can both be true, so it is a legal assignment, which must also satisfy all the clauses.

The reduction works both ways, so integer programming must be hard. Notice the following properties, which hold true in general for NP-complete:

- The reduction preserved the structure of the problem. It did not *solve* the problem, just put it into a different format.

- The possible IP instances that can result from this transformation are only a small subset of all possible IP instances. However, since some of them are hard, the general problem must be hard.

- The transformation captures the essence of *why* IP is hard. It has nothing to do with having big coefficients or big ranges on variables; since restricting them to 0/1 is enough. It has nothing to do with having inequalties with large numbers of variables. Integer programming is hard because satisfying a set of constraints is hard. A careful study of the properties needed for a reduction can tell us a lot about the problem.

6.4.2 Vertex Cover

Algorithmic graph theory proves to be a fertile ground for hard problems. The prototypical NP-complete graph problem is vertex cover, previously defined in Section 6.2.2 as follows:

Input: A graph $G = (V, E)$ and integer $k \leq |V|$.
Output: Is there a subset S of at most k vertices such that every $e \in E$ has at least one vertex in S?

Demonstrating the hardness of vertex cover proves more difficult than the previous reductions we have seen, because the structure of the two relevant problems is very different. A reduction from 3-satisfiability to vertex cover has to construct a graph G and bound k from the variables and clauses of the satisfiability instance.

First, we translate the variables of the 3-SAT problem. For each Boolean variable v_i, we create two vertices v_i and $\bar{v}_i$ connected by an edge. To cover these edges, at least n vertices will be needed, since no two of the edges will share a vertex.

Second, we translate the clauses of the 3-SAT problem. For each of the c clauses, we create three new vertices, one for each literal in each clause. The three vertices of each clause will be connected so as to form c triangles. At least two vertices per triangle must be included in any vertex cover of these triangles.

Finally, we will connect these two sets of components together. Each literal in the vertex "gadgets" is connected to corresponding vertices in the clause gadgets (triangles) that share the same literal. From a 3-SAT instance with n variables and c clauses, this constructs a graph with $2n + 3c$ vertices. The complete reduction for the 3-SAT problem $\{\{v_1, \bar{v}_3, \bar{v}_4\}, \{\bar{v}_1, v_2, \bar{v}_4\}\}$ is shown in Figure 6-3.

This graph has been designed to have a vertex cover of size $n + 2c$ if and only if the original expression is satisfiable. By the earlier analysis, any vertex cover must have at least $n + 2c$ vertices, since adding extra edges to the graph can only increase the size of the vertex cover. To show that our reduction is correct, we must demonstrate that:

- *Every satisfying truth assignment gives a vertex cover* – Given a satisfying truth assignment for the clauses, select the n vertices from the vertex gadgets that correspond to true literals to be members of the vertex cover. Since this is a satisfying truth assignment, a true literal from each clause will have covered one of the three cross edges connecting each clause triangle to a vertex gadget. Therefore, by selecting the other two vertices of each clause triangle, we can also pick up the remaining cross edges and complete the cover.

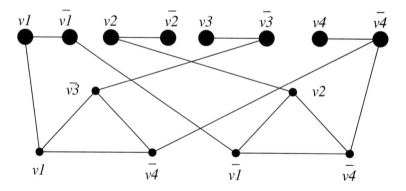

Figure 6-3. Reducing satisfiability instance $\{\{v_1, \bar{v}_3, \bar{v}_4\}, \{\bar{v}_1, v_2, \bar{v}_4\}\}$ to vertex cover

- *Every vertex cover gives a satisfying truth assignment* – Given any vertex cover C of size $n + 2c$, exactly n of the vertices must belong to the vertex gadgets. Let these first stage vertices define the truth assignment, while the $2c$ remaining cover vertices must be distributed at two per clause gadget; otherwise a clause gadget edge must go uncovered. These clause gadget vertices can cover only two of the three connecting cross edges per clause. Therefore, if C gives a vertex cover, at least one cross edge per clause must be covered, meaning that the corresponding truth assignment satisfies.

This proof of the hardness of vertex cover, chained with the clique and independent set arguments of Section 6.2.2, gives us a library of hard graph problems that we can use to make future hardness proofs easier.

6.5 Other NP-Complete Problems

Clique, vertex cover, and integer programming are just three of the literally hundreds of problems that have been shown to be NP-complete. It is important to be aware of which kinds of problems tend to be hard, so you recognize them when you see them in applications, and also to provide a suitable class of candidates for future reductions. Some, but by no means all, of the hard problems from the catalog include:

- *Integer partition* – Can you partition n integers into two subsets such that the sums of the subsets are equal? See Section 8.2.10 for details.
- *Bin packing* – How many bins of a given size do you need to hold n items of variable size? See Section 8.6.9 for details.
- *Chromatic number* – How many colors do you need to color a graph such that no neighboring vertices are of the same color? See Section 8.5.7 for details.
- *Bandwidth* – Which permutation p of the vertices of a graph minimizes the length of the longest edge when the vertices are ordered on a line, i.e. $\max_{(i,j) \in E} |p(i) - p(j)|$? See Section 8.2.2 for details.

A few other catalog problems exist in a limbo state, where it is not known whether the problem has a fast algorithm or is NP-complete. The most prominent of these are graph isomorphism (see Section 8.5.9) and primality testing (see Section 8.2.8). That this limbo list is so short is quite a tribute to the state of the art in algorithm design and the power of NP-completeness. For almost every important problem for which we do not know a fast algorithm, we have a good solid reason for why one doesn't exist.

The same should hold true for the problems you encounter in your work. One way or another they should be resolved as being either hard or polynomial. Leaving them in a limbo state is a sure sign of a bush-league algorithm designer.

It takes experience to be able to sense whether a problem is likely to be hard or not. Perhaps the quickest way to gain this experience is through careful study of the catalog. Note that slightly changing the wording of a problem can make the difference between it being polynomial or NP-complete. Finding the shortest path in a graph is easy, while finding the longest path in a graph is hard. Constructing a tour that visits all the edges once in a graph is easy, while constructing a tour that visits all the vertices once is hard.

The first thing to do when you suspect a problem might be NP-complete is look in Garey and Johnson's book *Computers and Intractability* [GJ79], which contains a list of several hundred problems known to be NP-complete. Likely you will find the problem you are interested in.

6.6 The Art of Proving Hardness

Proving that problems are hard is a skill. But once you get the hang of it, reductions can be surprisingly straightforward and pleasurable to do. Indeed, the dirty little secret of NP-completeness proofs is that they are usually easier to create than explain, in the same way that it is often easier to rewrite old code than it is to understand and modify it.

I offer the following advice to those seeking to prove the hardness of a given problem:

- *Make your source problem as simple (i.e. restricted) as possible.*
  Never try to use the general traveling salesman problem (TSP) as a source problem. Better, use Hamiltonian cycle, i.e. TSP where all the weights are 1 or ∞. Even better, use Hamiltonian path instead of cycle, so you don't have to worry about closing up the path. Best of all, use Hamiltonian path on directed planar graphs where each vertex has total degree 3. All of these problems are equally hard, and the more you can restrict the problem that you are reducing, the less work your reduction has to do.

  As another example, never try to use full satisfiability to prove hardness. Start with 3-satisfiability. In fact, you don't even have to use full 3-satisfiability. Instead, consider *planar 3-satisfiability*, where there is a way to draw the clauses as a graph in the plane such you can connect all instances of the same literal together without edges crossing. This property tends to be useful in proving the hardness of geometric problems. All these problems are equally hard, and hence NP-completeness reductions using any of them are equally convincing.

- *Make your target problem as hard as possible.*
  Don't be afraid to add extra constraints or freedoms in order to make your problem more general. Perhaps your undirected graph problem can be generalized into a directed graph problem and can hence be easier to prove hard. Once you have a proof of hardness for the general problem, you can then go back and try to simplify the target.

- *Select the right source problem for the right reason.*

  Selecting the right source problem makes a big difference in how difficult it is to prove a problem hard. Although theoretically any particular problem is as good as another, this is the first and easiest place to go wrong. When faced with trying to prove that a problem is hard, some people fish around through dozens of problems, looking for the one that seems the best fit. These people are amateurs; odds are they never will recognize what they are looking for when they see it.

  I use four and only four problems as candidates for my hard source problem. Limiting them to four means that I can know a lot about each of these problems, such as which variants of these problems are hard and which are soft. My favorite problems are:

    - *3-SAT*: The old reliable. When none of the three problems below seem appropriate, I go back to the original source.
    - *Integer partition*: This is the one and only choice for problems whose hardness seems to require using large numbers.
    - *Vertex cover*: This is the answer for any graph problems whose hardness depends upon *selection*. Chromatic number, clique, and independent set all involve trying to select the correct subset of vertices or edges.
    - *Hamiltonian path*: This is my choice for any graph problem whose hardness depends upon *ordering*. If you are trying to route or schedule something, Hamiltonian path is likely your lever into the problem.

- *Amplify the penalties for making the undesired selection.*

  Many people are too timid in their thinking about hardness proofs. You are trying to translate one problem into another, while keeping the problems as close to their original identities as as possible. The easiest way to do this is to be bold with your penalties, to punish anyone for trying to deviate from your intended solution. Your thinking should be, "if you select this element, then you have to pick up this huge set that prevents you from finding an optimal solution." The sharper the consequences for doing what is undesired, the easier it is to prove if and only if.

- *Think strategically at a high level, then build gadgets to enforce tactics.*

  You should be asking yourself the following types of questions. "How can I force that either A or B but not both are chosen?" "How can I force that A is taken before B?" "How can I clean up the things I did not select?" After you have an idea of what you want your gadgets to do, you can worry about how to actually craft them.

- *When you get stuck, alternate between looking for an algorithm or a reduction.*

  Sometimes the reason you cannot prove hardness is that there exists an efficient algorithm to solve your problem! Techniques such as dynamic programming or reducing to polynomial-time graph problems such as matching or network flow sometimes yield surprising polynomial algorithms. Whenever you can't prove hardness, it likely pays to alter your opinion occasionally to keep yourself honest.

6.7 War Story: Hard Against the Clock

My class's attention span was running down like sand through an hourglass. Eyes were starting to glaze even in the front row. Breathing had become soft and regular in the middle of the room. Heads were tilted back and eyes shut in the rear.

There were fifteen minutes left to go in my lecture on NP-completeness, and I couldn't really blame them. They had already seen several reductions like the ones presented here, but NP-completeness reductions are easier to create than to understand or explain. They had to watch one being created in order to appreciate this.

I reached for my trusty copy of Garey and Johnson's book [GJ79], which contains a list of over four hundred different known NP-complete problems in an appendix in the back.

"Enough of this!" I announced loudly enough to startle those in the back row. "NP-completeness proofs are routine enough that we can construct them on demand. I need a volunteer with a finger. Can anyone help me?"

A few students in the front held up their hands. A few students in the back held up their fingers. I opted for one from the front row.

"Select a problem at random from the back of this book. I can prove the hardness of any of these problems in the now twelve minutes remaining in this class. Stick your finger in and read me a problem."

I had definitely gotten their attention. But I could have done that by offering to juggle chainsaws. Now I had to deliver results without cutting myself into ribbons.

The student picked out a problem. "OK, prove that *Inequivalence of Programs with Assignments* is hard," she said.

"Huh? I've never heard of that problem before. What is it? Read me the entire description of the problem so I can write it on the board." The problem was as follows:

Input: A finite set X of variables, two programs P_1 and P_2, each a sequence of assignments of the form

$$x_0 \leftarrow \text{if } (x_1 = x_2) \text{ then } x_3 \text{ else } x_4$$

where the x_i are in X; and a value set V.

Output: Is there an initial assignment of a value from V to each variable in X such that the two programs yield different final values for some variable in X?

I looked at my watch. Ten minutes to go. But now everything was on the table. I was faced with a language problem. The input was two programs with variables, and I had to test to see whether they always do the same thing.

"First things first. We need to select a source problem for our reduction. Do we start with integer partition? 3-satisfiability? Vertex cover or Hamiltonian path?"

Since I had an audience, I tried thinking out loud. "Our target is not a graph problem or a numerical problem, so let's start thinking about the old reliable: 3-satisfiability. There seem to be some similarities. 3-SAT has variables. This thing has variables. To be more like 3-SAT, we could try limiting the variables in this problem so they only take on two values, i.e. $V = \{true, false\}$. Yes. That seems convenient."

My watch said nine minutes left. "So, class, which way does the reduction go. 3-SAT to language or language to 3-SAT?"

The front row correctly murmured, "3-SAT to language."

"Right. So we have to translate our set of clauses into two programs. How can we do that? We can try to split the clauses into two sets and write separate programs for each of them. But how do we split them? I don't see any natural way how to do it, because eliminating any single clause from the problem might suddenly make an unsatisfiable formula satisfiable, thus completely changing the answer. Instead, let's try something else. We can translate all the clauses into one program, and then make the second program be trivial. For example, the second program might ignore the input and always outputs either only true or only false. This sounds better. *Much* better."

I was still talking out loud to myself, which wasn't that unusual. But I had people listening to me, which was.

"Now, how can we turn a set of clauses into a program? We want to know whether the set of clauses can be satisfied, or if there is an assignment of the variables such that it is true. Suppose we constructed a program to evaluate whether $c_1 = (x_1, \bar{x}_2, x_3)$ is satisfied. We can do it like this …."

It took me a few minutes worth of scratching before I found the right program. I assumed that I had access to constants for true and false, which seemed reasonable in the sense that it shouldn't make the problem algorithmically harder. Once my proof worked, I could later worry about removing the extra assumption if desired.

$$c_1 = \text{if } (x_1 = true) \text{ then } true \text{ else } false$$
$$c_1 = \text{if } (x_2 = false) \text{ then } true \text{ else } c_1$$
$$c_1 = \text{if } (x_3 = true) \text{ then } true \text{ else } c_1$$

"Great. Now I have a way to evaluate the truth of each clause. I can do the same thing to evaluate whether all the clauses are satisfied."

$$sat = \text{if } (c_1 = true) \text{ then } true \text{ else } false$$
$$sat = \text{if } (c_2 = true) \text{ then } sat \text{ else } false$$
$$\vdots$$
$$sat = \text{if } (c_n = true) \text{ then } sat \text{ else } false$$

Now the back of the classroom was getting excited. They were starting to see a ray of hope that they would get to leave on time. There were two minutes left in class.

"Great. So now we have a program that can evaluate to be true if and only if there is a way to assign the variables so as to satisfy the set of clauses. We need a second program to finish the job. What about *sat* = false ? Yes, that is all we need. Our language problem asks whether the two programs always output the same thing, regardless of the possible variable assignments. If the clauses are satisfiable, that means that there must be an assignment of the variables such that the long program would output true. Testing whether the programs are equivalent is exactly the same as asking if the clauses are satisfiable."

I lifted my arms in triumph. "And so, the problem is neat, sweet, and NP-complete." I got the last word out just before the bell rang.

This exercise was so much fun that I repeated it the next time I taught the course, using a different randomly selected problem. The audio from my ultimately successful attempt to prove hardness is accessible from the CD-ROM, if you want to hear what the creation of an NP-completeness proof sounds like.

6.8 Approximation Algorithms

For the practical person, demonstrating that a problem is NP-complete is never the end of the line. Presumably, there was a reason why you wanted to solve it in the first place. That reason for wanting the solve it will not have gone away on being told that there is no polynomial-time algorithm. You still seek a program that solves the problem of interest. All you know is that you won't find one that quickly solves the problem to optimality in the worst case. You still have the following options:

- *Algorithms fast in the average case* – Examples of such algorithms include backtracking algorithms with substantial pruning.

- *Heuristics* – Heuristic methods like simulated annealing or greedy approaches can be used to find a solution with no requirement that it be the best one.

- *Approximation algorithms* – The theory of NP-completeness does not stipulate that it is hard to get close to the answer, only that it is hard to get *the* optimal answer. With clever, problem-specific heuristics, we can often get provably close to the optimal answer.

Approximation algorithms return solutions with a guarantee attached, namely that the optimal solution can never be much better than this given solution. Thus you can never go too far wrong in using an approximation algorithm. No matter what your input instance is and how lucky you are, you are doomed to do all right. Further, approximation algorithms realizing provably good bounds often are conceptually simple, very fast, and easy to program.

One thing that is usually not clear, however, is how well the solution from an approximation algorithm compares to what you might get from a heuristic that gives you no guarantees. The answer could be worse or it could be better. Leaving your money in a savings account in a bank guarantees you 3% interest without risk. Still, you likely will do much better putting your money in stocks than in the bank, even though performance is not guaranteed.

One way to get the best of approximation algorithms and heuristics is to run both of them on the problem instance and pick the solution giving the better answer. This way, you get a solution that comes with a guarantee and a second chance to do even better. When it comes to heuristics for hard problems, sometimes you can have it both ways.

6.8.1 Approximating Vertex Cover

As we have seen before, finding the minimum vertex cover of a graph is NP-complete. However, a very simple procedure can efficiently find a cover that is at most twice as large as the optimal cover:

```
VertexCover(G = (V, E))
        while (E ≠ ∅) do:
                Select an arbitrary edge (u, v) ≤ E
                Add both u and v to the vertex cover
                Delete all edges from E that are incident on either u or v.
```

It should be apparent that this procedure always produces a vertex cover, since each edge is only deleted immediately after an incident vertex has been added to the cover. More interesting is the claim that any vertex cover must use at least half as many vertices as this one. Why? Consider just the edges selected by the algorithm. No two of these edges can share a vertex. Therefore, any cover of just these edges must include at least one vertex per edge, which makes it at least half the size of the greedy cover.

There are several interesting things to notice about this algorithm:

- *Although the procedure is simple, it is not stupid* – Many seemingly smarter heuristics can give a far worse performance in the worst case. For example, why not modify the procedure above to select only one of the two vertices for the cover instead of both? After all, the selected edge will be equally well covered by only one vertex. However, consider the star-shaped graph of Figure 6-4. This heuristic will produce a two-vertex cover, while the single vertex heuristic can return a cover as large as $n - 1$ vertices, should we get unlucky and repeatedly select the leaf instead of the center as the cover vertex.

- *Greedy isn't always the answer* – Perhaps the most natural heuristic for this problem would repeatedly select and delete the vertex of highest remaining degree for the vertex cover. After all, this vertex will cover the largest number of possible edges. However, in the case of ties or near ties, this heuristic can

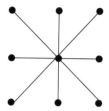

Figure 6–4. Neglecting to pick the center vertex leads to a terrible vertex cover

go seriously astray and in the worst case can yield a cover that is $\Theta(\lg n)$ times optimal.

- *Making a heuristic more complicated does not necessarily make it better* – It is easy to complicate heuristics by adding more special cases or details. For example, the procedure above does not specify which edge should be selected next. It might seem reasonable always to select the edge whose endpoints have highest degree. However, this does not improve the worst-case bound and just makes it more difficult to analyze.

- *A postprocessing cleanup step can't hurt* – The flip side of designing simple heuristics is that they can often be modified to yield better-in-practice solutions without weakening the approximation bound. For example, a postprocessing step that deletes any unnecessary vertex from the cover can only improve things in practice, even though it won't help the worst-case bound.

The important property of approximation algorithms is relating the size of the solution produced directly to a lower bound on the optimal solution. Instead of thinking about how well we might do, we have to think about the worst case i.e. how badly we might perform.

6.8.2   The Euclidean Traveling Salesman

In most natural applications of the traveling salesman problem, direct routes are inherently shorter than indirect routes. For example, if the edge weights of the graph are "as the crow flies", straight-line distances between pairs of cities, the shortest path from x to y will always be to fly directly.

The edge weights induced by Euclidean geometry satisfy the triangle inequality, which insists that $d(u,w) \le d(u,v) + d(v,w)$ for all triples of vertices u, v, and w. The reasonableness of this condition is shown in Figure 6-5. Note that the cost of airfares is an example of a distance function that violates the triangle inequality, since it is sometimes cheaper to fly through an intermediate city than to fly to the destination directly. TSP remains hard when the distances are Euclidean distances in the plane.

Whenever a graph obeys the triangle inequality, we can approximate the optimal traveling salesman tour using minimum spanning trees. First, observe that the weight of a minimum spanning tree is a lower bound on the cost of

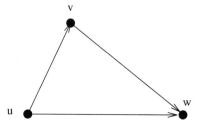

Figure 6–5. The triangle inequality typically holds in geometric and weighted graph problems.

the optimal tour. Why? Deleting any edge from a tour leaves a path, the total weight of which must be no greater than that of the original tour. This path has no cycles, and hence is a tree, which means its weight is at least that of the minimum spanning tree. Thus the minimum spanning tree cost gives a lower bound on the optimal tour.

Consider now what happens in performing a depth-first traversal of a spanning tree. Suppose we walk through each tree edge as we process it in a depth-first search. We will visit each edge twice, once going down the tree when exploring it and once going up after exploring the entire subtree. For example, in the depth-first search of Figure 6-6, we visit the vertices in order $1 - 2 - 1 - 3 - 5 - 8 - 5 - 9 - 5 - 3 - 6 - 3 - 1 - 4 - 7 - 10 - 7 - 11 - 7 - 4 - 1$, thus using every tree edge exactly twice. Therefore, this tour has weight twice that of the minimum spanning tree, and hence at most twice optimal.

However, vertices will be repeated on this depth-first search tour. To remove the extra vertices, at each step we can take a shortest path to the next unvisited vertex. The shortcut tour for the tree above is $1 - 2 - 3 - 5 - 8 - 9 - 6 - 4 - 7 - 10 - 11 - 1$. Because we have replaced a chain of edges by a single

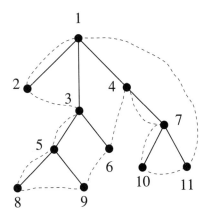

Figure 6–6. A depth-first traversal of a spanning tree, with the shortcut tour

direct edge, the triangle inequality ensures that the tour can only get shorter. Thus the shortcut tour is within weight twice that of optimal. More complicated but better approximation algorithms for Euclidean TSP are mentioned in Section 8.5.4. No approximation algorithms exist for TSPs that do not satisfy the triangle inequality.

6.9   Exercises

6-1. Prove that the vertex cover problem (does there exist a subset S of k vertices in a graph G such that every edge in G is incident upon at least one vertex in S?) remains NP-complete even when all the vertices in the graph are restricted to have even degree.

6-2. An instance of the *set cover* problem consists of a set X of n elements, a family F of subsets of X, and an integer k. The question is, do there exist k subsets from F whose union is X?

For example, if $X = \{1, 2, 3, 4\}$ and $F = \{\{1, 2\}, \{2, 3\}, \{4\}, \{2, 4\}\}$, there does not exist a solution for $k = 2$ but there does for $k = 3$ (for example, $\{1, 2\}, \{2, 3\}, \{4\}$).

Prove that set cover is NP-complete with a reduction from vertex cover.

6-3. The *baseball card collector problem* is as follows. Given packets $P_1, \ldots, P_m$, each of which contains a subset of that year's baseball cards, is it possible to collect all the year's cards by buying $\leq k$ packets?

For example, if the players are $\{Aaron, Mays, Ruth, Skiena\}$ and the packets are

$$\{\{Aaron, Mays\}, \{Mays, Ruth\}, \{Skiena\}, \{Mays, Skiena\}\},$$

there does not exist a solution for $k = 2$ but there does for $k = 3$, such as

$$\{Aaron, Mays\}, \{Mays, Ruth\}, \{Skiena\}$$

Prove that the baseball card collector problem is NP-hard using a reduction from vertex cover.

6-4. (*) An *Eulerian cycle* is a tour that visits every edge in a graph exactly once. An *Eulerian subgraph* is a subset of the edges and vertices of a graph that has an Eulerian cycle. Prove that the problem of finding the number of edges in the largest Eulerian subgraph of a graph is NP-hard. (Hint: the Hamiltonian circuit problem is NP-hard even if each vertex in the graph is incident upon exactly three edges.)

6-5. The *low degree spanning tree problem* is as follows. Given a graph G and an integer k, does G contain a spanning tree such that all vertices in the tree have degree *at most k* (obviously, only tree edges count towards the degree)? For example, in the following graph, there is no spanning tree such that all vertices have degree less than three.

(a) Prove that the low degree spanning tree problem is NP-hard with a reduction from Hamiltonian *path*.

(b) Now consider the *high degree spanning tree problem*, which is as follows. Given a graph G and an integer k, does G contain a spanning tree whose highest degree vertex is *at least* k? In the previous example, there exists a spanning tree of highest degree 8. Give an efficient algorithm to solve the high degree spanning tree problem, and an analysis of its time complexity.

6-6. (*) The problem of testing whether a graph G contains a Hamiltonian path is NP-hard, where a Hamiltonian *path* P is a path that visits each vertex exactly once. There does not have to be an edge in G from the ending vertex to the starting vertex of P, unlike in the Hamiltonian cycle problem.

Given a directed acyclic graph G (a DAG), give an $O(n + m)$-time algorithm to test whether or not it contains a Hamiltonian *path*. (Hint: think about topological sorting and DFS.)

6-7. (**) The 2-SAT problem is, given a Boolean formula in 2-conjunctive normal form (CNF), decide whether the formula is satisfiable. 2-SAT is like 3-SAT, except that each clause can have only two literals. For example, the following formula is in 2-CNF:

$$(x_1 \vee x_2) \wedge (\bar{x}_2 \vee x_3) \wedge (x_1 \vee \bar{x}_3)$$

Give a polynomial-time algorithm to solve 2-SAT.

6-8. (*) It is an open question whether the decision problem "Is integer n a composite number, in other words, not prime?" can be computed in time polynomial in the size of the input. Why doesn't the following algorithm suffice to prove it is in P, since it runs in $O(n)$ time?

```
PrimalityTesting(n)
    composite := false
    for i := 2 to n − 1 do
        if (n mod i) = 0 then
            composite := true
```

Implementation Challenges

6-1. Implement a translator that translates satisfiability instances into equivalent 3-SAT instances.

6-2. (*) Design and implement a backtracking algorithm to test whether a set of formulae are satisfiable. What criteria can you use to prune this search?

6-3. (*) Implement the vertex cover to satisfiability reduction, and run the resulting clauses through a satisfiability testing code. Does this seem like a practical way to compute things?

How to Design Algorithms

Designing the right algorithm for a given application is a difficult job. It requires a major creative act, taking a problem and pulling a solution out of the ether. This is much more difficult than taking someone else's idea and modifying it or tweaking it to make it a little better. The space of choices you can make in algorithm design is enormous, enough to leave you plenty of freedom to hang yourself.

This book is designed to make you a better algorithm designer. The techniques presented in Part I of this book provide the basic ideas underlying all combinatorial algorithms. The problem catalog of Part II will help you with modeling your application and point you in the right direction of an algorithm or implementation. However, being a successful algorithm designer requires more than book knowledge; it requires a certain attitude, the right problem-solving approach. It is difficult to teach this mindset in a book; yet getting it is essential to become a successful designer.

The key to algorithm design (or any other problem-solving task) is to proceed by asking yourself a sequence of questions to guide your thought process. What if we do this? What if we do that? Should you get stuck on the problem, the best thing to do is move onto the next question. In any group brainstorming session, the most useful person in the room is the one who keeps asking, "Why can't we do it this way?" not the person who later tells them why. Because eventually she will stumble on an approach that can't be shot down.

Towards this end, we provide below a sequence of questions to guide your search for the right algorithm for your problem. To use it effectively, you must not only ask the questions, but answer them. The key is working through the answers carefully, by writing them down in a log. The correct answer to, "Can I do it this way?" is never "no," but "no, because . . . ." By clearly articulating your reasoning as to why something doesn't work, you can check if it really holds up or whether you have just glossed over a possibility that you didn't want to think hard enough about. You will be surprised how often the reason you can't find a convincing explanation for something is because your conclusion is wrong.

An important distinction to keep aware of during any design process is the difference between *strategy* and *tactics*. Strategy represents the quest for the big picture, the framework around which we construct our path to the goal. Tactics are used to win the minor battles we must fight along the way. In problem solving, it is important to check repeatedly whether you are thinking on the right level. If you do not have a global strategy of how you are going to attack your problem, it is pointless to worry about the tactics. An example of a strategic question is, "How best can I model my application as a graph algorithm problem?" A tactical question might be, "Should I use an adjacency list or adjacency matrix data structure to represent my graph?" Of course, such tactical decisions are critical to the ultimate quality of the solution, but they can be properly evaluated only in light of a successful strategy.

When faced with a design problem, too many people freeze up in their thinking. After reading or hearing the problem, they sit down and realize that they *don't know what to do next*. They stare into space, then panic, and finally end up settling for the first thing that comes to mind. Avoid this fate. Follow the sequence of questions provided below and in most of the catalog problem sections. We'll *tell* you what to do next!

Obviously, the more experience you have with algorithm design techniques such as dynamic programming, graph algorithms, intractability, and data structures, the more successful you will be at working through the list of questions. Part I of this book has been designed to strengthen this technical background. However, it pays to work through these questions regardless of how strong your technical skills are. The earliest and most important questions on the list focus on obtaining a detailed understanding of the problem and do not require specific expertise.

This list of questions was inspired by a passage in that wonderful book about the space program *The Right Stuff* [Wol79]. It concerned the radio transmissions from test pilots just before their planes crashed. One might have expected that they would panic, so that ground control would hear the pilot yelling *Ahhhhh-hhhhhh —*, terminated only by the sound of smacking into a mountain. Instead, the pilots ran through a list of what their possible actions could be. *I've tried the flaps. I've checked the engine. Still got two wings. I've reset the —*. They had "the Right Stuff." Because of this, they sometimes managed to miss the mountain.

I hope this book and list will provide you with "the Right Stuff" to be an algorithm designer. And I hope it prevents you from smacking into any mountains along the way.

1. Do I really understand the problem?

    (a) What exactly does the input consist of?
    (b) What exactly are the desired results or output?
    (c) Can I construct an example input small enough to solve by hand? What happens when I try to solve it?
    (d) How important is it to my application that I always find an exact, optimal answer? Can I settle for something that is usually pretty good?

(e) How large will a typical instance of my problem be? Will I be working on 10 items? 1,000 items? 1,000,000 items?

(f) How important is speed in my application? Must the problem be solved within one second? One minute? One hour? One day?

(g) How much time and effort can I invest in implementing my algorithm? Will I be limited to simple algorithms that can be coded up in a day, or do I have the freedom to experiment with a couple of approaches and see which is best?

(h) Am I trying to solve a numerical problem? A graph algorithm problem? A geometric problem? A string problem? A set problem? Might my problem be formulated in more than one way? Which formulation seems easiest?

2. Can I find a simple algorithm or heuristic for the problem?

(a) Can I find an algorithm to solve my problem *correctly* by searching through all subsets or arrangements and picking the best one?

    i. If so, why am I sure that this algorithm always gives the correct answer?

    ii. How do I measure the quality of a solution once I construct it?

    iii. Does this simple, slow solution run in polynomial or exponential time? Is my problem small enough that this brute-force solution will suffice?

    iv. If I can't find a slow, *guaranteed* correct algorithm, why am I certain that my problem is sufficiently well-defined to have a correct solution?

(b) Can I solve my problem by repeatedly trying some simple rule, like picking the biggest item first? The smallest item first? A random item first?

    i. If so, on what types of inputs does this heuristic work well? Do these correspond to the data that might arise in my application?

    ii. On what types of inputs does this heuristic work badly? If no such examples can be found, can I show that it always works well?

    iii. How fast does my heuristic come up with an answer? Does it have a simple implementation?

3. Is my problem in the catalog of algorithmic problems in the back of this book?

(a) If it is, what is known about the problem? Is there an implementation available that I can use?

(b) If I don't see my problem, did I look in the right place? Did I browse through all the pictures? Did I look in the index under all possible keywords?

(c) Are there relevant resources available on the World-Wide Web? Did I do a Lycos, Alta Vista, or Yahoo search? Did I go to the WWW page associated with this book, http://www.cs.sunysb.edu/~algorith?

4. Are there special cases of the problem that I know how to solve exactly?

   (a) Can I solve the problem efficiently when I ignore some of the input parameters?

   (b) What happens when I set some of the input parameters to trivial values, such as 0 or 1? Does the problem become easier to solve?

   (c) Can I simplify the problem to the point where I *can* solve it efficiently? Is the problem now trivial or still interesting?

   (d) Once I know how to solve a certain special case, why can't this be generalized to a wider class of inputs?

   (e) Is my problem a special case of a more general problem in the catalog?

5. Which of the standard algorithm design paradigms are most relevant to my problem?

   (a) Is there a set of items that can be sorted by size or some key? Does this sorted order make it easier to find the answer?

   (b) Is there a way to split the problem in two smaller problems, perhaps by doing a binary search? How about partitioning the elements into big and small, or left and right? Does this suggest a divide-and-conquer algorithm?

   (c) Do the input objects or desired solution have a natural left-to-right order, such as characters in a string, elements of a permutation, or the leaves of a tree? If so, can I use dynamic programming to exploit this order?

   (d) Are there certain operations being repeatedly done on the same data, such as searching it for some element, or finding the largest/smallest remaining element? If so, can I use a data structure to speed up these queries? What about a dictionary/hash table or a heap/priority queue?

   (e) Can I use random sampling to select which object to pick next? What about constructing many random configurations and picking the best one? Can I use some kind of directed randomness like simulated annealing in order to zoom in on the best solution?

   (f) Can I formulate my problem as a linear program? How about an integer program?

   (g) Does my problem seem something like satisfiability, the traveling salesman problem, or some other NP-complete problem? If so, might the problem be NP-complete and thus not have an efficient algorithm? Is it in the problem list in the back of Garey and Johnson [GJ79]?

6. Am I still stumped?

   (a) Am I willing to spend money to hire an expert to tell me what to do? If so, check out the professional consulting services mentioned in Section 9.5.

   (b) Why don't I go back to the beginning and work through these questions again? Did any of my answers change during my latest trip through the list?

Problem solving is not a science, but part art and part skill. It is one of the skills most worth developing. My favorite book on problem solving remains Pólya's *How to Solve It* [P57], which features a catalog of problem solving techniques that are fascinating to browse through, both before and after you have a problem.

P
A
R
T

II

Resources

8

A Catalog of Algorithmic Problems

This is a catalog of algorithmic problems that arise commonly in practice. It describes what is known about them and gives suggestions about how best to proceed if the problem arises in your application.

What is the best way to use this catalog? First, think a little about your problem. If you recall the name of your problem, look up the catalog entry in the index or table of contents and start reading. Read through the *entire* entry, since it contains pointers to other relevant problems that might be yours. If you don't find what you are looking for, leaf through the catalog, looking at the pictures and problem names to see if something strikes a chord. Don't be afraid to use the index, for every problem in the book is listed there under several possible keywords and applications. If you *still* don't find something relevant, your problem is either not suitable for attack by combinatorial algorithms or else you don't fully understand it. In either case, go back to step one.

The catalog entries contain a variety of different types of information that have never been collected in one place before. Different fields in each entry present information of practical and historical interest.

To make this catalog more easily accessible, we introduce each problem with a pair of graphics representing the problem instance or input on the left and the result of solving the problem on this instance on the right. We have invested considerable thought in selecting stylized images and examples that illustrate desired behaviors, more than just definitions. For example, the minimum spanning tree example illustrates how points can be clustered using minimum spanning trees. We hope that people without a handle on algorithmic terminology can flip through the pictures and identify which problems might be relevant to them. We augment these pictures with more formal written input and problem descriptions in order to eliminate any ambiguity inherent in a purely pictorial representation.

Once you have identified your problem of interest, the discussion section tells you what you should do about it. We describe applications where the problem is likely to arise and special issues associated with data from them. We discuss the kind of results you can hope for or expect and, most importantly, what you should do to get them. For each problem, we outline a quick-and-dirty algorithm and pointers to algorithms to try next if the first attempt is not sufficient. We also identify other, related problems in the catalog.

For most if not all of the problems presented, we identify readily available software implementations, which are discussed in the implementation field of each entry. Many of these routines are quite good, and they can perhaps be plugged directly into your application. Others will be incomplete or inadequate for production use, but they hopefully can provide a good model for your own implementation. In general, the implementations are listed in order of descending usefulness, but we will explicitly recommend the best one available for each problem if a clear winner exists. More detailed information for many of these implementations appears in Chapter 9. Essentially all of the implementations are available on the enclosed CD-ROM and via the WWW site associated with this book, reachable at http://www.cs.sunysb.edu/~algorith.

Finally, in deliberately smaller print, we discuss the history of each problem and present results of primarily theoretical interest. We have attempted to report the best results known for each problem and point out empirical comparisons of algorithms if they exist. This should be of interest to students and researchers, and also to practitioners for whom our recommended solutions prove inadequate and who need to know if anything better is possible.

Caveats

This is a catalog of algorithmic problems. It is not a cookbook. It cannot be because there are too many possible recipes and too many possible variations on what people want to eat. My goal is to aim you in the right direction so that you can solve your own problems. I point out the issues you are likely to encounter along the way, problems that you are going to have to work out for yourself. In particular:

- For each problem, I suggest algorithms and directions to attack it. These recommendations are based on my experiences and study and aimed towards what I see as typical applications. I felt it was more important to make concrete recommendations for the masses rather than to try to cover all possible situations. If you don't want to follow my advice, don't follow my advice. But before you ignore me, make sure you understand the reasoning behind my recommendations and can articulate a reason why your application violates my assumptions.

- The implementations I recommend are not necessarily complete solutions to your problem. I point to an implementation whenever I feel it might be more useful to someone than just a textbook description of the algorithm. Some programs are useful only as models for you to write your own codes. Others are embedded in large systems and so might be too painful to extract and run on their own. Assume that all of them contain bugs. Many are quite serious, so beware.

- Please respect the licensing conditions for any implementations you use commercially. Many of these codes are not in the public domain and have restrictions. See Section 9.1 for a further discussion of this issue.

- I would be interested in hearing about your experiences with my recommendations, both positive and negative. I would be especially interested in learning about any other implementations that you know about. Feel free to drop me a line at algorith@cs.sunysb.edu.

8.1 Data Structures

Data structures are not really algorithms that you can find and plug into your application. Instead, they are the fundamental constructs for you to build your program around. Becoming fluent in what data structures can do for you is essential to get full value from them.

Because of this, this section is slightly out of sync with the rest of the catalog. Perhaps the most useful aspect of it will be the pointers to implementations of various data structures. Many of these data structures are nontrivial to implement well, so the programs we point to will likely be useful as models even if they do not do exactly what you need. Certain fundamental data structures, like kd-trees and suffix trees, are not as well known as they should be. Hopefully, this catalog will serve to better publicize them.

There is a large number of books on elementary data structures available. Our favorites include:

- *Gonnet and Baeza-Yates* [GBY91] – The book is a comprehensive reference to fundamental searching, sorting, and text searching algorithms. It features over 2,000 references and implementations in C and Pascal. These programs are now available by ftp/WWW. See Section 9.1.6.2 for more details.

- *Weiss* [Wei92] – A nice text, emphasizing data structures more than algorithms. Comes in Pascal, C++, and Ada editions.

- *Wood* [Woo93] – A thorough and accessible treatment of modern data structures, including suffix trees and geometric data structures. Pascal implementations for many of the data structures are provided.

Mehlhorn and Tsakalidis [MT90b] provide a detailed and up-to-date survey of research in data structures. The student who took only an elementary course in data structures is likely to be impressed by the volume and quality of recent work on the subject.

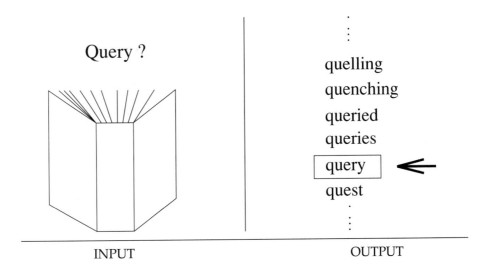

Query ?

quelling
quenching
queried
queries
query   ←
quest

INPUT   OUTPUT

8.1.1 Dictionaries

Input description: A set of n records, each identified by one or more key fields.

Problem description: Build and maintain a data structure to efficiently locate, insert, or delete the record associated with any query key q.

Discussion: The abstract data type "dictionary" is one of the most important structures in computer science. Dozens of different data structures have been proposed for implementing dictionaries including hash tables, skip lists, and balanced/unbalanced binary search trees – so choosing the right one can be tricky. Depending on the application, it is also a decision that can significantly impact performance. *In practice, it is more important to avoid using a bad data structure than to identify the single best option available.*

An essential piece of advice is to carefully isolate the implementation of the dictionary data structure from its interface. Use explicit calls to subroutines that initialize, search, and modify the data structure, rather than embedding them within the code. This leads to a much cleaner program, but it also makes it easy to try different dictionary implementations to see how they impact performance. Do not obsess about the cost of the procedure call overhead inherent in such an abstraction. If your application is so time-critical that such overhead can impact performance, then it is even more essential that you be able to easily experiment with different implementations of your dictionary.

In choosing the right data structure for your dictionary, ask yourself the following questions:

- *How many items will you typically have in your data structure?* – Will you know this number in advance? Are you looking at a problem small enough that the simple data structure will be best, or will it be so large that we must worry about using too much memory or swapping?

- *Do you know the relative number of insertions, deletions, and search queries?* – Will there be any modifications to the data structure after it is first constructed, or will it be static from that point on?

- *Do you have an understanding of the relative frequency with which different keys will be accessed?* – Can we assume that the access pattern will be uniform and random, or will it exhibit a skewed access distribution (i.e. certain elements are much more popular than others) or a sense of locality (i.e. elements are likely to be repeatedly accessed in clusters, instead of at fairly random intervals). Usually, the world is both skewed and clustered.

- *Is it critical that individual operations be fast, or only that the total amount of work done over the entire program be minimized?* – When response time is critical, such as in a program controlling a heart-lung machine, you can't wait too long between steps. When you have a program that is doing a lot of queries over the database, such as identifying all sex offenders who happen to be Republicans, it is not so critical that you pick out any particular congressman quickly as that you get them all with the minimum total effort.

Once you understand what your needs are, try to identify the best data structure from the list below:

- *Unsorted linked lists or arrays* – For small data sets, say up to 10 to 20 items, an unsorted array is probably the easiest and most efficient data structure to maintain. They are easier to work with than linked lists, and if the dictionary will be kept this small, you cannot possibly save a significant amount of space over allocating a full array. If your dictionary will be too much larger, the search time will kill you in either case.

  A particularly interesting and useful variant is a *self-organizing list*. Whenever a key is accessed or inserted, always move it to head of the list. Thus if the key is accessed again in the near future, it will be near the front and so require only a short search to find it. Since most applications exhibit both uneven access frequencies and locality of reference, the average search time for a successful search in a self-organizing list is typically much better than in a sorted or unsorted list. Of course, self-organizing data structures can be built from arrays as well as linked lists.

- *Sorted linked lists or arrays* – Maintaining a sorted linked list is usually not worth the effort (unless you are trying to eliminate duplicates), since we cannot perform binary search in such a data structure. A sorted array will be appropriate if and only if there are not many insertions or deletions. When the array gets so large that it doesn't fit in real memory, think B-trees instead.

- *Hash tables* – For applications involving a moderate-to-large number of keys (say between 100 and 1,000,000), a hash table with bucketing is probably the right way to go. In a hash table, we use a function that maps keys (be they strings, numbers, or whatever) to integers between 0 and $m - 1$. We maintain an array of m buckets, each typically implemented using an unsorted linked list. For a given key, the hash function immediately identifies which bucket will contain it. If we use a hash function that spreads the keys out nicely and a sufficiently large hash table, each bucket should contain very few items, thus making linear search acceptable. Insertion and deletion from a hash table reduce to insertion and deletion from the bucket/list.

A well-tuned hash table will likely outperform a sorted array in most applications. However, several design decisions go into creating a well-tuned hash table:

– *How big should the table be?* Typically, *m* should about the same as the maximum number of items you expect to put in the table. Make sure that *m* is a prime number, so as to minimize the dangers of a bad hash function.

– *What hash function should I use?* For strings, something like

$$F(S) = \sum_{i=1}^{|S|} char(S_i) \times \alpha^{i-1} \bmod m$$

should work, where α is the size of the alphabet and *char(x)* is the function that maps each character *x* to its ASCII character code. For long strings, 8 to 10 characters should be sufficient to hash upon, provided they are unlikely to be padded blanks or some other invariant. Use Horner's rule to implement this hash function computation efficiently, as discussed in Section 8.2.9.

Regardless of which hash function you decide to use, print statistics on the distribution of keys per bucket to see how uniform it *really* is. Odds are the first hash function you try will not prove to be the best. Botching up the hash function is an excellent way to slow down any application.

• *Binary search trees* – Binary search trees are elegant data structures that support fast insertions, deletions, and queries. The big distinction between different types of trees is whether they are explicitly rebalanced after insertion or deletion, and how this rebalancing takes place. In *random search trees*, no rebalancing takes place and we simply insert a node at the leaf position where we can find it. Although search trees perform well under random insertions, most applications are not really random. Indeed, unbalanced search trees constructed by inserting keys in sorted order are a disaster, performing like a linked list.

Balanced search trees use local *rotation* operations to restructure search trees, moving more distant nodes closer to the root while maintaining the in-order search structure of the tree. Among balanced search trees, AVL and 2/3 trees are now passé, and *red-black trees* seem to be more popular. A particularly interesting self-organizing data structure is the *splay tree*, which uses rotations to move any accessed key to the root. Frequently used or recently accessed nodes thus sit near the top of the tree, allowing fast search.

Bottom line: Which binary search tree is best for your application? Probably the balanced tree for which you have the best implementation readily available. See the choices below. Which flavor of balanced tree is probably not as important as how good the programmer was who coded it.

- *B-trees* – For data sets so large that they will not fit in main memory (say more than 1,000,000 items) your best bet will be some flavor of a B-tree. As soon as the data structure gets outside of main memory, the search time to access a particular location on a disk or CD-ROM can kill you, since this is several orders of magnitude slower than accessing RAM.

  The idea behind a B-tree is to collapse several levels of a binary search tree into a single large node, so that we can make the equivalent of several search steps before another disk access is needed. We can thereafter reference enormous numbers of keys using only a few disk accesses. To get the full benefit from using a B-tree, it is important to understand explicitly how the secondary storage device and virtual memory interact, through constants such as page size and virtual/real address space.

  Even for modest-sized data sets, unexpectedly poor performance of a data structure may be due to excessive swapping, so listen to your disk to help decide whether you should be using a B-tree.

- *Skip lists* – These are somewhat of a cult data structure. Their primary benefits seem to be ease of implementation relative to balanced trees. If you are using a canned tree implementation, and thus not coding it yourself, this benefit is eliminated. I wouldn't bother with them.

Implementations: LEDA (see Section 9.1.1) provides an extremely complete collection of dictionary data structures in C++, including hashing, perfect hashing, B-trees, red-black trees, random search trees, and skip lists. Given all of these choices, their default dictionary implementation is a randomized search tree [AS89], presumably reflecting which structure they expect to be most efficient in practice.

XTango (see Section 9.1.5) is an algorithm animation system for UNIX and X-windows that includes animations of such dictionary data structures as AVL trees, binary search trees, hashing, red-black trees, and treaps (randomized search trees). Many of these are interesting and quite informative to watch. Further, the C source code for each animation is included.

The 1996 DIMACS implementation challenge focused on elementary data structures like dictionaries. The world's best available implementations were likely to be identified during the course of the challenge, and they are accessible from http://dimacs.rutgers.edu/ .

Bare bones implementations in C and Pascal of a dizzying variety of dictionary data structures appear in [GBY91], among them several variations on hashing and binary search trees, and optimal binary search tree construction. See Section 9.1.6.2 for details.

Implementation-oriented treatments of a variety of dictionary data structures appear in [BR95], including hashing, splay trees, red-black trees, and what looks like a thorough implementation of B-trees. Code in C for these data structures is included in the text and is available on disk for a modest fee.

Notes: Mehlhorn and Tsakalidis [MT90b] give a thorough survey of the state of the art in modern data structures. Knuth [Knu73a] provides a detailed analysis and exposition on fundamental dictionary data structures but misses such modern data structures as red-black and splay trees. Gonnet and Baeza-Yates [GBY91] provide implementations (in C and Pascal), detailed references, and experimental results for a wide variety of dictionary data structures. We defer to these sources to avoid giving original references for each of the data structures described above.

Good expositions on red-black trees [GS78] include [BR95, CLR90, Woo93]. Good expositions on splay trees [ST85] include [Tar83, Woo93]. Good expositions on B-trees [BM72] include [BR95, CLR90]. Good expositions on hashing includes [Meh84, Woo93].

Several modern data structures, such as splay trees, have been studied via *amortized analysis*, where we bound the total amount of time used by any sequence of operations. In an amortized analysis, we show that if a single operation is very expensive, this is because we have already benefited from enough cheap operations before it to pay off the higher cost. A data structure realizing an amortized complexity of $O(f(n))$ is less desirable than one whose worst-case complexity is $O(f(n))$ (since a very bad operation might still occur) but better than one with an average-case complexity $O(f(n))$, since the amortized bound will achieve this average on any input.

Newer dictionary data structures that explicitly incorporate randomization into the construction include randomized search trees [AS89] and skip lists [Pug90].

Related Problems: Sorting (see page 236), searching (see page 240).

October 30

December 7

July 4

January 1

February 2

December 25

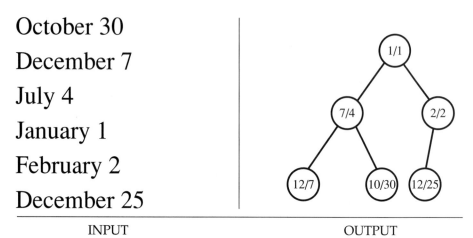

INPUT   OUTPUT

8.1.2 Priority Queues

Input description: A set of records with numerically or otherwise totally ordered keys.

Problem description: Build and maintain a data structure for quickly inserting and deleting records, while enabling quick access to the *smallest* or *largest* key in the set.

Discussion: Priority queues are useful data structures in simulations, particularly for maintaining a set of future events ordered by time so that we can quickly retrieve what the next thing to happen is. They are called "priority" queues because they enable you to retrieve items not by the insertion time (as in a stack or queue), nor by a key match (as in a dictionary), but by which item has the highest priority of retrieval.

If your application performs no insertions after the first query, there is no need for an explicit priority queue. Simply sort the records by priority and proceed from top to bottom, maintaining a pointer to the last record deleted. This situation occurs in Kruskal's minimum spanning tree algorithm, or when simulating a completely scripted set of events.

However, if you are mixing insertions, deletions, and queries, you need a real priority queue. The following questions will help select the right one:

- *Besides access to the smallest element, what other operations will you need?* – Will you be searching for arbitrary keys, or just searching for the smallest? Will you be deleting arbitrary elements from the data, or just repeatedly deleting the top or smallest element?

- *Will you know the maximum size of your data structure in advance, or might an arbitrary number of items be inserted into it?* – The issue here is whether you can preallocate space for the data structure.

- *Will you be changing the priority of elements already in the queue, or simply inserting and removing them?* – Changing the priority of elements implies that we must be able to look up elements in the queue based on their key, in addition to being able to retrieve the largest element.

Depending upon the answers, you have the following basic priority queue choices:

- *Sorted array or list* – In a sorted array, it is very efficient to find and (by decrementing the top index) delete the smallest element. However, maintaining sortedness makes the insertion of new elements slow. Sorted arrays are suitable when there will be few insertions into the priority queue.

- *Binary heaps* – This simple, elegant data structure supports both insertion and extract-min in $O(\lg n)$ time each. Heaps maintain an implicit binary tree structure in an array, such that the key of the root of any subtree is less than that of all its descendents. Thus the minimum key is always at the root of the heap. New keys can be inserted by placing them at an open leaf and percolating the element upwards until it sits at its proper place in the partial order.

  Binary heaps are the right answer whenever you know an upper bound on the number of items in your priority queue, since you must specify the array size at creation time.

- *Bounded height priority queue* – This array-based data structure permits constant-time insertion and find-min operations whenever the range of possible key values is limited. It is the hero of Section 2.6. Suppose we know that all key values will be integers between 1 and n. We can set up an array of n linked lists, such that the ith list serves as a bucket containing all items with key i. We will maintain a pointer *top* to the smallest nonempty list. To insert an item with key k into the priority queue, add it to the kth bucket and set $top = \min(top, k)$. To extract the minimum, report the first item from bucket *top*, delete it, and move *top* down if the bucket is now empty.

  Bounded height priority queues are very useful in maintaining the vertices of a graph sorted by degree, which is a fundamental operation in graph algorithms. Still, they are not as widely known as they should be. They are usually the right priority queue for any small, discrete range of keys.

- *Binary search trees* – Binary search trees make effective priority queues, since the smallest element is always the leftmost leaf, while the largest element is always the rightmost leaf. The min (max) is found by simply tracing down left (right) pointers until the next pointer is nil. Binary tree heaps prove most appropriate when you also need to search a dictionary of the values, or if you have an unbounded key range and do not know the maximum priority queue size in advance.

- *Fibonacci and pairing heaps* – These complicated priority queues are designed to speed up *decrease-key* operations, where the priority of an item already in the priority queue is reduced. This arises, for example, in shortest path computations whenever we discover a shorter route to a vertex v than we had previously established. Thus v has a higher priority of being accessed next.

Properly implemented and used, they lead to better performance on very large computations. Still, they are sufficiently complicated that you shouldn't mess with them unless you really know what you are doing.

Implementations: LEDA (see Section 9.1.1) provides a complete collection of priority queues in C++, including Fibonacci heaps, pairing heaps, Emde-Boas trees, and bounded height priority queues. Fibonacci heaps are their default implementation.

SimPack/Sim++ is a library of routines for implementing discrete event simulations, built by Robert Cubert and Paul Fishwick of the University of Florida. Priority queues are integral to such simulations, and Sim++ contains implementations of linked, binary, leftist, and calendar heaps [Bro88]. If you need a priority queue to control a simulation, check out http://www.cis.ufl.edu/~fishwick/ simpack/simpack.html. An associated book [Fis95] describes model design using SimPack.

Bare bones implementations in C and Pascal of the basic priority queue data structures appear in [GBY91]. Most notable is the inclusion of implementations of exotic priority queues such as P-trees and pagodas. See Section 9.1.6.2 for further details.

XTango (see Section 9.1.5) is an algorithm animation system for UNIX and X-windows, that includes animations of such advanced priority queue data structures as binomial and Fibonacci heaps, as well as a spiffy animation of heapsort.

Many textbooks provide implementations of simple priority queues, including [MS91] (see Section 9.1.6.4). Algorithm 561 [Kah80] of the Collected Algorithms of the ACM is a Fortran implementation of a heap (see Section 9.1.2).

Notes: Good expositions on efficient heap construction algorithms include [Baa88, Ben86, CLR90, Man89, MT90b]. See [GBY91] for a description of several exotic priority queues. Empirical comparisons between priority queue data structures include [Jon86].

Bounded height priority queues are useful data structures in practice, but they do not have good worst-case performance bounds when arbitrary insertions and deletions are permitted. However, von Emde Boas priority queues [vEBKZ77] support $O(\lg \lg n)$ insertion, deletion, search, max, and min operations where each key is an element from 1 to n.

Fibonacci heaps [FT87] support insert and decrease-key operations in $O(1)$ amortized time, with $O(\lg n)$ amortized time extract-min and delete operations. The constant-time decrease-key operation leads to faster implementations of classical algorithms for shortest-paths, weighted bipartite-matching, and minimum-spanning-tree. In practice, Fibonacci heaps are nontrivial to implement and have large constant factors associated with them. However, pairing heaps have been proposed to realize the same bounds with less overhead. Experiments with pairing heaps are reported in [SV87].

Heaps define a partial order that can be built using a linear number of comparisons. The familiar linear-time merging algorithm for heap construction is due to Floyd [Flo64]. In the worst case, $1.625n$ comparisons suffice [GM86] and $1.5n - O(\lg n)$ comparisons are necessary [CC92].

Related Problems: Dictionaries (see page 175), sorting (see page 236), shortest path (see page 279).

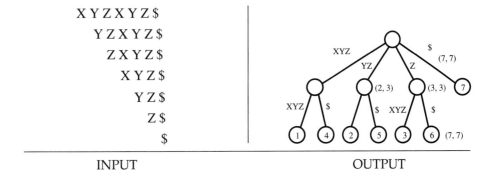

X Y Z X Y Z $

Y Z X Y Z $

Z X Y Z $

X Y Z $

Y Z $

Z $

$

INPUT   OUTPUT

8.1.3 Suffix Trees and Arrays

Input description: A reference string S.

Problem description: Build a data structure for quickly finding all places where an arbitrary query string q is a substring of S.

Discussion: Suffix trees and arrays are phenomenally useful data structures for solving string problems efficiently and with elegance. If you need to speed up a string processing algorithm from $O(n^2)$ to linear time, proper use of suffix trees is quite likely the answer. Indeed, suffix trees are the hero of the war story reported in Section 2.8.

In its simplest instantiation, a suffix tree is simply a *trie* of the n strings that are suffixes of an n-character string S. A trie is a tree structure, where each node represents one character, and the root represents the null string. Thus each path from the root represents a string, described by the characters labeling the nodes traversed. Any finite set of words defines a trie, and two words with common prefixes will branch off from each other at the first distinguishing character. Each leaf represents the end of a string. Figure 8-1 illustrates a simple trie.

Tries are useful for testing whether a given query string q is in the set. Starting with the first character of q, we traverse the trie along the branch defined by the next character of q. If this branch does not exist in the trie, then q cannot be one

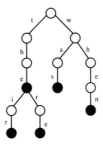

Figure 8–1. A trie on strings *the, their, there, was,* and *when*

of the set of strings. Otherwise we find q in $|q|$ character comparisons *regardless* of how many strings are in the trie. Tries are very simple to build (repeatedly insert new strings) and very fast to search, although they can be expensive in terms of memory.

A *suffix tree* is simply a trie of all the proper suffixes of S. The suffix tree enables you to quickly test whether q is a substring of S, because any substring of S is the prefix of some suffix (got it?). The search time is again linear in the length of q.

The catch is that constructing a full suffix tree in this manner can require $O(n^2)$ time and, even worse, $O(n^2)$ space, since the average length of the n suffices is $n/2$ and there is likely to be relatively little overlap representing shared prefixes. However, linear space suffices to represent a full suffix tree by being clever. Observe that most of the nodes in a trie-based suffix tree occur on simple paths between branch nodes in the tree. Each of these simple paths corresponds to a substring of the original string. By storing the original string in an array and collapsing each such path into a single node described by the starting and ending array indices representing the substring, we have all the information of the full suffix tree in only $O(n)$ space. The output figure for this section displays a collapsed suffix tree in all its glory.

Even better, there exist linear-time algorithms to construct this collapsed tree that make clever use of pointers to minimize construction time. The additional pointers used to facilitate construction can also be used to speed up many applications of suffix trees.

But what can you do with suffix trees? Consider the following applications. For more details see the books by Gusfield [Gus97] or Crochemore and Rytter [CR94]:

- *Find all occurrences of q as a substring of S* – Just as with a trie, we can walk down from the root to find the node n_q associated with q. The positions of all occurrences of q in S are represented by the descendents of n_q, which can be identified using a depth-first search from n_q. For collapsed suffix trees, this takes $O(|q| + k)$ time if there are k occurrences of q in S.

- *Longest substring common to a set T of strings $S_1, \ldots, S_k$* – Build a single collapsed suffix tree containing all suffixes of all strings, with each leaf labeled with its original string. In the course of doing a depth-first search on this tree, we can label each node with both the length of its common prefix and the number of distinct strings from T that are children of it. Thus the best node can be selected in linear time.

- *Find the longest palindrome in S* – A *palindrome* is a string that reads the same if the order of characters is reversed, such as *madam*. To find the longest palindrome in a string S, build a single suffix tree containing all suffixes of S and the reversal of S, with each leaf identified by its starting position. A palindrome is defined by any node in this tree that has forward and reversed children from the same position.

Since the linear time suffix tree construction algorithm is tricky, I recommend either starting from an existing implementation or using a simple, potentially quadratic-time incremental-insertion algorithm to build a compressed suffix tree. Another good option is to use suffix arrays, discussed below.

Suffix arrays do most of what suffix trees do, while typically using four times less memory than suffix trees. They are also easier to implement. A suffix array is basically just an array that contains all the n suffixes of S in sorted order. Thus a binary search of this array for string q suffices to locate the prefix of a suffix that matches q, permitting efficient substring search in $O(\lg n)$ string comparisons. In fact, only $\lg n + |q|$ *character* comparisons need be performed on any query, since we can identify the next character that must be tested in the binary search. For example, if the lower range of the search is *cowabunga* and the upper range is *cowslip*, all keys in between must share the same first three letters, so only the fourth character of any intermediate key must be tested against q.

The space savings of suffix arrays result because as with compressed suffix trees, it suffices to store pointers into the original string instead of explicitly copying the strings. Suffix arrays use less memory than suffix trees by eliminating the need for explicit pointers between suffixes since these are implicit in the binary search. In practice, suffix arrays are typically as fast or faster to search than suffix trees. Some care must be taken to construct suffix arrays efficiently, however, since there are $O(n^2)$ characters in the strings being sorted. A common solution is to first build a suffix *tree*, then perform an in-order traversal of it to read the strings off in sorted order!

Implementations: Ting Chen's and Dimitris Margaritis's C language implementations of suffix trees, reported in the war story of Section 2.8, are available on the algorithm repository WWW site: http://www.cs.sunysb.edu/~algorith.

Bare bones implementations in C of digital and Patricia trie data structures and suffix arrays appear in [GBY91]. See Section 9.1.6.2 for details.

Notes: Tries were first proposed by Fredkin [Fre62], the name coming from the central letters of the word "retrieval". A survey of basic trie data structures with extensive references appears in [GBY91]. Expositions on tries include [AHU83].

Efficient algorithms for suffix tree construction are due to Weiner [Wei73], Mc-Creight [McC76], and Ukkonen [Ukk92]. Good expositions on these algorithms include Crochmore and Wytter [CR94] and Gusfield [Gus97]. Textbooks include [Woo93] and [AHU74], where they are called position trees. Several applications of suffix trees to efficient string algorithms are discussed in [Apo85].

Suffix arrays were invented by Manber and Myers [MM90], although an equivalent idea called *Pat trees* due to Gonnet and Baeza-Yates appears in [GBY91].

The power of suffix trees can be further augmented by using a data structure for computing the *least common ancestor* of any pair of nodes x, y in a tree in *constant* time, after linear-time preprocessing of the tree. The original data structure is due to Harel and Tarjan [HT84], but it was significantly simplified by Schieber and Vishkin [SV88]. Expositions include Gusfield [Gus97]. The least common ancestor (LCA) of two nodes in a suffix tree or trie defines the node representing the longest common prefix of the two associated strings. Being able to answer such queries in constant time is amazing, and useful as a building block for many other algorithms. The correctness of the LCA data structure is difficult to see; however, it is implementable and can perform well in practice.

Related Problems: string matching (see page 403), longest common substring (see page 422).

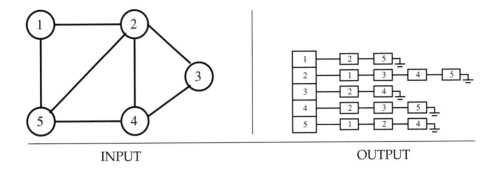

<table>
<tr><td>INPUT</td><td>OUTPUT</td></tr>
</table>

8.1.4 Graph Data Structures

Input description: A graph G.

Problem description: Give a flexible, efficient data structure to represent G.

Discussion: While there are several possible variations, the two basic data structures for graphs are *adjacency matrices* and *adjacency lists*. What these data structures actually are is discussed in Section 4.2. The issues in deciding which one to use include:

- *How big will your graph be?* – How many vertices will it have, both typically and in the worse case? Ditto for the number of edges? If your graph has 100 vertices, your adjacency matrix contains 10,000 entries. If your graph has 1,000 vertices, your adjacency matrix contains 1,000,000 entries. If your graph has 10,000 vertices, your adjacency matrix contains 100,000,000 entries – so forget about it. Adjacency matrices work only for small or very dense graphs.

- *How dense will your graph be?* – If the graph is very dense, meaning that a large fraction of the vertex pairs define edges, there is probably no compelling reason to use adjacency lists, since you will be doomed to using $\Theta(n^2)$ space, anyway.

- *Which algorithms will you be implementing?* – Certain algorithms are easier on adjacency matrices (such as all-pairs shortest path) and others on adjacency lists (such as most DFS-based algorithms). Adjacency matrices win for algorithms that repeatedly ask, "Is (i, j) in G?" However, most graph algorithms can be modified to eliminate such queries.

- *Will you be modifying the graph over the course of your application, and if so, how?* – Repeated edge insertions and (particularly) deletions argue for adjacency matrices, or perhaps for fancier versions of adjacency lists such as binary search trees. However, more likely than modifying the topology of graph is modifying the *attributes* of a vertex or edge of the graph, such as size, weight, or color. Attributes are best handled as extra fields in the vertex or edge records of adjacency lists.

Building a good general-purpose graph type is surprisingly tricky and difficult. For this reason, we suggest that you check out existing implementations (particularly LEDA) before hacking up your own. Note that it costs only time linear in the size of the larger data structure to convert between adjacency matrices and adjacency lists. This conversion is unlikely to be the bottleneck in any application, if you decide you want to use both data structures and have the space to store them. This usually isn't necessary but might prove simplest if you are confused about the alternatives.

Planar graphs are those that can be drawn in the plane so that no two edges cross. Many graphs arising in applications are planar by definition, such as maps of countries, while others are planar by happenstance, like any tree. Planar graphs are always sparse, since any n-vertex planar graph can have at most $3n - 6$ edges, so they should usually be represented by adjacency lists. If the planar drawing (or *embedding*) of the graph is fundamental to what is being computed, planar graphs are best represented geometrically. See Section 8.4.12 for algorithms for constructing planar embeddings from graphs, and Section 8.6.15 for algorithms maintaining graphs implicit in the arrangements of geometric objects like lines and polygons.

Hypergraphs are generalized graphs where each edge may link subsets of more than two vertices. For example, suppose we want to represent who is on which Congressional committee. The vertices of our hypergraph would be the individual congressmen, while each hyperedge would represent one committee. Such arbitrary collections of subsets of a set are naturally thought of as hypergraphs. Two basic data structures for hypergraphs are:

- *Incidence matrices*, which are analogous to adjacency matrices and require $n \times m$ space, where m is the number of hyperedges. Each row corresponds to a vertex, and each column to an edge, with a nonzero entry in $M[i, j]$ iff vertex i is incident to edge j. For standard graphs, there are two nonzero entries in each column. The degree of each vertex governs the number of nonzero entries in each row.

- *Bipartite incidence structures*, which are analogous to adjacency lists, and hence suited for sparse hypergraphs. There is a vertex of the incidence structure associated with each edge and vertex of the hypergraphs, and an edge (i, j) in the incidence structure if vertex i of the hypergraph is in edge j of the hypergraph. Adjacency lists are typically used to represent this incidence structure. This bipartite incidence structure also provides a natural way to visualize the hypergraph, by drawing the associated bipartite graph.

Special efforts must be taken to represent very large graphs efficiently. However, interesting problems have been solved on graphs with millions of edges and vertices. The first step is to make your data structure as lean as possible, by packing your adjacency matrix as a bit vector (see Section 8.1.5) or removing extra pointers from your adjacency list representation. For example, in a static graph (no edge insertions or deletions) each edge list can be replaced by a packed array of vertex identifiers, thus eliminating pointers and saving potentially half the space.

At some point it may become necessary to switch to a hierarchical representation of the graph, where the vertices are clustered into subgraphs that are compressed into single vertices. Two approaches exist for making such a hierarchical decomposition. The first breaks things into components in a natural or application-specific way. For example, knowing that your graph is a map of roads and cities suggests a natural decomposition – partition the map into districts, towns, counties, and states. The other approach runs a graph partition algorithm as in Section 8.5.6. If you are performing the decomposition for space or paging reasons, a natural decomposition will likely do a better job than some naive heuristic for an NP-complete problem. Further, if your graph is really unmanageably large, you cannot afford to do a very good job of algorithmically partitioning it. You should first verify that standard data structures fail on your problem before attempting such heroic measures.

Implementations: LEDA (see Section 9.1.1) provides the best implemented graph data type currently available in C++. If at all possible, you should use it. If not, you should at least study the methods it provides for graph manipulation, so as to see how the right level of abstract graph type makes implementing algorithms very clean and easy. Although a general graph implementation like LEDA may be 2 to 5 times slower and considerably less space efficient than a stripped-down special-purpose implementation, you have to be a pretty good programmer to realize this performance improvement. Further, this speed is likely to come at the expense of simplicity and clarity.

GraphEd [Him94], written in C by Michael Himsolt, is a powerful graph editor that provides an interface for application modules and a wide variety of graph algorithms. If your application demands interaction and visualization more than sophisticated algorithmics, GraphEd might be the right place to start, although it can be buggy. GraphEd can be obtained by anonymous ftp from forwiss.uni-passau.de (132.231.20.10) in directory /pub/local/graphed. See Section 8.4.10 for more details on GraphEd and other graph drawing systems.

The Stanford Graphbase (see Section 9.1.3) provides a simple but flexible graph data structure in CWEB, a literate version of the C language. It is instructive to see what Knuth does and does not place in his basic data structure, although we recommend LEDA as a better basis for further development.

LINK is an environment for combinatorial computing that provides special support for hypergraphs, including the visualization of hypergraphs. Although written in C++, it provides a Scheme language interface for interacting with the graphs. LINK is available from http://dimacs.rutgers.edu/Projects/LINK.html.

An elementary implementation of a "lazy" adjacency matrix in Pascal, which does not have to be initialized, appears in [MS91]. See Section 9.1.6.4.

Simple graph data structures in Mathematica are provided by Combinatorica [Ski90], with a library of algorithms and display routines. See Section 9.1.4.

Notes: It was not until the linear-time algorithms of Hopcroft and Tarjan [HT73b, Tar72] that the advantages of adjacency list data structures for graphs became apparent. The basic adjacency list and matrix data structures are presented in essentially all books on algorithms or data structures, including [CLR90, AHU83, Tar83].

An interesting question concerns minimizing the number of bits needed to represent arbitrary graphs on n vertices, particularly if certain operations must be supported efficiently. Such issues are discussed in [vL90b].

Dynamic graph algorithms are essentially data structures that maintain quick access to an invariant (such as minimum spanning tree or connectivity) under edge insertion and deletion. *Sparsification* [EGIN92] is a general and interesting approach to constructing dynamic graph algorithms. See [ACI92] for an experimental study on the practicality of dynamic graph algorithms.

Hierarchically-defined graphs often arise in VLSI design problems, because designers make extensive use of cell libraries [Len90]. Algorithms specifically for hierarchically-defined graphs include planarity testing [Len89], connectivity [LW88], and minimum spanning trees [Len87a].

The theory of hypergraphs is presented by Berge [Ber89].

Related Problems: Set data structures (see page 191), graph partition (see page 326).

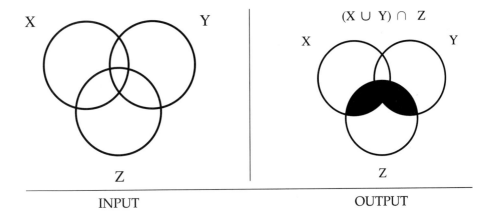

INPUT   OUTPUT

8.1.5 Set Data Structures

Input description: A universe of items $U = \{u_1, \ldots, u_n\}$ and a collection of subsets $S = \{S_1, \ldots, S_m\}$, where $S_i \subset U$.

Problem description: Represent each subset so as to efficiently (1) test whether $u_i \in S_j$, (2) find the union or intersection of S_i and S_j, and (3) insert or delete members of S.

Discussion: In mathematical terms, a set is an unordered collection of objects drawn from a fixed universal set. However, it is usually useful for implementation to represent each set in a single *canonical order*, typically sorted, so as to speed up or simplify various operations. Sorted order turns the problem of finding the union or intersection of two subsets into a linear-time operation – just sweep from left to right and see what you are missing. It also makes possible element searching in sublinear time. Finally, printing the elements of a set in a canonical order paradoxically reminds us that order really doesn't matter.

We distinguish sets from two other kinds of objects: strings and dictionaries. If there is no fixed-size universal set, a collection of objects is best thought of as a *dictionary*, as discussed in Section 8.1.1. If the order does matter in a subset, i.e. if $\{A, B, C\}$ is not the same as $\{B, C, A\}$, then your structure is more profitably thought of as a string, so see Sections 8.1.3 and 8.7.

When each subset has cardinality exactly two, they form edges in a graph whose vertices are the universal set. A system of subsets with no restrictions on the cardinality of its members is called a *hypergraph*. It often can be profitable to consider whether your problem has a graph-theoretical analogy, like connected components or shortest path in a hypergraph.

Your primary alternatives for representing arbitrary systems of subsets are:

- *Bit vectors* – If your universal set U contains n items, an n-bit vector or array can represent any subset $S \subset U$. Bit i will be 1 if $i \in S$, otherwise bit i is 0. Since only one bit is used per element, bit vectors can be very space efficient for surprisingly large values of $|U|$. Element insertion and deletion simply flips the appropriate bit. Intersection and union are done by "and-ing" or "or-ing" the bits together. The only real drawback of a bit vector is that for sparse subsets, it takes $O(n)$ time to explicitly identify all members of S.

- *Containers or dictionaries* – A subset can also be represented using a linked list, array, binary tree, or dictionary containing exactly the elements in the subset. No notion of a fixed universal set is needed for such a data structure. For sparse subsets, dictionaries can be more space and time efficient than bit vectors and easier to work with and program. For efficient union and intersection operations, it pays to keep the elements in each subset sorted, so a linear-time traversal through both subsets identifies all duplicates.

In many applications, the subsets are all pairwise disjoint, meaning that each element is in exactly one subset. For example, consider maintaining the connected components of a graph or the party affiliations of politicians. Each vertex/hack is in exactly one component/party. Such a system of subsets is called a *set partition*. Algorithms for constructing partitions of a given set are provided in Section 8.3.6.

For data structures, the primary issue is maintaining a given set partition as things change over time, perhaps as edges are added or party members defect. The queries we are interested in include "which set is a particular item in?" and "are two items in the same set?" as we modify the set by (1) changing one item, (2) merging or unioning two sets, or (3) breaking a set apart. Your primary options are:

- *Dictionary with subset attribute* – If each item in a binary tree has associated a field recording the name of the subset it is in, set identification queries and single element modifications can be performed in the time it takes to search in the dictionary, typically $O(\lg n)$. However, operations like performing the union of two subsets take time proportional to (at least) the sizes of the subsets, since each element must have its name changed. The need to perform such union operations quickly is the motivation for the ...

- *Union-Find Data Structure* – Suppose we represent a subset using a rooted tree, where each node points to its *parent* instead of its children. Further, let the name of the subset be the name of the item that is the root. Finding out which subset we are in is simple, for we keep traversing up the parent pointers until we hit the root. Unioning two subsets is also easy. Just make the root of one of two trees point to the other, so now *all* elements have the same root and thus the same subset name.

    Certain details remain, such as which subset should be the ultimate root of a union, but these are described in most every algorithms text. Union-Find is a fast, extremely simple data structure that every programmer should know about. It does not support breaking up subsets created by unions, but usually this is not an issue.

Neither of these options provides access to all of the items in a particular subset without traversing all the items in the set. However, both can be appropriately augmented with extra pointers if it is important that this operation be fast.

Implementations: LEDA (see Section 9.1.1) provides dictionary data structures to maintain sets and the union-find data structure to maintain set partitions, all in C++.

LINK is an environment for combinatorial computing that provides special support for hypergraphs, including visualization of hypergraphs. Although written in C++, it provides an additional Scheme language interface for interacting with the graphs. LINK is available from http://dimacs.rutgers.edu/Projects/LINK.html.

Many textbooks contain implementations of the union-find data structure, including [MS91] (see Section 9.1.6.4). An implementation of union-find underlies any implementation of Kruskal's minimum spanning tree algorithm. Section 8.4.3 contains a selection of minimum spanning tree codes.

Notes: Optimal algorithms for such set operations as intersection and union were presented in [Rei72]. Good expositions on set data structures include [AHU83].

Galil and Italiano [GI91] survey data structures for disjoint set union. Expositions on the union-find data structure appear in most algorithm texts, including [CLR90, MS91]. The upper bound of $O(m\alpha(m, n))$ on m union-find operations on an n-element set is due to Tarjan [Tar75], as is a matching lower bound on a restricted model of computation [Tar79]. The inverse Ackerman function $\alpha(m, n)$ grows notoriously slowly, so this performance is close to linear. Expositions on the Ackerman bound include [AHU74]. An interesting connection between the worst-case of union-find and the length of Davenport-Schintzel sequences, a combinatorial structure that arises in computational geometry, is established in [SA95].

The *power set* of a set S is the collection of all $2^{|S|}$ subsets of S. Explicit manipulation of power sets quickly becomes intractable due to their size. Implicit representations of power sets in symbolic form becomes necessary for nontrivial computations. See [BCGR92] for algorithms and computational experience with symbolic power set representations.

Related Problems: Generating subsets (see page 250), generating partitions (see page 253), set cover (see page 398), minimum spanning tree (see page 275).

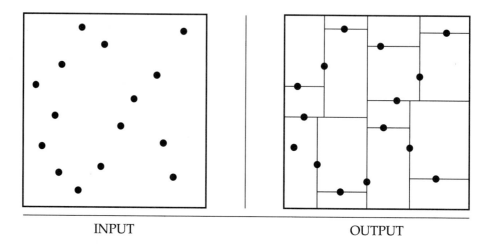

INPUT   OUTPUT

8.1.6   Kd-Trees

Input description: A set S of n points in k dimensions.

Problem description: Construct a tree that partitions the space by half-planes such that each point is contained in its own box-shaped region.

Discussion: Although many different flavors of kd-trees have been devised, their purpose is always to hierarchically decompose space into a relatively small number of cells such that no cell contains too many input objects. This provides a fast way to access any input object by position. We traverse down the hierarchy until we find the cell containing the object and then scan through the few objects in the cell to identify the right one.

Typical algorithms construct kd-trees by partitioning point sets. Each node in the tree is defined by a plane through one of the dimensions that partitions the set of points into left/right (or up/down) sets, each with half the points of the parent node. These children are again partitioned into equal halves, using planes through a different dimension. Partitioning stops after $\lg n$ levels, with each point in its own leaf cell. Alternate kd-tree construction algorithms insert points incrementally and divide the appropriate cell, although such trees can become seriously unbalanced.

The cutting planes along any path from the root to another node defines a unique box-shaped region of space, and each subsequent plane cuts this box into two boxes. Each box-shaped region is defined by $2k$ planes, where k is the number of dimensions. Indeed, the 'kd' in *kd-tree* is short for k-dimensional tree. In any search performed using a kd-tree, we maintain the current region defined by the intersection of these half-spaces as we move down the tree.

Different flavors of kd-trees differ in exactly how the splitting plane is selected. Options include:

- *Cycling through the dimensions* – partition first on d_1, then $d_2, \ldots, d_k$ before cycling back to d_1.

- *Cut along the largest dimension* – select the partition dimension so as to make the resulting boxes as square or cube-like as possible. Selecting a plane to partition the points in half does not mean selecting a splitter in the middle of the box-shaped regions, since all the points may be in the left side of the box.
- *Quadtrees or Octtrees* – Instead of partitioning with single planes, use all axis-parallel planes that pass through a given partition point. In two dimensions, this means creating four child cells, in 3D this means eight child cells. Quadtrees seem particularly popular on image data, where leaf cells imply that all pixels in the region have the same color.

Nonorthogonal (i.e. not axis-parallel) cutting planes have also been used, although they make maintaining the cell boundaries more complicated.

Ideally, our partitions evenly split both the space (ensuring nice, fat, regular regions) and the set of points (ensuring a log height tree) evenly, but this can be impossible for a given point set. The advantages of fat cells become clear in many applications of kd-trees:

- *Point location* – To identify which cell a query point q lies in, we start at the root and test which side of the partition plane contains q. By repeating this process on the appropriate child node, we travel the tree to find the leaf cell containing q in time proportional to its height. See Section 8.6.7 for more on point location.
- *Nearest neighbor search* – To find the point in S closest to a query point q, we perform point location to find the cell c containing q. Since c is bordered by some point p, we can compute the distance $d(p,q)$ from p to q. Point p is likely very close to q, but it might not be the single closest neighbor. Why? Suppose q lies right at the boundary of a cell. Then q's nearest neighbor might lie just to the left of the boundary in another cell. Thus we must traverse all cells that lie within a distance of $d(p,q)$ of cell c and verify that none of them contain closer points. With nice, fat cells, very few cells should need to be tested. See Section 8.6.5 for more on nearest neighbor search.
- *Range search* – Which points lie within a query box or region? Starting from the root, check to see whether the query region intersects or contains the cell defining the current node. If it does, check the children; if not, none of the leaf cells below this node can possibly be of interest. We quickly prune away the irrelevant portions of the space. See Section 8.6.6 for more on range search.
- *Partial key search* – Suppose we want to find a point p in S, but we do not have full information about p. Say we are looking for someone of age 35 and height 5'8" but of unknown weight in a 3d-tree with dimensions age, weight, and height. Starting from the root, we can identify the correct decendant for all but the weight dimension. To be sure we find the right point, we must search *both children* of this node. We are better off the more fields we know, but such partial key search can be substantially faster than checking all points against the key.

Kd-trees are most effective data structures for small and moderate numbers of dimensions, say from 2 up to maybe 20 dimensions. As the dimensionality increases, they lose effectiveness, primarily because the ratio of the volume of a unit sphere in k-dimensions shrinks exponentially compared to a unit cube in k-dimensions. Thus exponentially many cells will have to be searched within a given radius of a query point, say for nearest-neighbor search. Also, the number of neighbors for any cell grows to $2k$ and eventually become unmanageable.

The bottom line is that you should try to avoid working in high-dimensional spaces, perhaps by discarding the least important dimensions.

Implementations: *Ranger* is a tool for visualizing and experimenting with nearest neighbor and orthogonal range queries in high-dimensional data sets, using multidimensional search trees. Four different search data structures are supported by *Ranger*: naive kd-trees, median kd-trees, nonorthogonal kd-trees, and the vantage point tree. For each of these, *Ranger* supports queries in up to 25 dimensions under any Minkowski metric. It includes generators for a variety of point distributions in arbitrary dimensions. Finally, *Ranger* provides a number of features to aid in visualizing multidimensional data, best illustrated by the accompanying video [MS93]. To identify the most appropriate projection at a glance, *Ranger* provides a $k \times k$ matrix of all two-dimensional projections of the data set. *Ranger* is written in C, runs on Silicon Graphics and HP workstations, and is available from the algorithm repository.

The 1996 DIMACS implementation challenge focuses on data structures for higher-dimensional data sets. The world's best kd-tree implementations were likely to be identified in the course of the challenge, and they are accessible from http://dimacs.rutgers.edu/.

Bare bones implementations in C of kd-tree and quadtree data structures appear in [GBY91]. See Section 9.1.6.2 for details on how to ftp them.

Notes: The best reference on kd-trees and other spatial data structures are two volumes by Samet [Sam90a, Sam90b], in which all major variants are developed in substantial detail.

Bentley [Ben75] is generally credited with developing kd-trees, although they have the typically murky history associated with most folk data structures. The most reliable history is likely from Samet [Sam90b].

An exposition on kd-trees for orthogonal range queries in two dimensions appears in [PS85]. Expositions of grid files and other spatial data structures include [NH93].

Algorithms that quickly produce a point provably close to the query point are a recent development in higher-dimensional nearest neighbor search. A sparse weighted-graph structure is built from the data set, and the nearest neighbor is found by starting at a random point and walking greedily in the graph towards the query point. The closest point found during several random trials is declared the winner. Similar data structures hold promise for other problems in high-dimensional spaces. See [AM93, AMN+94].

Related Problems: Nearest-neighbor search (see page 361), point location (see page 367), range search (see page 364).

8.2 Numerical Problems

If most problems you encounter are numerical in nature, there is a good chance that you are reading the wrong book. *Numerical Recipes* [PFTV86] gives a terrific overview to the fundamental problems in numerical computing, including linear algebra, numerical integration, statistics, and differential equations. Different flavors of the book include source code for all the algorithms in C, Pascal, and Fortran. Their coverage is skimpier on the combinatorial/numerical problems we consider in this section, but you should be aware of that book.

Numerical algorithms tend to be different beasts than combinatorial algorithms, for at least two distinct reasons:

- *Issues of Precision and Error* – Numerical algorithms typically perform repeated floating-point computations, which accumulate error at each operation until, eventually, the results are meaningless. An amusing example [SK93] concerns the Vancouver Stock Exchange, which over a twenty-two month period accumulated sufficient round-off error to reduce its index from the correct value of 1098.982 to 574.081.

  A simple and dependable way to test for round-off errors in numerical programs is to run them both at single and double precision, and then think hard whenever there is a disagreement.

- *Extensive Libraries of Codes* – Large, high-quality libraries of numerical routines have existed since the 1960s, which is still not the case for combinatorial algorithms. This is true for several reasons, including (1) the early emergence of Fortran as a standard for numerical computation, (2) the nature of numerical computations to remain independent rather than be embedded within large applications, and (3) the existence of large scientific communities needing general numerical libraries.

  Regardless of why, you should exploit this software base. There is probably no reason to implement algorithms for any of the problems in this section instead of stealing existing codes. Searching Netlib (see Section 9.1.2) is always a good place to start.

Most scientist's and engineer's ideas about algorithms derive from Fortran programming and numerical methods, while computer scientists grew up programming with pointers and recursion, and so are comfortable with the more sophisticated data structures required for combinatorial algorithms. Both sides can and should learn from each other, since several problems such as pattern recognition can be modeled either numerically or combinatorially.

There is a vast literature on numerical algorithms. In addition to *Numerical Recipes*, recommended books include:

- *Skeel and Keiper* [SK93] – A readable and interesting treatment of basic numerical methods, avoiding overly detailed algorithm descriptions through its use of the computer algebra system Mathematica. I like it.

- *Pizer and Wallace* [PW83] – A numerical analysis book written for computer scientists, not engineers. The organization is by issues instead of problems. A different but interesting perspective.

- *Cheney and Kincaid* [CK80] – A traditional Fortran-based numerical analysis text, with discussions of optimization and Monte Carlo methods in addition to such standard topics as root-finding, numerical integration, linear systems, splines, and differential equations.

- *Buchanan and Turner* [BT92] – Thorough language-independent treatment of all standard topics, including parallel algorithms. Most comprehensive of the texts described here.

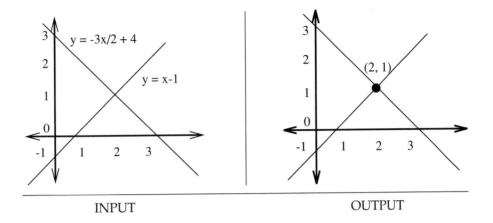

INPUT   OUTPUT

8.2.1 Solving Linear Equations

Input description: An $m \times n$ matrix A, and an $m \times 1$ vector b, together representing m linear equations on n variables.

Problem description: What is the vector x such that $A \cdot x = b$?

Discussion: The need to solve linear systems arises in an estimated 75% of all scientific computing problems [DB74]. For example, applying Kirchhoff's laws to analyze electric circuits generates a system of equations, the solution of which gives currents through each branch of the circuit. Analysis of the forces acting on a mechanical truss generates a similar set of equations. Even finding the point of intersection between two or more lines reduces to solving a (small) linear system.

Not all systems of equations have solutions; consider the equations $2x+3y = 5$ and $2x+3y = 6$. Some systems of equations have multiple solutions; consider the equations $2x+3y = 5$ and $4x+6y = 10$. Such *degenerate* systems of equations are called *singular*, and they can be recognized by testing whether the determinant of the coefficient matrix is zero.

Solving linear systems is a problem of such scientific and commercial importance that excellent codes are readily available. There is likely no good reason to implement your own solver, even though the basic algorithm (Gaussian elimination) is one we learned in high school. This is especially true if you are working with large systems.

Gaussian elimination is based on the fact that the solution to a system of linear equations is invariant under scaling (multiplying both sides by a constant; i.e. if $x = y$, then $2x = 2y$) and adding equations (i.e. the solution to the equations $x = y$ and $w = z$ is the same as the solution to $x = y$ and $x+w = y+z$). Gaussian elimination scales and adds equations so as to eliminate each variable from all but one equation, leaving the system in such a state that the solution can just be read off from the equations.

The time complexity of Gaussian elimination on an $n \times n$ system of equations is $O(n^3)$, since for the ith variable we add a scaled copy of the n-term ith row to each of the $n - 1$ other equations. On this problem, however, constants matter. Algorithms that only partially reduce the coefficient matrix and then backsubstitute to get the answer use 50% fewer floating-point operations than the naive algorithm.

Issues to worry about include:

- *Are roundoff errors and numerical stability affecting my solution?* – Implementing Gaussian elimination would be quite straightforward except for round-off errors, which accumulate with each row operation and can quickly wreak havoc on the solution, particularly with matrices that are *almost* singular.

  To eliminate the danger of numerical errors, it pays to substitute the solution back into each of the original equations and test how close they are to the desired value. *Iterative methods* for solving linear systems refine initial solutions to obtain more accurate answers – good linear systems packages will include such routines.

  The key to minimizing roundoff errors in Gaussian elimination is selecting the right equations and variables to pivot on, and to scale the equations so as to eliminate large coefficients. This is as much an art as a science, which is why you should use one of the many well-crafted library routines described below.

- *Which routine in the library should I use?* – Selecting the right code is also somewhat of an art. If you are taking your advice from this book, you should start with the general linear system solvers. Hopefully they will suffice for your needs. But search through the manual for more efficient procedures for solving special types of linear systems. If your matrix happens to be one of these special types, the solution time can reduce from cubic to quadratic or even linear.

- *Is my system sparse?* – The key to recognizing that you might have a special-case linear system is establishing how many matrix elements you really need to describe A. If there are only a few nonzero elements, your matrix is *sparse* and you are in luck. If these few nonzero elements are clustered near the diagonal, your matrix is *banded* and you are in even more luck. Algorithms for reducing the bandwidth of a matrix are discussed in Section 8.2.2. Many other regular patterns of sparse matrices can also be exploited, so see the manual of your solver or a better book on numerical analysis for details.

- *Will I be solving many systems with the same coefficient matrix?* – In certain applications, such as least-squares curve fitting and differential equations, we have to solve $A \cdot x = b$ repeatedly with different b vectors. For efficiency, we seek to preprocess A to make this easier. The lower-upper or *LU-decomposition* of A creates lower- and upper-triangular matrices L and U such that $L \cdot U = A$. We can use this decomposition to solve $A \cdot x = b$, since

$$A \cdot x = (L \cdot U) \cdot x = L \cdot (U \cdot x) = b$$

This is efficient since backsubstitution solves a triangular system of equations in quadratic time. Solving $L \cdot y = b$ and then $U \cdot x = y$ gives the solution x

using two $O(n^2)$ steps instead of one $O(n^3)$ step, once the LU-decomposition has been found in $O(n^3)$ time.

The problem of solving linear systems is equivalent to that of matrix inversion, since $Ax = B \leftrightarrow A^{-1}Ax = A^{-1}B$, where $I = A^{-1}A$ is the identity matrix. However, avoid it, since matrix inversion proves to be three times slower than Gaussian elimination. LU-decompositions prove useful in inverting matrices as well as computing determinants (see Section 8.2.4).

Implementations: The library of choice for solving linear systems is apparently LAPACK, a descendant of LINPACK [DMBS79]. Both of these Fortran codes, as well as many others, are available from Netlib. See Section 9.1.2.

Algorithm 533 [She78], Algorithm 576 [BS81], and Algorithm 578 [DNRT81] of the Collected Algorithms of the ACM are Fortran codes for Gaussian elimination. Algorithm 533 is designed for sparse systems, algorithm 576 to minimize roundoff errors, and algorithm 578 to optimize virtual memory performance. Algorithm 645 [NW86] is a Fortran code for testing matrix inversion programs. See Section 9.1.2 for details on fetching these programs.

Numerical Recipes [PFTV86] provides routines for solving linear systems. However, there is no compelling reason to use these ahead of the free codes described in this section.

C++ implementations of $O(n^3)$ algorithms to solve linear equations and invert matrices are embedded in LEDA (see Section 9.1.1).

Notes: Good expositions on algorithms for Gaussian elimination and LU-decomposition include [CLR90] and a host of numerical analysis texts [BT92, CK80, SK93].

Parallel algorithms for linear systems are discussed in [Ort88]. Solving linear systems is one of relatively few problems where parallel architectures are widely used in practice.

Matrix inversion, and hence linear systems solving, can be done in matrix multiplication time using Strassen's algorithm plus a reduction. Good expositions on the equivalence of these problems include [AHU74, CLR90].

Certain types of nonsparse systems can be solved efficiently via special algorithms. In particular, *Toeplitz* matrices are constructed so that all the elements along a particular diagonal are identical, and *Vandermonde* matrices are defined by an n-element vector x where $A[i, j - 1] = x_i^j$.

Related Problems: Matrix multiplication (see page 204), determinant/permanent (see page 207).

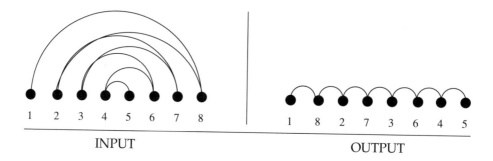

INPUT   OUTPUT

8.2.2 Bandwidth Reduction

Input description: A graph $G = (V, E)$, representing an $n \times n$ matrix M of zero and nonzero elements.

Problem description: Which permutation p of the vertices of V minimizes the length of the longest edge when the vertices are ordered on a line, i.e. minimizes $\max_{(i,j) \in E} |p(i) - p(j)|$?

Discussion: Bandwidth reduction lurks as a hidden but important problem for both graphs and matrices, and it is important to see how it arises so as to properly recognize it. Applied to matrices, it permutes the rows and columns of a sparse matrix so as to minimize the distance b of any nonzero entry from the center diagonal. This is important in solving linear systems, because Gaussian elimination (see Section 8.2.1) can be performed in $O(nb^2)$ on matrices of bandwidth b. This is a big win over the general $O(n^3)$ algorithm if $b << n$.

Bandwidth minimization on graphs arises in more subtle ways. Arranging a set of n circuit components in a line on a circuit board so as to minimize the length of the longest wire (and hence time delay) is a bandwidth problem, where each vertex of our graph corresponds to a circuit component and there is an edge for every wire linking two components. Alternatively, consider a hypertext application where we must store large objects (say images) on a magnetic tape. From each image there is a set of possible images we can go to next (i.e. the hyperlinks). To minimize the search time, we seek to place linked images near each other on the tape. This is exactly the bandwidth problem. More general formulations, such as rectangular circuit layouts and magnetic disks, inherit the same hardness and classes of heuristics from the linear versions.

Unfortunately, bandwidth minimization is NP-complete. It stays NP-complete even if the input graph is a tree whose maximum vertex degree is 3, which is an unusually strong condition. Further, there is no known approximation algorithm for bandwidth reduction, even for trees. Thus our only options are brute-force search or ad hoc heuristics.

Fortunately, ad hoc heuristics have been well-studied in the numerical analysis community, and production-quality implementations of the best heuristics are available. These are based on performing a breadth-first search from a given vertex v, where v is placed at the leftmost point of the ordering. All of the vertices that are distance 1 from v are placed to its immediate right, followed by all the vertices at distance 2, and so forth until we reach the vertex furthest from v. We then continue the breadth-first search from the vertex immediately to the right of v until all vertices in G are accounted for. The popular heuristics differ according to how many different start vertices are considered and how equidistant vertices are ordered among themselves. However, breaking ties with low-degree vertices over to the left seems to be a good idea.

Implementations of the most popular heuristics, the Cuthill-McKee and Gibbs-Poole-Stockmeyer algorithms, are discussed in the implementation section. The worst case of the Gibbs-Poole-Stockmeyer algorithm is $O(n^3)$, which would wash out any possible savings in solving linear systems, but its performance in practice is close to linear.

Brute-force search programs can find the exact minimum bandwidth work by backtracking through the set of $n!$ possible permutations of vertices. Considerable pruning can be achieved to reduce the search space by starting with a good heuristic bandwidth solution and alternately adding vertices to the left- and rightmost open slots in the partial permutation. The first edge connecting a vertex on the left to a vertex on the right will likely define an edge whose length is greater than our best example to date, thus leading to fast pruning. In our experience, graphs of size $n = 30$ or so can be solved to optimality. See the discussion on backtracking in Section 5.1. However, for almost any application such an exact solution will not be worth the expense of finding it.

Implementations: Fortran language implementations of both the Cuthill-McKee algorithm [CGPS76, Gib76, CM69] and the Gibbs-Poole-Stockmeyer algorithm [Lew82, GPS76] are available from Netlib. See Section 9.1.2. Empirical evaluations of these and other algorithms on a test suite of 30 matrices are discussed in [Eve79b], showing Gibbs-Poole-Stockmeyer to be the consistent winner.

Brute-force implementations written by Ingmar Bitter and Dario Vlah in C and C++ as Stony Brook class projects and capable of solving instances of size $n = 30$ to optimality are provided on the algorithm repository http://www.cs.sunysb.edu/~algorith.

Notes: An excellent survey on graph-theoretic and algorithmic results on the bandwidth problem to 1981 appears in [CCDG82]. Ad hoc heuristics have been widely studied, a tribute to its importance in numerical computation. Everstine [Eve79b] cites no less than 49 different bandwidth reduction algorithms!

The hardness of the bandwidth problem was first established by Papadimitriou [Pap76b], and its hardness on trees of maximum degree 3 in [GGJK78]. There are algorithms that run in polynomial time for fixed bandwidth k [Sax80]. An exposition on the hardness of the linear arrangement problem appears in [Eve79a].

Related Problems: Solving linear equations (see page 199), topological sorting (see page 273).

$$\begin{bmatrix} 2 & 3 \\ 3 & 4 \\ 4 & 5 \end{bmatrix} \begin{bmatrix} 2 & 3 & 4 \\ 3 & 4 & 5 \end{bmatrix} \qquad \begin{bmatrix} 13 & 18 & 23 \\ 18 & 25 & 32 \\ 23 & 32 & 41 \end{bmatrix}$$

INPUT   OUTPUT

8.2.3 Matrix Multiplication

Input description: An $x \times y$ matrix A and a $y \times z$ matrix B.

Problem description: The $x \times z$ matrix $A \times B$.

Discussion: Although matrix multiplication is an important problem in linear algebra, its main significance for combinatorial algorithms is its equivalence to a variety of other problems, such as transitive closure and reduction, solving linear systems, and matrix inversion. Thus a faster algorithm for matrix multiplication implies faster algorithms for all of these problems. Matrix multiplication arises in its own right in computing the results of such coordinate transformations as scaling, rotation, and translation for robotics and computer graphics.

The straightforward algorithm to compute the product of $x \times y$ matrix A and $y \times z$ matrix B runs in $O(xyz)$ time and is tough to beat in practice:

> for $i = 1$ to x do
>   for $j = 1$ to z
> $M[i,j] = \sum_{k=1}^{y} A[i,k] \cdot A[k,j]$

In two multiplying bandwidth-b matrices, where all nonzero elements of A and B lie within b elements of the main diagonals, a speedup to $O(xbz)$ is possible, since zero elements will not contribute to the product.

Asymptotically faster algorithms for matrix multiplication exist, based on clever divide-and-conquer recurrences. However, these prove difficult to program and require very large matrices to beat the trivial algorithm. In particular, some empirical results show that Strassen's $O(n^{2.81})$ algorithm is unlikely to beat the straightforward algorithm for $n \leq 100$, and it is less numerically stable to boot. Other studies have been more encouraging, claiming that the crossover point is as low as $n \leq 32$. Still, I consider it unlikely that you will speed up any serious application by implementing Strassen's algorithm.

There is a better way to save computation when you are multiplying a chain of more than two matrices together. Recall that multiplying an $x \times y$ matrix by a $y \times z$ matrix creates an $x \times z$ matrix. Thus multiplying a chain of matrices from left to right might create large intermediate matrices, each taking a lot of time to compute. Matrix multiplication is not commutative, but it is associative, so we can parenthesize the chain in whatever manner we deem best without changing the final product. A standard dynamic programming algorithm can be used to construct the optimal parenthesization. Whether it pays to do this optimization will depend upon whether your matrices are large enough or your chain is multiplied often enough to justify it. Note that we are optimizing over the sizes of the dimensions in the chain, not the actual matrices themselves. If all your matrices are the same dimensions, you are out of luck, since no such optimization is possible.

Matrix multiplication has a particularly interesting interpretation in counting the number of paths between two vertices in a graph. Let A be the adjacency matrix of a graph G, meaning $A[i, j] = 1$ if there is an edge between i and j. Otherwise, $A[i, j] = 0$. Now consider the square of this matrix, $A^2 = A \times A$. If $A^2[i, j] \geq 1$, this means that there must be a k such that $A[i, k] = A[k, j] = 1$, so i to k to j is a path of length 2 in G. More generally, $A^k[i, j]$ counts the number of paths of length exactly k between i and j. This count includes nonsimple paths, where vertices are repeated, such as i to k to i.

Implementations: The quick and dirty $O(n^3)$ algorithm will be your best bet unless your matrices are very large. For example, [CLR90] suggests that $n > 45$ before you have a hope of winning. Experimental results suggest that $n > 100$ is more realistic [CR76], with Bailey [BLS91] finding a crossover point of $n = 128$ for Cray systems. Strassen's algorithm is difficult to implement efficiently because of the data structures required to maintain the array partitions. That said, an implementation of Strassen's algorithm in Mathematica by Stan Wagon is offered "without promise of efficiency" on the algorithm repository WWW site.

The linear algebra library of choice is LAPACK, a descendant of LINPACK [DMBS79], which includes several routines for matrix multiplication. These Fortran codes are available from Netlib as discussed in Section 9.1.2.

Algorithm 601 [McN83] of the Collected Algorithms of the ACM is a sparse matrix package written in Fortran that includes routines to multiply any combination of sparse and dense matrices. See Section 9.1.2 for details.

XTango (see Section 9.1.5) is an algorithm animation system for UNIX and X-windows that includes an animation of the $O(n^3)$ matrix multiplication algorithm. A C++, $O(n^3)$ implementation of matrix multiplication is embedded in LEDA (see Section 9.1.1).

Notes: Winograd's algorithm for fast matrix multiplication reduces the number of multiplications by a factor of two over the straightforward algorithm. It is implementable, although the additional bookkeeping required makes it doubtful whether it is a win. Expositions on Winograd's algorithm [Win68] include [CLR90, Man89, Win80].

In my opinion, the history of theoretical algorithm design began when Strassen published his $O(n^{2.81})$-time matrix multiplication algorithm. For the first time, improving an algorithm in the asymptotic sense became a respected goal in its own right. Good expositions on Strassen's algorithm [Str69] include [Baa88, CLR90, Cra94]. Progressive improvements to Strassen's algorithm have gotten progressively less practical. The current best result for matrix multiplication is Coppersmith and Winograd's [CW87] $O(n^{2.376})$ algorithm, while the conjecture is that $\Theta(n^2)$ suffices.

The interest in the squares of graphs goes beyond counting paths. Fleischner [Fle74] proved that the square of any biconnected graph has a Hamiltonian cycle. See [LS95] for results on finding the square roots of graphs, i.e. finding A given A^2.

The problem of Boolean matrix multiplication can be reduced to that of general matrix multiplication [CLR90]. The four-Russians algorithm for Boolean matrix multiplication [ADKF70] uses preprocessing to construct all subsets of $\lg n$ rows for fast retreival in performing the actual multiplication, yielding a complexity of $O(n^3/\lg n)$. Additional preprocessing can improve this to $O(n^3/\lg^2 n)$ [Ryt85]. An exposition on the four-Russians algorithm, including this speedup, appears in [Man89].

Good expositions of the matrix-chain algorithm include [Baa88, CLR90], where it is a standard example of dynamic programming.

Related Problems: Solving linear equations (see page 199), shortest path (see page 279).

$$\det \begin{bmatrix} 2 & 1 & 3 \\ 4 & -2 & 10 \\ 5 & -3 & 13 \end{bmatrix}$$

$$2 * \det \begin{bmatrix} -2 & 10 \\ -3 & 13 \end{bmatrix} +$$

$$-1 * \det \begin{bmatrix} 4 & 10 \\ 5 & 13 \end{bmatrix} +$$

$$3 * \det \begin{bmatrix} 4 & -2 \\ 5 & -3 \end{bmatrix} = 0$$

INPUT   OUTPUT

8.2.4 Determinants and Permanents

Input description: An $n \times n$ matrix M.

Problem description: What is the determinant $|M|$ or the permanent $perm(M)$ of the matrix m?

Discussion: Determinants of matrices provide a clean and useful abstraction in linear algebra that can used to solve a variety of problems:

- Testing whether a matrix is *singular*, meaning that the matrix does not have an inverse. A matrix M is singular iff $|M| = 0$.
- Testing whether a set of d points lies on a plane in fewer than d dimensions. If so, the system of equations they define is singular, so $|M| = 0$.
- Testing whether a point lies to the left or right of a line or plane. This problem reduces to testing whether the sign of a determinant is positive or negative, as discussed in Section 8.6.1.
- Computing the area or volume of a triangle, tetrahedron, or other simplicial complex. These quantities are a function of the magnitude of the determinant, as discussed in Section 8.6.1.

The determinant of a matrix M is defined as the sum over all $n!$ possible permutations π_i of the n columns of M:

$$|M| = \sum_{i=1}^{n!} (-1)^{sign(\pi_i)} \prod_{j=1}^{n} M[j, \pi_j]$$

where $sign(\pi_i)$ is the number of pairs of elements out of order (called *inversions*) in permutation π_i.

A direct implementation of this definition yields an $O(n!)$ algorithm, as does the cofactor expansion method we learned in high school. However, better algorithms are available to evaluate determinants based on LU-decomposition. They are discussed in Section 8.2.1. The determinant of M is simply the product of the diagonal elements of the LU-decomposition of M, which can be found in $O(n^3)$ time.

A closely related function, called the *permanent*, arises often in combinatorial problems. For example, the permanent of the adjacency matrix of a graph G counts the number of perfect matchings in G. The permanent of a matrix M is defined by

$$perm(M) = \sum_{i=1}^{n!} \prod_{j=1}^{n} M[j, \pi_j]$$

differing from the determinant only in that all products are positive.

Surprisingly, it is NP-hard to compute the permanent, even though the determinant can easily be computed in $O(n^3)$ time. The fundamental difference is that $det(AB) = det(A) \times det(B)$, while $perm(AB) \neq perm(A) \times perm(B)$. Fortunately, there are permanent algorithms that prove to be considerably faster than the $O(n!)$ definition, running in $O(n^2 2^n)$ time. Thus finding the permanent of a 20×20 matrix is not out of the realm of possibility.

Implementations: The linear algebra package LINPACK contains a variety of Fortran routines for computing determinants, optimized for different data types and matrix structures. It can be obtained from Netlib, as discussed in Section 9.1.2. A C++ program to compute determinants in $O(n^3)$ time is embedded in LEDA (see Section 9.1.1).

Nijenhuis and Wilf [NW78] provide an efficient Fortran routine to compute the permanent of a matrix. See Section 9.1.6.3.

Notes: Cramer's rule reduces the problems of matrix inversion and solving linear systems to that of computing determinants. However, algorithms based on LU-determination are faster. See [BM53] for an exposition on Cramer's rule.

Determinants can be computed in $o(n^3)$ time using fast matrix multiplication, as shown in [AHU83]. Section 8.2.3 discusses such algorithms. A fast algorithm for computing the sign of the determinant, an important problem for performing robust geometric computations, is due to Clarkson [Cla92].

The problem of computing the permanent was shown to be #P-complete by Valiant [Val79], where #P is the class of problems solvable on a "counting" machine in polynomial time. A counting machine returns the number of distinct solutions to a problem. Counting the number of Hamiltonian cycles in a graph is a #P-complete problem that is trivially NP-hard (and presumably harder), since any count greater than zero proves that the graph is Hamiltonian. Counting problems can be #P-complete even if the corresponding decision problem can be solved in polynomial time, as shown by the permanent and perfect matchings.

Minc [Min78] is the definitive work on permanents. A variant of an $O(n^2 2^n)$-time algorithm due to Ryser for computing the permanent is presented in [NW78]. Recently, probabilistic algorithms have been developed for estimating the permanent [FJ95].

Related Problems: Solving linear systems (see page 199), matching (see page 287), geometric primitives (see page 347).

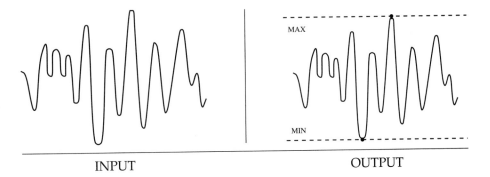

INPUT   OUTPUT

8.2.5 Constrained and Unconstrained Optimization

Input description: A function $f(x_1, \ldots, x_n)$.

Problem description: What point $p = (p_1, \ldots, p_n)$ maximizes (or minimizes) the function f?

Discussion: Most of this book concerns algorithms that optimize one thing or another. This section considers the general problem of optimizing functions where, due to lack of structure or knowledge, we are unable to exploit the problem-specific algorithms seen elsewhere in this book.

Optimization arises whenever there is an objective function that must be tuned for optimal performance. Suppose we are building a program to identify good stocks to invest in. We have available certain financial data to analyze, such as the price-earnings ratio, the interest and inflation rates, and the stock price, all as a function of time t. The key question is how much weight we should give to each of these factors, where these weights correspond to coefficents of a formula:

$$\text{stock-goodness}(t) = c_1 \times \text{price}(t) + c_2 \times \text{interest}(t) + c_3 \times \text{PE-ratio}(t)$$

$$+ c_4 \times \text{inflation}(t)$$

We seek the numerical values c_1, c_2, c_3, c_4 whose stock-goodness function does the best job of evaluating stocks. Similar issues arise in tuning evaluation functions for game playing programs such as chess.

Unconstrained optimization problems also arise in scientific computation. Physical systems from protein structures to particles naturally seek to minimize their "energy functions." Thus programs that attempt to simulate nature often define energy potential functions for the possible configurations of objects and then take as the ultimate configuration the one that minimizes this potential.

Global optimization problems tend to be hard, and there are lots of ways to go about them. Ask the following questions to steer yourself in the right direction:

- *Am I doing constrained or unconstrained optimization?* – In unconstrained optimization, there are no limitations on the values of the parameters other than

that they maximize the value of f. Often, however, there are costs or constraints on these parameters. These constraints make certain points illegal,
points that might otherwise be the global optimum. Constrained optimization problems typically require mathematical programming approaches like
linear programming, discussed in Section 8.2.6.

- *Is the function I am trying to optimize described by a formula or data?* – If the
function that you seek to optimize is presented as an algebraic formula (such
as the minimum of $f(n) = n^2 - 6n + 2^{n+1}$), the solution is to analytically take its
derivative $f'(n)$ and see for which points p' we have $f'(p') = 0$. These points
are either local maxima or minima, which can be distinguished by taking
a second derivative or just plugging back into f and seeing what happens.
Symbolic computation systems such as Mathematica and Maple are fairly
effective at computing such derivatives, although using computer algebra
systems effectively is somewhat of a black art. They are definitely worth a try,
however, and you can always use them to plot a picture of your function to
get a better idea of what you are dealing with.

- *How expensive is it to compute the function at a given point?* – If the function
f is not presented as a formula, what to do depends upon what is given.
Typically, we have a program or subroutine that evaluates f at a given point,
and so can request the value of any given point on demand. By calling this
function, we can poke around and try to guess the maxima. Our freedom to
search in such a situation depends upon how efficiently we can evaluate f.
If f is just a complicated formula, evaluation will be very fast. But suppose
that f represents the effect of the coefficients $x_1, \ldots, x_n$ on the performance of
the board evaluation function in a computer chess program, such that x_1 is
how much a pawn is worth, x_2 is how much a bishop is worth, and so forth.
To evaluate a set of coefficients as a board evaluator, we must play a bunch
of games with it or test it on a library of known positions. Clearly, this is
time-consuming, so we must be frugal in the number of evaluations of f we
use.

- *How many dimensions do we have? How many do we need?* – The difficulty in
finding a global maximum increases rapidly with the number of dimensions
(or parameters). For this reason, it often pays to reduce the dimension by
ignoring some of the parameters. This runs counter to intuition, for the naive
programmer is likely to incorporate as many variables as possible into their
evaluation function. It is just too hard to tweak such a complicated function.
Much better is to start with the 3 to 5 seemingly most important variables and
do a good job optimizing the coefficients for these.

- *How smooth is my function?* The main difficulty of global optimization is getting
trapped in local optima. Consider the problem of finding the highest point
in a mountain range. If there is only one mountain and it is nicely shaped,
we can find the top by just walking in whatever direction is up. However, if
there are many false summits or other mountains in the area, it is difficult to
convince ourselves whether we are really at the highest point. *Smoothness* is
the property that enables us to quickly find the local optimum from a given

point. We assume smoothness in seeking the peak of the mountain by walking up. If the height at any given point was a completely random function, there would be no way we could find the optimum height short of sampling every single point.

Efficient algorithms for unconstrained global optimization use derivatives and partial derivatives to find local optima, to point out the direction in which moving from the current point does the most to increase or decrease the function. Such derivatives can sometimes be computed analytically, or they can be estimated numerically by taking the difference between values of nearby points. A variety of *steepest descent* and *conjugate gradient* methods to find local optima have been developed, similar in many ways to numerical root-finding algorithms.

It is a good idea to try out several different methods on any given optimization problem. For this reason, we recommend experimenting with the implementations below before attempting to implement your own method. Clear descriptions of these algorithms are provided in several numerical algorithms books, in particular [PFTV86].

For constrained optimization, finding points that satisfy all the constraints is often the difficult problem. One approach is to use a method for unconstrained optimization, but add a penalty according to how many constraints are violated. Determining the right penalty function is problem-specific, but it often makes sense to vary the penalties as optimization proceeds. At the end, the penalties should be very high to ensure that all constraints are satisfied.

Simulated annealing is a fairly robust and simple approach to constrained optimization, particularly when we are optimizing over combinatorial structures (permutations, graphs, subsets) instead of continuous functions. Techniques for simulated annealing are described in Section 5.5.1.

Implementations: Several of the *Collected Algorithms of the ACM* are Fortran codes for unconstrained optimization, most notably Algorithm 566 [MGH81], Algorithm 702 [SF92], and Algorithm 734 [Buc94]. Algorithm 744 [Rab95] does unconstrained optimization in Lisp. They are available from Netlib (see Section 9.1.2). Also check out the selection at GAMS, the NIST *Guide to Available Mathematical Software*, at http://gams.nist.gov.

NEOS (Network-Enabled Optimization System) provides a unique service, the opportunity to solve your problem on computers and software at Argonne National Laboratory, over the WWW. Linear programming and unconstrained optimization are both supported. This is worth checking out at http://www.mcs.anl.gov/home/otc/Server/ when you need a solution instead of a program.

General purpose simulated annealing implementations are available and probably are the best place to start experimenting with this technique for constrained optimization. Particularly popular is Adaptive Simulated Annealing (ASA), written in C and retrievable via anonymous ftp from ftp.alumni.caltech.edu [131.215.139.234] in the /pub/ingber directory. To get on the ASA mailing list send e-mail to asa-request@alumni.caltech.edu.

Genocop, by Zbigniew Michalewicz [Mic92], is a genetic algorithm-based program for constrained and unconstrained optimization, written in C. I tend to be quite skeptical of genetic algorithms (see Section 5.5.3), but many people find them irresistible. Genocop is available from ftp://ftp.uncc.edu/coe/evol/ for noncommercial purposes.

Notes: Steepest-descent methods for unconstrained optimization are discussed in most books on numerical methods, including [PFTV86, BT92]. Unconstrained optimization is the topic of several books, including [Bre73, Fle80].

Simulated annealing was devised by Kirkpatrick et. al. [KGV83] as a modern variation of the Metropolis algorithm [MRRT53]. Both use Monte Carlo techniques to compute the minimum energy state of a system. Good expositions on simulated annealing include [AK89].

Genetic algorithms were developed and popularized by Holland [Hol75, Hol92]. Expositions on genetic algorithms include [Gol89, Koz92, Mic92]. Tabu search [Glo89a, Glo89b, Glo90] is yet another heuristic search procedure with a devoted following.

Related Problems: Linear programming (see page 213), satisfiability (see page 266).

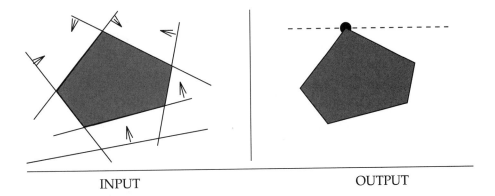

INPUT   OUTPUT

8.2.6 Linear Programming

Input description: A set S of n linear inequalities on m variables $S_i := \sum_{j=1}^{m} c_{ij} \cdot x_j \geq b_i$, $1 \leq i \leq n$, and a linear optimization function $f(X) = \sum_{j=1}^{m} c_j \cdot x_j$.

Problem description: Which variable assignment X' maximizes the objective function f while satisfying all inequalities S?

Discussion: Linear programming is the most important problem in mathematical optimization and operations research. Applications include:

- *Resource allocation* – We seek to invest a given amount of money so as to maximize our return. Our possible options, payoffs, and expenses can usually be expressed as a system of linear inequalities, such that we seek to maximize our possible profit given the various constraints. Very large linear programming problems are routinely solved by airlines and other corporations.

- *Approximating the solution of inconsistent equations* – A set of m linear equations on n variables x_i, $1 \leq i \leq n$, is overdetermined if $m > n$. Such overdetermined systems are often *inconsistent*, meaning that no assignment of variables simultaneously solves all the equations. To find the variable assignment that best fits the equations, we can replace each variable x_i by $x_i' + \epsilon_i$ and solve the new system as a linear program, minimizing the sum of the error terms.

- *Graph algorithms* – Many of the standard graph problems described in this book, such as shortest paths, bipartite matching, and network flow, can all be solved as special cases of linear programming. Most of the rest, including traveling salesman, set cover, and knapsack, can be solved using integer linear programming.

The standard algorithm for linear programming is called the *simplex method*. Each constraint in a linear programming problem acts like a knife that carves away a region from the space of possible solutions. We seek the point within the remaining region that maximizes (or minimizes) $f(X)$. By appropriately rotating the solution space, the optimal point can always be made to be the highest point in the region. Since the region (simplex) formed by the intersection of a set of linear constraints is convex, we can find the highest point by starting from any vertex of the region and walking to a higher neighboring vertex. When there is no higher neighbor, we are at the highest point.

While the basic simplex algorithm is not too difficult to program, there is a considerable art to producing an efficient implementation capable of solving large linear programs. For example, large programs tend to be sparse (meaning that most inequalities use few variables), so sophisticated data structures must be used. There are issues of numerical stability and robustness, as well as which neighbor we should walk to next (so called *pivoting rules*). Finally, there exist sophisticated *interior-point* methods, which cut through the interior of the simplex instead of walking along the outside, that beat simplex in many applications.

The bottom line on linear programming is this: you are much better off using an existing LP code than writing your own. Further, you are much better off paying money than surfing the net. Linear programming is one algorithmic problem of such economic importance that commercial implementations are far superior to free versions.

Issues that arise in linear programming include:

- *Do any variables have integrality constraints?* – It is impossible to send 6.54 airplanes from New York to Washington each business day, even if that value maximizes profit according to your model. Such variables often have natural integrality constraints. A linear program is called an *integer program* when all its variables have integrality constraints, or a *mixed integer progam* if some of them do.

  Unfortunately, it is NP-complete to solve integer or mixed programs to optimality. However, there are techniques for integer programming that work reasonably well in practice. *Cutting plane techniques* solves the problem first as a linear program, and then adds extra constraints to enforce integrality around the optimal solution point before solving it again. After a sufficient number of iterations, the optimum point of the resulting linear program matches that of the original integer program. As with most exponential-time algorithms, run times for integer programming depend upon the difficulty of the problem instance and are unpredictable. If they do not make progress quickly, they are unlikely to make much progress over longer periods of time. Therefore, if you have multiple implementations available, it may well pay to try the same problem using different codes in the hopes that one can complete in a reasonable amount of time.

- *What if my optimization function or constraints are not linear?* – In least-squares curve fitting, we seek the line that best approximates a set of points by minimizing the sum of squares of the distance between each point and the line. In formulating this as a mathematical program, the natural objective function is no longer linear, but quadratic. Unfortunately, *quadratic programming* is NP-complete, even without integer variables.

  There are three possible courses of action when you must solve a nonlinear program. The best is to see if you can model it in some other way, as is the case with least-squares fitting. The second is to try to track down special codes for quadratic programming, which do exist. Finally, you can model your problem as a constrained or unconstrained optimization problem and try to solve it with the codes discussed in Section 8.2.5.

- *What if my model does not match the input format of my LP solver?* – Many linear programming implementations accept models only in so-called *standard form*, where all variables are constrained to be nonnegative, the object function must be minimized, and all constraints must be equalities (instead of inequalities). Do not fear. There exist standard transformations to map arbitrary LP models into standard form. To convert a maximization problem to a minimization one, simply multiply each coefficient of the objective function by -1. The remaining problems can be solved by adding *slack variables* to the model. See any textbook on linear programming for details.

Implementations: A very useful resource on solving linear programs is the USENET frequently asked question (FAQ) list, maintained by John W. Gregory. In particular, it provides a list of available codes with descriptions of experiences. Check out the plaintext version at ftp://rtfm.mit.edu/pub/usenet/sci.answers/linear-programming-faq or a slicker WWW version at http://www.skypoint.com/~ashbury/linear-programming-faq.html.

The noncommercial code of choice appears to be *lp_solve*, written in ANSI C by Michel Berkelaar, who has solved problems as large as 30,000 variables and 50,000 constraints. Lp_solve can also handle (smaller) integer and mixed-integer problems. It is available by anonymous ftp from ftp://ftp.es.ele.tue.nl/pub/lp_solve but is not in the public domain. A user community for lp_solve exists, which has ported it to a variety of different platforms.

NEOS (Network-Enabled Optimization System) provides a unique service, an opportunity to solve your problem on computers and software at Argonne National Laboratory via the WWW. Linear programming and unconstrained optimization are both supported. This is worth checking out at http://www.mcs.anl.gov/home/otc/Server/ if you need an answer instead of a program.

If you are serious about solving large linear programs, you likely need a commercial implementation. The book [MW93] provides an overview of commercial linear programming systems, online at http://www.mcs.anl.gov/home/otc/Guide/SoftwareGuide/index.html. Surveys of commercial LP codes appear in [SR95, Sha93] and in the linear programming FAQ. I have heard good things from various people about CPLEX and AMPL, but do your own research before spending money.

For low-dimensional linear programming problems, computational geometry algorithms can outperform more general LP codes. See ftp://icemcfd.com/pub/linprog.a for a C language implementation of Seidel's randomized incremental LP algorithm, by Mike Hohmeyer.

Algorithm 551 [Abd80] and Algorithm 552 [BR80] of the *Collected Algorithms of the ACM* are simplex-based codes for solving overdetermined systems of linear equations, in Fortran. See Section 9.1.2 for details.

Pascal implementations of the revised and dual simplex methods for linear programming, as well as cutting plane and explicit enumeration algorithms for integer programming, are provided in [SDK83]. See Section 9.1.6.1. These are likely to work only for small problems.

Sedgewick [Sed92] provides a bare bones implementation of the simplex algorithm in C++. See Section 9.1.6.6 for details.

Notes: Good expositions on the simplex and ellipsoid algorithms for linear programming include [PS82, Chv83]. Expositions on low-dimensional linear programming include [PS85]. For an implementation-oriented exposition on linear and integer programming, with references to experimental work, see [SDK83].

The need for optimization via linear programming arose in logistics problems in World War II. The simplex algorithm was invented by George Danzig in 1947 [Dan63]. Klee and Minty [KM72] proved that the simplex algorithm is exponential in worst case, but it is very efficient in practice. Khachian's ellipsoid algorithm [Kha79] proved that linear programming was polynomial in 1979. Karmarkar's algorithm [Kar84] is an interior-point method that has proven to be both a theoretical and practical improvement of the ellipsoid algorithm, as well as a challenge for the simplex method.

Linear programming is P-complete under log-space reductions [DLR79]. This makes it unlikely to have an NC parallel algorithm, where a problem is in NC iff it can be solved on a PRAM in polylogarithmic time using a polynomial number of processors. Any problem that is P-complete under log-space reduction cannot be in NC unless P=NC. See [GHR95] for a thorough exposition of the theory of P-completeness, including an extensive list of P-complete problems.

Related Problems: Constrained and unconstrained optimization (see page 209), network flow (see page 297).

HTHTHHTHHT

HHHTTHHTTT

HHTTTTHTHT

HTHHHTTHHT

HTTHHTTHTH

INPUT   OUTPUT

8.2.7 Random Number Generation

Input description: Nothing, or perhaps a seed.

Problem description: Generate a sequence of random integers.

Discussion: Random number generation forms the foundation behind such standard algorithmic techniques as simulated annealing and Monte Carlo integration. Discrete event simulations, used to model everything from transportation systems to casino poker, all run on streams of random numbers. Initial passwords and cryptographic keys are typically generated randomly. New developments in randomized algorithms for graph and geometric problems are revolutionizing these fields and establishing randomization as one of the fundamental ideas of computer science.

Unfortunately, generating random numbers is a task that looks a lot easier than it really is, primarily because it is fundamentally impossible to produce truly random numbers on any deterministic device. Von Neumann [Neu63] said it best: "Anyone who considers arithmetical methods of producing random digits is, of course, in a state of sin." All we can hope for are *pseudorandom* numbers, a stream of numbers that appear as if they were generated randomly.

There can be serious consequences to using a bad random number generator. For example, the security of an Internet password scheme was recently invalidated with the discovery that its keys were produced using a random number generator of such small period that brute-force search quickly exhausted all possible passwords. The accuracy of simulations is regularly compromised or invalidated by poor random number generation. Bottom line: This is an area where people shouldn't mess around, but they do. Issues to think about include:

- *Should my program use the same "random" numbers each time it runs?* – A poker game that deals you the exact same hand each time you play quickly loses interest. One common solution is to use the lower-order bits of the machine clock as a seed or source for random numbers, so that each time the program runs it does something different.

Such methods are perhaps adequate for games, but not for serious simulations. There are liable to be periodicities in the distribution of random numbers whenever calls are made in a loop. Also, debugging is seriously complicated by the fact that the results are not repeatable. If the program crashes, you cannot go back and discover why. One possible compromise is to use a deterministic pseudorandom number generator, but write the current seed to a file between runs. During debugging, this file can be overwritten with a fixed initial value or *seed*.

- *How good is my compiler's built-in random number generator?* – If you need uniformly generated random numbers, and you are not going to bet the farm on the accuracy of your simulation, my recommendation is simply to use what your compiler provides. Your best opportunity to mess it up is with a bad choice of starting seed, so read the manual for its recommendations.

  If you *are* going to bet the farm on the quality of your simulation, you had better test your random number generator. Be aware that it is very difficult to eyeball the results and decide whether the output is really random. This is because people have very skewed ideas of how random sources should behave and often see patterns that don't really exist. To evaluate a random number generator, several different tests should be used and the statistical significance of the results established. Such tests are implemented in *plab* and *DIEHARD* (discussed below) and explained in [Knu81].

- *What if I have to implement my own random number generator?* – The algorithm of choice is the *linear congruential generator*. It is fast, simple, and (if instantiated with the right constants) gives reasonable pseudorandom numbers. The *n*th random number R_n is a function of the $(n - 1)$st random number:

$$R_n = (aR_{n-1} + c) \bmod m$$

In theory, linear congruential generators work the same way roulette wheels do. The long path of the ball around and around the wheel (captured by $aR_{n-1} + c$) ends in one of a relatively small number of bins, the choice of which is extremely sensitive to the length of the path (captured by the truncation of the mod m).

A substantial theory has been developed to select the constants a, c, m, and R_0. The period length is largely a function of the modulus m, which is typically constrained by the word length of the machine. A presumably safe choice for a 32-bit machine would be $R_0 = 0$, $a = 1366$, $c = 150889$, and $m = 714025$. Don't get creative and change any of the constants, unless you use the theory or run tests to determine the quality of the resulting sequence.

- *What if I don't want large, uniformly distributed random integers?* – The linear congruential generator produces a uniformly distributed sequence of large integers, which can be scaled to produce other uniform distributions. For uniformly distributed real numbers between 0 and 1, use R_i/m. Note that 1 cannot be realized this way, although 0 can. If you want uniformly distributed integers between l and h, use $\lfloor l + (h - l + 1)R_i/m \rfloor$.

Generating random numbers according to a given nonuniform distribution can be a tricky business. The most reliable way to do this correctly is the acceptance-rejection method. Suppose we bound the desired probability distribution function or geometric region to sample from a box and then select a random point p from the box. This point can be selected by p by generating the x and y coordinates independently, at random. If this p is within the distribution, or region, we can return p as selected at random. If p is in the portion of the box outside the region of interest, we throw it away and repeat with another random point. Essentially, we throw darts at random and report those that hit the target.

This method is correct, but it can be slow. If the volume of the region of interest is small relative to that of the box, most of our darts will miss the target. Efficient generators for Gaussian and other special distributions are described in the references and implementations below.

Be cautious about inventing your own technique, however, since it can be tricky to obtain the right probability distribution. For example, an *incorrect* way to select points uniformly from a circle of radius r would be to generate polar coordinates and select an angle from 0 to 2π and a displacement between 0 and r, both uniformly at random. In such a scheme, half the generated points will lie within $r/2$ of the radius, when only one-fourth of them should be! This is a substantial enough difference to seriously skew the results, while being subtle enough that it might escape detection.

- *How long should I run my Monte Carlo simulation to get the best results?* – It makes sense that the longer you run a simulation, the more accurately the results will approximate the limiting distribution, thus increasing accuracy. However, this is only true until you exceed the *period*, or cycle length, of your random number generator. At that point, your sequence of random numbers repeats itself, and further runs generate no additional information. Check the period length of your generator before you jack up the length of your simulation. You are liable to be very surprised by what you learn.

Implementations: An excellent WWW page on random number generation and stochastic simulation is available at http://random.mat.sbg.ac.at/others/. It includes pointers to papers and literally dozens of implementations of random number generators. From there are accessible *pLab* [Lee94] and *DIEHARD*, systems for testing the quality of random number generators.

The Stanford Graphbase (see Section 9.1.3) contains a machine-independent random number generator based on the recurrence $a_n = (a_{n-24} - a_{n-55}) \bmod 2^{31}$. With the proper initialization, this generator has a period of at least $2^{55} - 1$.

Algorithm 488 [Bre74], Algorithm 599 [AKD83], and Algorithm 712 [Lev92] of the *Collected Algorithms of the ACM* are Fortran codes for generating random numbers according to several probability distributions, including normal, exponential, and Poisson distributions. They are available from Netlib (see Section 9.1.2).

Sim++ is a library of routines for implementing discrete event simulations, built by Robert Cubert and Paul Fishwick, of the University of Florida. It contains random number generators for a variety of different distributions, including uniform, exponential, and normal. Check out http://www.cis.ufl.edu/~fishwick/simpack/simpack.html if you need a random number generator to control a simulation. Fishwick's book [Fis95] describes model design using SimPack.

LEDA (see Section 9.1.1) provides a comprehensive random source in C++ for generating random bits, integers, and double precision reals. Sedgewick [Sed92] provides simple implementations of linear and additive congruential generators in C++. See Section 9.1.6.6 for details.

XTango (see Section 9.1.5) is an algorithm animation system for UNIX and X-windows, which includes an animation illustrating the uniformity of random number generation.

Notes: Knuth [Knu81] has a thorough and interesting discussion of random number generation, which I heartily recommend. He presents the theory behind several methods, including the middle square and shift-register methods we have not described here, as well as a detailed discussion of statistical tests for validating random number generators. Another good source is [PFTV86] – our recommended constants for the linear congruential generator are drawn from here. Comparisons of different random number generators in practice include [PM88].

Tables of random numbers appear in most mathematical handbooks, as relics from the days before there was ready access to computers. Most notable is [RC55], which provides one million random digits.

The deep relationship between randomness, information, and compressibility is explored within the theory of Kolmogorov complexity, which measures the complexity of a string by its compressibility. Truly random strings are incompressible. The string of seemingly random digits of π cannot be random under this definition, since the entire sequence is defined by any program implementing a series expansion for π. Li and Vitáni [LV93] provide a thorough introduction to the theory of Kolmogorov complexity.

Related Problems: Constrained and unconstrained optimization (see page 209), generating permutations (see page 246), generating subsets (see page 250), generating partitions (see page 253).

		179424673
8338169264555846052842102071		2038074743
	*	22801763489

8338169264555846052842102071

INPUT   OUTPUT

8.2.8 Factoring and Primality Testing

Input description: An integer n.

Problem description: Is n a prime number, and if not what are the factors of n?

Discussion: The dual problems of factoring integers and testing primality have surprisingly many applications for a problem long suspected of being only of mathematical interest. The security of the RSA public-key cryptography system (see Section 8.7.6) is based on the computational intractability of factoring large integers. As a more modest application, hash table performance typically improves when the table size is a prime number. To get this benefit, an initialization routine must identify a prime near the desired table size. Finally, prime numbers are just interesting to play with. It is no coincidence that programs to generate large primes often reside in the games directory of UNIX systems.

Although factoring and primality testing are related problems, algorithmically they are quite different. There exist algorithms that can demonstrate that an integer is *composite* (i.e. not prime) without actually giving the factors. To convince yourself of the plausibility of this, note that you can demonstrate the compositeness of any nontrivial integer whose last digit is 0, 2, 4, 5, 6, or 8 without doing the actual division.

The simplest algorithm for both of these problems is brute-force trial division. To factor n, compute the remainder of n/i for all $1 < i \leq \sqrt{n}$. The prime factorization of n will contain at least one instance of every i such that $n/i = \lfloor n/i \rfloor$, unless n is prime. Make sure you handle the multiplicities correctly and account for any primes larger than $\sqrt{n}$.

Such algorithms can be sped up by using a precomputed table of small primes to avoid testing all possible i. Surprisingly large numbers of primes can be represented in surprisingly little space by using bit vectors (see Section 8.1.5). A bit vector of all odd numbers less than 1,000,000 fits in under 64 kilobytes. Even tighter encodings become possible by eliminating all multiples of three and other small primes.

Considerably faster factoring algorithms exist, whose correctness depends upon more substantial number theory. The fastest known algorithm, the *number field sieve*, uses randomness to construct a system of congruences, the solution of which usually gives a factor of the integer. Integers with as many at 128 digits have been factored using this method, although such feats require enormous amounts of computation.

Randomized algorithms make it much easier to test whether an integer is prime. Fermat's little theorem states that $a^{n-1} = 1 \pmod{n}$ for all a, when n is prime. Suppose we pick a random value of $1 \leq a \leq n$ and compute the residue of $a^{n-1} \pmod{n}$. If this residue is not 1, we have just proven that n cannot be prime. Such randomized primality tests are very efficient. PGP (see Section 8.7.6) finds 300+ digit primes using hundreds of these tests in minutes, for use as cryptographic keys.

Although the primes are scattered in a seemingly random way throughout the integers, there is some regularity to their distribution. The *prime number theorem* states that the number of primes less than n, commonly denoted by $\pi(n)$, is approximately $n/\ln n$. Further, there are never large gaps between primes, so in general, one would expect to examine about $\ln n$ integers if one wanted to find the first prime larger than n. This distribution and the fast randomized primality test explain how PGP can find such large primes so quickly.

Implementations: My first choice for factoring or primality testing applications would be PARI, a system capable of handling complex number-theoretic problems on integers with up to 300,000 decimal digits, as well as reals, rationals, complex numbers, polynomials, and matrices. It is written mainly in C, with assembly code for inner loops on major architectures, and includes more than 200 special predefined mathematical functions. PARI can be used as a library, but it also possesses a calculator mode that gives instant access to all the types and functions. The main advantage of PARI is its speed. On a Unix platform, it is between 5 to 100 times faster than Maple or Mathematica, depending on the applications. PARI is available for PC, Amiga, Macintosh, and most Unix platforms by anonymous ftp at ftp://megrez.ceremab.u-bordeaux.fr/pub/pari/.

A Mathematica implementation by Ilan Vardi of Lenstra's elliptic curve method of factorization is available in Packages/NumberTheory/FactorInteger-ECM.m of the standard Mathematica distribution and MathSource. It is designed to find prime factors of up to about 18 digits in reasonable time, extending Mathematica's ability to factor all numbers of up to 40 digits. It is faster when factoring the product of small primes.

Notes: Bach and Shallit's book [BS96] is the most comprehensive reference on computational number theory, while Adleman's excellent survey [Adl94a] describes the state of the art, as well as open problems. Good expositions on modern algorithms for factoring and primality testing include [CLR90].

The Miller-Rabin [Mil76, Rab80] randomized primality testing algorithm eliminates problems with Carmichael numbers, which are composite integers that always satisfy Fermat's theorem. The best algorithms for integer factorization include the quadratic-sieve [Pom84] and the elliptic-curve methods [Len87b].

Mechanical sieving devices provided the fastest way to factor integers surprisingly far into the computing era. See [SWM95] for a fascinating account of one such device, built during World War I. Hand-cranked, it proved the primality of $2^{31} - 1$ in fifteen minutes of sieving time.

An important problem in computational complexity theory is whether $P = NP \cap co-NP$. The decision problem "is n a composite number?" is perhaps the best candidate for a counterexample. By exhibiting the factors of n, it is trivially in NP. It can be shown to be in co-NP, since every prime has a short proof of its primality [Pra75]. However, there is no evidence it is in P. For more information on complexity classes, see [GJ79, Joh90].

A group headed by Arjen Lenstra has regularly broken records for general-purpose integer factoring, using an Internet-distributed implementation of the quadratic sieve factoring method. The June 1993 factorization of RSA-120 took approximately 830 MIP-years of computation. The April 1994 factorization of RSA-129, famous for appearing in the original RSA paper [RSA78], was factored in eight months using over 1,600 computers. This was particularly noteworthy because in [RSA78] they had originally predicted such a factorization would take 40 quadrillion years using 1970s technology.

Related Problems: Cryptography (see page 414), high precision arithmetic (see page 224).

49578291287491495151508905425869578	2
74367437231237242727263358138804367	3

INPUT	OUTPUT

8.2.9   Arbitrary-Precision Arithmetic

Input description: Two very large integers, x and y.

Problem description: What is $x + y$, $x - y$, $x \times y$, and x/y?

Discussion: Any programming language whose level rises above basic assembler supports single- and perhaps double-precision integer/real addition, subtraction, multiplication, and division. But what if we wanted to represent the national debt of the United States in pennies? One trillion dollars worth of pennies requires 15 decimal digits, which is far more than can fit into a 32-bit integer.

In other applications *much* larger integers are needed. The RSA algorithm for public-key cryptography requires integer keys of at least 100 digits to achieve any level of security, and 1000 digits are recommended. Experimenting with number-theoretic conjectures for fun or research always requires playing with large numbers. I once solved a minor open problem [GKP89] by performing an exact computation on the integer $\binom{5906}{2953} \approx 9.93285 \times 10^{1775}$.

What should you do when you need large integers?

- *Am I solving a problem instance requiring large integers, or do I have an embedded application?* – If you just need the answer to a specific problem with large integers, such as in the number theory application above, you would be well advised to consider using a computer algebra system like Maple or Mathematica. These use arbitrary-precision arithmetic as a default and use nice Lisp-like programming languages as a front end, together often reducing your problem to a 5 to 10 line program.

  If instead you have an embedded application requiring high-precision arithmetic, you would be well advised to use an existing library. In addition to the four basic operations, you are likely to get additional functions for computing things like greatest common divisor in the bargain. See the implementations below for details.

- *Do I need high- or arbitrary-precision arithmetic?* – Is there an upper bound on how big your integers can get, or do you really need *arbitrary*-precision, i.e. unbounded. This determines whether you can use a fixed-length array to represent your integers as opposed to a linked-list of digits. The array is likely to be simpler and will not be a constraint in most applications.

- *What base should I do arithmetic in?* – It is conceptually simplest to implement your own high-precision arithmetic package in decimal and represent each integer as a string of base-10 digits, at one digit per node. However, it is far more efficient to use a higher base, ideally the square root of the largest integer supported fully by hardware arithmetic.

Why? The higher the base, the fewer digits we need to represent the number (compare 64 decimal with 1000000 binary). Since hardware addition usually takes one clock cycle independent of the actual numbers, best performance is achieved using the highest base. The reason for limiting us to $\sqrt{maxint}$ is that in performing high-precision multiplication, we will multiply two of these "digits" together and need to avoid overflow.

The only complication of using a larger base is that integers must be converted to and from base-10 for input and output, but the conversion is easily performed once all four high-precision arithmetical operations are supported.

- *How low-level are you willing to get for fast computation?* Hardware addition is much faster than a subroutine call, so you are going to take a significant hit on speed whenever your package is used where low-precision arithmetic suffices. High-precision arithmetic is one of few problems in this book where inner loops in assembly language can be the right idea to speed things up. Finally, using bit-level masking and shift operations instead of arithmetical operations can be a win.

The algorithm of choice for each of the five basic arithmetic operations is as follows:

- *Addition* – The basic schoolhouse method of lining up the decimal points and then adding the digits from right to left with carries works in time linear in the number of digits. More sophisticated carry-look-ahead parallel algorithms are available for low-level hardware implementation. One may hope that they are used on your chip for low-precision addition.

- *Subtraction* – Depending upon the sign bits of the numbers, subtraction can be a special case of addition: $(A - (-B)) = (A + B)$. The tricky part of subtraction is performing the borrow. This can be simplified by always subtracting from the number with the larger absolute value and adjusting the signs afterwards, so we can be certain there will always be something to borrow from.

- *Multiplication* – The simplest method of repeated addition will take exponential time on large integers, so stay away. The digit-by-digit schoolhouse method is reasonable to program and will work much better, presumably well enough for your application. On very large integers, Karatsuba's $O(n^{1.59})$ divide-and-conquer algorithm (cited in the notes) wins. Dan Grayson, author of Mathematica's arbitrary-precision arithmetic, found that the switch-over happened at well under 100 digits. Even faster on very large integers is an algorithm based on Fourier transforms. A discussion of such algorithms appears in Section 8.2.11.

- *Division* – Repeated subtraction will take exponential time, so the easiest reasonable algorithm to use is the long-division method you hated in school. This is a far more complicated algorithm than needed for the other operations, requiring arbitrary-precision multiplication and subtraction as subroutines, as well as trial and error to determine the correct digit to use at each position of the quotient.

In fact, integer division can be reduced to integer multiplication, although in a nontrivial way, so if you are implementing asymptotically fast multiplication, you can reuse that effort in long division. See the references below for details.

- *Exponentiation* – We can compute a^b in the obvious manner using $b - 1$ multiplications, but a better way is to exploit the fact that $a^b = a^{b/2} \times a^{b/2}$. By repeatedly squaring the results of our partial product, we can escape using $O(\lg n)$ multiplications, a big win when b is large. See Section 3.6 for a discussion of this algorithm.

High- but not arbitrary-precision arithmetic can be conveniently performed using the Chinese remainder theorem and modular arithmetic. The Chinese remainder theorem states that an integer between 1 and $P = \prod_{i=1}^{k} p_i$ is uniquely determined by its set of residues mod p_i, where each p_i, p_j are relatively prime integers. Addition, subtraction, and multiplication (but not division) can be supported using such residue systems, with the advantage that large integers can be manipulated without complicated data structures.

Many of these algorithms for computations on long integers can be directly applied to computations on polynomials. See the references for more details. A particularly useful algorithm is Horner's rule for fast polynomial evaluation. When $P(x) = \sum_{i=0}^{n} c_i \cdot x^i$ is blindly evaluated term by term, $O(n^2)$ multiplications will be performed. Much better is observing that $P(x) = c_0 + x(c_1 + x(c_2 + x(c_3 + \ldots)))$, the evaluation of which uses only a linear number of operations.

Implementations: All major commercial computer algebra systems incorporate high-precision arithmetic, including Maple, Mathematica, Axiom, and Macsyma. If you have access to one of these, this is your best option for a quick, nonembedded application. The rest of this section focuses on source code available for embedded applications.

PARI, developed by Henri Cohen and his colleagues in France, is a system capable of handling complex number-theoretic problems on integers with up to 300,000 decimal digits, as well as reals, rationals, complex numbers, polynomials, and matrices. It is probably the most powerful free software available for number theory. Written mainly in C (with assembly-language code for speed-critical routines), it includes more than 200 special predefined mathematical functions. PARI can be used as a library, but it possesses also a powerful calculator mode that gives instant access to all the types and functions. The main advantage of PARI is its speed. On a Unix platform, it runs between 5 to 100 times faster than Maple or Mathematica, depending on the applications. PARI is available for PC, Amiga, Macintosh, and most Unix platforms by anonymous ftp at ftp://megrez.ceremab.u-bordeaux.fr/pub/pari/.

Algorithm 693 [Smi91] of the *Collected Algorithms of the ACM* is a Fortran implementation of floating-point, multiple-precision arithmetic. See Section 9.1.2.

An implementation of arbitrary-precision integer and rational arithmetic in C++ is embedded in LEDA (see Section 9.1.1), including GCD, square roots, and logarithms as well as the basic four operations. Sparc assembler code is used for certain time-critical functions.

Implementations in C of a high-precision calculator with all four elementary operations appear in [BR95]. The authors use base 10 for arithmetic and arrays of digits to represent long integers, with short integers as array indices, thus limiting computations to 32,768 digits. The code for these algorithms is printed in the text and available on disk for a modest fee.

Bare bones implementations in C of high-precision multiplication and in Pascal of such special functions as logarithm and arctangent appear in [GBY91]. See Section 9.1.6.2 for further details. Sedgewick [Sed92] provides a bare bones implementation of polynomial arithmetic in C++. See Section 9.1.6.6 for details.

Notes: Knuth [Knu81] is the primary reference on algorithms for all basic arithmetic operations, including implementations of them in the MIX assembly language. Bach and Shallit [BS96] provide a more recent treatment of computational number theory.

Expositions on the $O(n^{1.59})$-time divide-and-conquer algorithm for multiplication [KO63] include [AHU74, Man89]. An FFT-based algorithm multiplies two n-bit numbers in $O(n \lg n \lg \lg n)$ time and is due to Schönhage and Strassen [SS71]. Expositions include [AHU74, Knu81]. The reduction between integer division and multiplication is presented in [AHU74, Knu81].

Good expositions of algorithms for modular arithmetic and the Chinese remainder theorem include [AHU74, CLR90]. A good exposition of circuit-level algorithms for elementary arithmetic algorithms is [CLR90].

Euclid's algorithm for computing the greatest common divisor of two numbers is perhaps the oldest interesting algorithm. Expositions include [CLR90, Man89].

Related Problems: Factoring integers (see page 221), cryptography (see page 414).

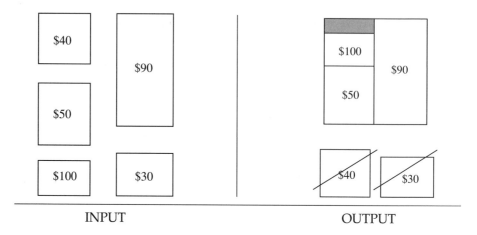

INPUT   OUTPUT

8.2.10   Knapsack Problem

Input description: A set of items $S = \{1, \ldots, n\}$, where item i has size s_i and value v_i. A knapsack capacity C.

Problem description: Find the subset $S' \subset S$ that maximizes the value of $\sum_{i \in S'} v_i$ given that $\sum_{i \in S'} s_i \leq C$; i.e. all the items fit in a knapsack of size C.

Discussion: The knapsack problem arises whenever there is resource allocation with financial constraints. Given a fixed budget, how do you select what things you should buy. Everything has a cost and value, so we seek the most value for a given cost. The term *knapsack problem* invokes the image of the backbacker who is constrained by a fixed-size knapsack and so must fill it only with the most useful items.

The typical formulation in practice is the *0/1 knapsack problem*, where each item must be put entirely in the knapsack or not included at all. Objects cannot be broken up arbitrarily, so its not fair taking one can of coke from a six-pack or opening the can to take just a sip. It is this 0/1 property that makes the knapsack problem hard, for a simple greedy algorithm finds the optimal selection whenever we are allowed to subdivide objects arbitrarily. For each item, we could compute its "price per pound", and take as much of the most expensive item until we have it all or the knapsack is full. Repeat with the next most expensive item, until the knapsack is full. Unfortunately, this 0/1 constraint is usually inherent in most applications.

Issues that arise in selecting the best algorithm include:

- *Does every item have the same cost/value or the same size?* – If each item is worth the same amount to us as any other item, say $1, I maximize my value by taking the greatest number of items. In this case the optimal solution is to sort the items in order of increasing size and insert them into the knapsack in this order until nothing fits. The problem is solved similarly when each object has the same size but the costs are different. These are the easy cases of knapsack.

- *Does each item have the same "price per pound"?* – In this case, our problem is equivalent to ignoring the price and just trying to minimize the amount of empty space left in the knapsack. Unfortunately, even this restricted version of the problem is NP-complete, and so we cannot expect an efficient algorithm that always solves the problem. Don't lose hope, however, because knapsack proves to be an "easy" hard problem, one that can usually be handled with the algorithms described below.

  An important special case of constant "price-per-pound" knapsack is the *integer partition* problem, presented in cartoon form in Figure 8-2. Here, we seek to partition the elements of S into two sets A and B such that $\sum_{a \in A} a = \sum_{b \in B} b$, or alternately make the difference as small as possible. Integer partition can be thought of as bin packing with two equal-sized bins or knapsack with a capacity of half the total weight, so all three problems are closely related and NP-complete.

  The constant 'price-per-pound' knapsack problem is often called the *subset sum* problem, because given a set of numbers, we seek a subset that adds up to a specific target number, i.e. the capacity of our knapsack.

- *Are all the sizes relatively small integers?* – When the sizes of the items and the knapsack capacity C are all integers, there is an efficient dynamic programming algorithm that finds the optimal solution in time $O(nC)$ and $O(C)$ space. Whether this is good for you depends upon how big C is. For $C \le 1,000$, this might be great, but not for $C \ge 10,000,000$.

  The algorithm works as follows: Let S' be a set of items, and let $C[i, S']$ be true if and only if there is a subset of S' whose size adds up exactly to i. For the empty set, $C[i, \emptyset]$ is false for $1 \le i \le C$. One by one we add a new item s_j to S' and update the affected values of $C[i, S']$. Observe that $C[i, S' \cup s_j]$ = true iff either $C[i, S']$ or $C[i - s_j, S']$ is true, since either we use s_j in our subset or we don't. By performing n sweeps through all C elements, one for each s_j, $1 \le j \le n$, and updating the array, we identify which sums of sizes can be realized. The knapsack solution is the largest realizable size. In order to reconstruct the winning subset, we must store the name of the item number that turned $C[i]$ from false to true, for each $1 \le i \le C$, and then scan backwards through the array.

  The dynamic programming formulation described above ignored the values of the items. To generalize the algorithm, add a field to each element of the array to store the value of the best subset to date summing up to i. We

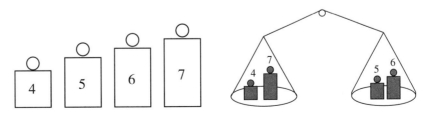

Figure 8–2. Integer partition is a variant of the Knapsack problem

now update not only when $C[i]$ turns from false to true, but when the sum of the cost of $C[i - s_j, S']$ plus the cost of s_j is better than the previous cost of $C[i]$.

- *What if I have multiple knapsacks?* – When there are multiple knapsacks, your problem is probably better thought of as a bin packing problem. See Section 8.6.9.

When the knapsack capacity gets too large for dynamic programming, exact solutions can be found using integer programming or backtracking. A 0/1 integer variable x_i is used to denote whether item i is present in the optimal subset. We maximize $\sum_{i=1}^{n} x_i \cdot v_i$ given the constraint that $\sum_{i=1}^{n} x_i \cdot s_i \le C$. Algorithms and implementations of integer and linear programming are discussed in Section 8.2.6.

When exact solutions prove too costly to compute, heuristics should be used. The simple greedy heuristic inserts items according to the maximum 'price per pound' rule, described above. Often this heuristic solution is close to optimal, but it can be arbitrarily bad depending upon the problem instance. The "price per pound" rule can also be used to reduce the size of the problem instance in exhaustive search-based algorithms by eliminating "cheap but heavy" objects from future consideration.

Another heuristic is based on *scaling*. Dynamic programming works well if the capacity of the knapsack is a reasonably small integer, say C_s. But what if we have a problem with capacity $C_m > C_s$? We scale down the sizes of all items by a factor of C_m/C_s, round the size down to an integer, and then use dynamic programming on the scaled items. Scaling works well in practice, especially when the range of sizes of items is not too large.

Implementations: Martello and Toth's book [MT90a] comes with a disk of Fortran implementations of a variety of knapsack algorithms. This is likely the best source of code currently available.

Algorithm 632 [MT85] of the *Collected Algorithms of the ACM* is a Fortran code for the 0/1 knapsack problem, with the twist that it supports multiple knapsacks. See Section 9.1.2.

Pascal implementations of several knapsack algorithms, including backtracking and a refined greedy algorithm, are provided in [SDK83]. See Section 9.1.6.1 for details.

Notes: Martello and Toth's book [MT90a] and survey article [MT87] are the standard references on the knapsack problem, including most theoretical and experimental results. An excellent exposition on integer programming approaches to knapsack problems appears in [SDK83]. See [FP75a] for a computational study of algorithms for 0-1 knapsack problems.

A polynomial-time approximation scheme is an algorithm that approximates the optimal solution of a problem in time polynomial in both its size and the approximation factor ϵ. This very strong condition implies a smooth tradeoff between running time and approximation quality. Good expositions on the polynomial-time approximation scheme [IK75] for knapsack and subset sum includes [Baa88, CLR90, GJ79, Man89].

The first algorithm for generalized public key encryption by Merkle and Hellman [MH78] was based on the hardness of the knapsack problem. See [Sch94] for an exposition.

Related Problems: Bin packing (see page 374), integer programming (see page 213).

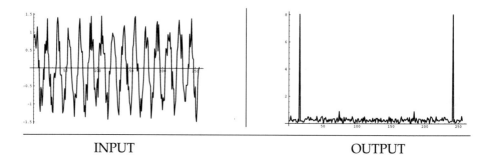

INPUT   OUTPUT

8.2.11 Discrete Fourier Transform

Input description: A sequence of n real or complex values h_i, $0 \leq i \leq n - 1$, sampled at uniform intervals from a function h.

Problem description: The discrete Fourier transform H of h,

$$H_m = \sum_{k=0}^{n-1} h_k e^{2\pi ikm/n},$$

for $0 \leq m \leq n - 1$.

Discussion: Although computer scientists tend to be relatively unfamiliar with Fourier transforms, electrical engineers and signal processors eat them for breakfast. Functionally, Fourier transforms provide a way to convert samples of a standard time-series into the "frequency domain". This provides a "dual" representation of the function, in which certain operations become easier than in the time domain. Applications of Fourier transforms include:

- *Filtering* – Taking the Fourier transform of a function is equivalent to representing it as the sum of sine functions. By eliminating undesirable high- and/or low-frequency components (i.e. dropping some of the sine functions) and taking an inverse Fourier transform to get us back into the time domain, we can filter an image to remove noise and other artifacts. For example, the sharp spike in the figure above describes the period of a single sine function that closely models the input data.

- *Image Compression* – A smoothed, filtered image contains less information than a noisy image, while retaining a similar appearance. Thus encoding the smoothed image will require fewer bits to represent than the original image. By eliminating the coefficients of sine functions that contribute relatively little to the image, we can further reduce the size of the image, at little cost in image fidelity.

- *Convolution and Deconvolution* – Fourier transforms can be used to efficiently compute convolutions of two sequences. A convolution is the pairwise product of elements from two different sequences, such as in multiplying two

n-variable polynomials f and g or multiplying two long integers. Implementing the product directly takes $O(n^2)$, while $O(n \lg n)$ suffices using the fast Fourier transform. Another example comes from image processing. Because a scanner measures the darkness of an image patch instead of a single point, the scanned input is always blurred. A reconstruction of the original signal can be obtained by deconvoluting the input signal with a Gaussian point-spread function.

- *Computing the correlation of functions* – The *correlation function* of two functions $f(t)$ and $g(t)$ is defined by

$$z(t) = \int_{-\infty}^{\infty} f(\tau)g(t + \tau)$$

and can be easily computed using Fourier transforms. Note that if the two functions are similar in shape but one is shifted relative to the other (such as $f(t) = \sin(t)$ and $g(t) = \cos(t)$), the value of $z(t_0)$ will be large at this shift offset t_0. As an application, suppose that we want to detect whether there are any funny periodicities in our random number generator. We can generate a large series of random numbers, turn it into a time series (the ith number at time i), and take the Fourier transform of this series. Any funny spikes will correspond to potential periodicities.

The discrete Fourier transform takes as input n complex numbers h_k, $0 \leq k \leq n - 1$, corresponding to equally spaced points in a time series, and outputs n complex numbers H_k, $0 \leq k \leq n - 1$, each describing a sine function of given frequency. The discrete Fourier transform is defined by

$$H_m = \sum_{k=0}^{n-1} h_k e^{-2\pi ikm/n}$$

and the inverse Fourier transform is defined by

$$h_m = \frac{1}{n} \sum_{k=0}^{n-1} H_k e^{2\pi ikm/n}$$

which enables us move easily between h and H.

Since the output of the discrete Fourier transform consists of n numbers, each of which can be computed using a formula on n numbers, they can be computed in $O(n^2)$ time. The fast Fourier transform (FFT) is an algorithm that computes the discrete Fourier transform in $O(n \log n)$. This is arguably the most important algorithm known, as measured by practical impact, for it opened the door to modern image processing. There are several different algorithms that call themselves FFTs, all of which are based on a divide-and-conquer approach. Essentially, the problem of computing the discrete Fourier transform on n points is reduced to computing two transforms on $n/2$ points each and is then applied recursively.

The FFT usually assumes that n is a power of two. If this is not the case for your data, you are usually better off padding your data with zeros to create $n = 2^k$ elements rather than hunting for a more general code.

Since many image processing systems have strong real-time constraints, FFTs are often implemented in hardware, or at least in assembly language tuned to the particular machine. Be aware of this possibility if the codes below prove too slow.

Implementations: FFTPACK is a package of Fortran subprograms for the fast Fourier transform of periodic and other symmetric sequences, written by P. Swartzrauber. It includes complex, real, sine, cosine, and quarter-wave transforms. A C language translation of the main routines is also provided. FFTPACK resides on Netlib (see Section 9.1.2) at http://www.netlib.org/fftpack.

Algorithm 545 [Fra79] of the *Collected Algorithms of the ACM* is an implementation of the fast Fourier transform optimizing virtual memory performance and written in Fortran. See Section 9.1.2 for further information.

XTango (see Section 9.1.5) is an algorithm animation system for UNIX and X-windows, which includes an interesting animation of the fast Fourier transform.

A Pascal implementation of the fast Fourier transform for $n = 2^k$ points appears in [MS91]. For more details, see Section 9.1.6.4. Sedgewick [Sed92] provides a bare bones implementation of the fast Fourier transform in C++. See Section 9.1.6.6 for details.

Notes: Brigham [Bri74] is an excellent introduction to Fourier transforms and the FFT and is strongly recommended, as is the exposition in [PFTV86]. Expositions in algorithms texts on the fast Fourier transform include [AHU74, Baa88, CLR90, Man89].

Credit for inventing the fast Fourier transform is usually given to Cooley and Tukey [CT65], although it is not completely deserved. See [Bri74] for a complete history.

An interesting divide-and-conquer algorithm for polynomial multiplication [KO63] does the job in $O(n^{1.59})$ time and is discussed in [AHU74, Man89]. An FFT-based algorithm that multiplies two n-bit numbers in $O(n \lg n \lg \lg n)$ time is due to Schönhage and Strassen [SS71] and is presented in [AHU74].

In recent years, wavelets have been proposed to replace Fourier transforms in filtering. See [Dau92] for an introduction to wavelets.

Related Problems: Data compression (see page 410), high-precision arithmetic (see page 224).

8.3 Combinatorial Problems

In this section, we consider several classic algorithmic problems of a purely combinatorial nature. These include sorting and permutation generation, both of which were among the first nonnumerical problems arising on electronic computers. Sorting, searching, and selection can all be classified in terms of operations on a partial order of keys. Sorting can be viewed as identifying or imposing the total order on the keys, while searching and selection involve identifying specific keys based on their position in the total order.

The rest of this section deals with other combinatorial objects, such as permutations, partitions, subsets, calendars, and schedules. We are particularly interested in algorithms that *rank* and *unrank* combinatorial objects, i.e. that map each distinct object to and from a unique integer. Once we have rank and unrank operations, many other tasks become simple, such as generating random objects (pick a random number and unrank) or listing all objects in order (iterate from 1 to n and unrank).

We conclude with the problem of generating graphs. Graph algorithms are more fully presented in subsequent sections.

Books on general combinatorial algorithms, in this restricted sense, include:

- *Nijenhuis and Wilf* [NW78] – This book specializes in algorithms for constructing basic combinatorial objects such as permutations, subsets, and partitions. Such algorithms are often very short but hard to locate and usually are surprisingly subtle. Fortran programs for all of the algorithms are provided, as well as a discussion of the theory behind each of them. See Section 9.1.6.3 for details.

- *Ruskey* [Rus97] – On its completion, this manuscript in preparation will become the standard reference on generating combinatorial objects. A preview is available via the WWW at http://sue.csc.uvic.ca/~cos/.

- *Knuth* [Knu73a, Knu73b] – The standard reference on searching and sorting, with significant material on combinatorial objects such as permutations.

- *Reingold, Nievergelt, Deo* [RND77] – A comprehensive algorithms text with a particularly thorough treatment of combinatorial generation and search.

- *Stanton and White* [SW86] – An undergraduate combinatorics text with algorithms for generating permutations, subsets, and set partitions. It contains relevant programs in Pascal.

- *Skiena* [Ski90] – This description of *Combinatorica*, a library of 230 Mathematica functions for generating combinatorial objects and graph theory (see Section 9.1.4) provides a distinctive view of how different algorithms can fit together. Its author is uniquely qualified to write a manual on algorithm design.

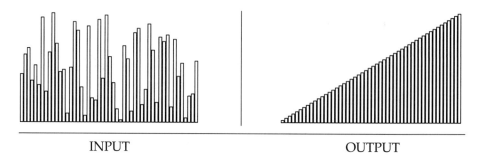

| INPUT | OUTPUT |

8.3.1 Sorting

Input description: A set of n items.

Problem description: Arrange the items in increasing order.

Discussion: Sorting is the fundamental algorithmic problem in computer science. Learning the different sorting algorithms is like learning scales for a musician. Sorting is the first step in solving a host of other algorithm problems, as discussed in Section 2.5. Indeed, *"when in doubt, sort"* is one of the first rules of algorithm design.

Sorting is also used to illustrate the standard paradigms of algorithm design. The result is that most programmers are familiar with many different sorting algorithms, which sows confusion as to which should be used for a given application. The following criteria can help you decide:

- *How many keys will you be sorting?* – For small amounts of data (say $n \leq 100$), it really doesn't matter much which of the quadratic-time algorithms you use. Insertion sort is faster, simpler, and less likely to be buggy than bubblesort. Shellsort is much faster than insertion sort, but it involves somewhat trickier programming and looking up good insert sequences in Knuth [Knu73b].

  If you have more than 100 items to sort, it is important to use an $O(n \lg n)$-time algorithm, like heapsort, quicksort, or mergesort. If you have more than 1,000,000 items to sort, you probably need an external-memory algorithm that minimizes disk access. Both types of algorithms are discussed below.

- *Will there be duplicate keys in the data?* – When all items have distinct keys, the sorted order is completely defined. However, when two items share the same key, something else must determine which one comes first. For many applications it doesn't matter, so any sorting algorithm is equally good. Often, ties are broken by sorting on a secondary key, like the first name or initial if the family names collide.

  Occasionally, ties need to be broken by their initial position in the data set. If the 5th and 27th items of the initial data set share the same key in such a case, the 5th item must be before the 27th in the final order. A *stable* sorting algorithm preserves the original ordering in case of ties. Most of the quadratic-time sorting algorithms are stable, while many of the $O(n \lg n)$ algorithms are not. If it is important that your sort be stable, it is probably better to explicitly use the initial position as a secondary key rather than trust the stability of your implementation.

- *What do you know about your data?* – In special applications, you can often exploit knowledge about your data to get it sorted faster or more easily. Of course, general sorting is a fast $O(n \lg n)$ algorithm, so if the time spent sorting is really the bottleneck in your application, you are a fortunate person indeed.

    - *Is the data already partially sorted?* If so, algorithms like insertion sort perform better than they otherwise would.

    - *Do you know the distribution of the keys?* If the keys are randomly or uniformly distributed, a *bucket* or *distribution sort* makes sense. Throw the keys into bins based on their first letter, and recur until each bin is small enough to sort by brute force. This is very efficient when the keys get evenly distributed into buckets. However, bucket sort would be bad news sorting names on the mailing list of the "Smith Society."

    - *Are your keys very long or hard to compare?* If your keys are long text strings, it might pay to use a radix or bucket sort instead of a standard comparison sort, because the time of each comparison can get expensive. A radix sort always takes time linear in the number of characters in the file, instead of $O(n \lg n)$ times the cost of comparing two keys.

    - *Is the range of possible keys very small?* If you want to sort a subset of $n/2$ distinct integers, each with a value from 1 to n, the fastest algorithm would be to initialize an n-element bit vector, turn on the bits corresponding to keys, then scan from left to right and report which bits are on.

- *Do I have to worry about disk accesses?* – In massive sorting problems, it may not be possible to keep all data in memory simultaneously. Such a problem is called *external sorting*, because one must use an external storage device. Traditionally, this meant tape drives, and Knuth [Knu73b] describes a variety of intricate algorithms for efficiently merging data from different tapes. Today, it usually means virtual memory and swapping. Any sorting algorithm will work with virtual memory, but most will spend all their time swapping.

    The simplest approach to external sorting loads the data into a B-tree (see Section 8.1.1) and then does an in-order traversal of the tree to read the keys off in sorted order. Other approaches are based on mergesort. Files containing portions of the data are sorted using a fast internal sort, and then these files are merged in stages using 2- or k-way merging. Complicated merging patterns based on the properties of the external storage device can be used to optimize performance.

- *How much time do you have to write and debug your routine?* – If I had under an hour to deliver a working routine, I would probably just use a simple selection sort. If I had an afternoon to build an efficient sort routine, I would probably use heapsort, for it delivers reliable performance without tuning.

If I was going to take the time required to build a fast system sort routine, I would carefully implement quicksort.

The best general-purpose sorting algorithm is quicksort (see Section 2.5), although it requires considerable tuning effort to achieve maximum performance. Indeed, you are probably better off using a library function instead of doing it yourself. A poorly written quicksort will likely run more slowly than a poorly written heapsort.

If you are determined to implement your own quicksort, use the following heuristics, which make a big difference in practice:

- *Use randomization* – By randomly permuting (see Section 8.3.4) the keys before sorting, you can eliminate the potential embarrassment of quadratic-time behavior on nearly-sorted data.

- *Median of three* – For your pivot element, use the median of the first, last, and middle elements of the array, to increase the likelihood of partitioning the array into roughly equal pieces. Some experiments suggest using a larger sample on big subarrays and a smaller sample on small ones.

- *Leave small subarrays for insertion sort* – Terminating the quicksort recursion and switching to insertion sort makes sense when the subarrays get small, say fewer than 20 elements. You should experiment to determine the best switchpoint for your implementation.

- *Do the smaller partition first* – Assuming that your compiler is smart enough to remove tail recursion, you can minimize runtime memory by processing the smaller partition before the larger one. Since successive stored calls are at most half as large as before, only $O(\lg n)$ stack space is needed.

Before you get started, see Bentley's article on building a faster quicksort [Ben92b].

Implementations: Pascal implementations of all the primary sorting algorithms are available from [MS91]. See Section 9.1.6.4 for details. Timing comparisons show an optimized version of quicksort to be the winner.

Bare bones implementations of all basic sorting algorithms, in C and Pascal, appear in [GBY91]. Most notable is the inclusion of implementations of external memory sorting algorithms. Sedgewick includes similar sort routine fragments in C++. See Section 9.1.6.6 for details.

XTango (see Section 9.1.5) is an algorithm animation system for UNIX and X-windows, which includes animations of all the basic sorting algorithms, including bubblesort, heapsort, mergesort, quicksort, radix sort, and shellsort. Many of these are quite interesting to watch. Indeed, sorting is the canonical problem for algorithm animation.

Algorithm 410 [Cha71] and Algorithm 505 [Jan76] of the *Collected Algorithms of the ACM* are Fortran codes for sorting. The latter is an implementation of Shellsort on linked lists. Both are available from Netlib (see Section 9.1.2).

C language implementations of Shellsort, quicksort, and heapsort appear in [BR95]. The code for these algorithms is printed in the text and available on disk for a modest fee.

A bare bones implementation of heapsort in Fortran from [NW78] can be obtained in Section 9.1.6.3. A bare bones implementation of heapsort in Mathematica from [Ski90] can be obtained in Section 9.1.4.

Notes: Knuth [Knu73b] is the best book that has been written on sorting. It is now almost twenty-five years old, and a revised edition is promised, but it remains fascinating reading. One area that has developed since Knuth is sorting under presortedness measures. A newer and noteworthy reference on sorting is [GBY91], which includes pointers to algorithms for partially sorted data and includes implementations in C and Pascal for all of the fundamental algorithms.

Expositions on the basic internal sorting algorithms appear in every algorithms text, including [AHU83, Baa88, CLR90, Man89]. Treatments of external sorting are rarer but include [AHU83]. Heapsort was first invented by Williams [Wil64]. Quicksort was invented by Hoare [Hoa62], with careful analysis and implementation by Sedgewick [Sed78]. Von Neumann is credited with having produced the first implementation of mergesort, on the EDVAC in 1945. See Knuth for a full discussion of the history of sorting, dating back to the days of punched-card tabulating machines.

Sorting has a well-known $\Omega(n \lg n)$ lower bound under the algebraic decision tree model [BO83]. Determining the exact number of comparisons required for sorting n elements, for small values of n, has generated considerable study. See [Aig88, Raw92] for expositions.

This lower-bound does not hold under different models of computation. Fredman and Willard [FW93] present an $O(n \sqrt{\lg n})$ algorithm for sorting under a model of computation that permits arithmetic operations on keys. Under a similar model, Thorup [Tho96] developed a priority queue supporting $O(\lg \lg n)$ operations, implying an $O(n \lg \lg n)$ sorting algorithm.

Related Problems: Dictionaries (see page 175), searching (see page 240), topological sorting (see page 273).

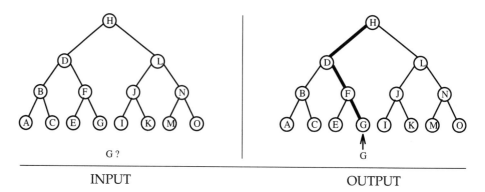

INPUT   OUTPUT

8.3.2 Searching

Input description: A set S of n keys, a query key q.

Problem description: Where is q in S?

Discussion: Searching means different things to different people. Searching for the global maximum or minimum of a function is the problem of *unconstrained optimization* and is discussed in Section 8.2.5. Chess playing programs search for the best move to make next by using alpha-beta minimax search, which is an exhaustive search of the possible moves using a variation of backtracking (see Section 5.1).

Here we consider the simple task of searching for a key in a list or in an array, which is a fundamental problem associated with information retrieval. Dictionary data structures maintain efficient access to sets of keys under insertion and deletion and are discussed in Section 8.1.1. Typical dictionaries include binary trees and hash tables.

We treat searching here as a problem distinct from dictionaries because simpler and more efficient solutions emerge when our primary interest is static searching. These little data structures can yield large performance improvements when properly employed in an innermost loop. Also, several of the ideas from list and array searching, such as binary search and self-organization, apply to other problems and justify our attention.

There are two basic approaches to array searching: sequential search and binary search. Both are simple, yet have interesting and subtle variations. In *sequential search*, we simply start from the front of our list or array of keys and compare each successive item against the key until we find a match or reach the end. In *binary search*, we start with a sorted array of keys. To search for key q, we compare q to the middle key $S_{n/2}$. If q is before $S_{n/2}$, it must reside in the top half of our set; if not, it must reside in the bottom half of our set. By repeating this process on the correct half, we find the key in a total of $\lceil \lg n \rceil$ comparisons. This is a big win over the $n/2$ comparisons we expect with sequential search. See Section 3.6 for more on binary search.

Sequential search is the simplest algorithm, and likely to be fastest on up to 10-20 elements. For such tiny problems, forget about binary search. Beyond 50-

100 elements, there is no question that binary search will be more efficient than sequential search, even factoring in the cost of the sorting (assuming multiple queries). Other issues do come into play, however, particularly in identifying the proper variant of the algorithm:

- *How much time can you spend programming?* – Binary search is a notoriously tricky algorithm to program correctly. It took seventeen years after its invention until the first *correct* version of binary search was published! Don't be afraid to start from one of the implementations described below. Test it completely by writing a driver that searches for every key in the set S as well as between the keys.

- *Are certain items accessed more often than other ones?* – Certain English words (such as "the") are much more likely to occur than others (such as "defenestrate"). We can reduce the number of comparisons in a sequential search by putting the most popular words at the top of the list and the least popular ones at the bottom. Further, nonuniform access is typically the rule, not the exception. Many real-world distributions, such as word use in English, are more accurately modeled by *Zipf's law*. Under *Zipf's law*, the ith most frequently accessed key is selected with probability $(i - 1)/i$ times the probability of the $(i - 1)$st most popular key, for all $1 \le i \le n$.

  However, preordering the list to exploit a skewed access pattern requires knowing the access pattern in advance. For many applications, it can be difficult to obtain such information. Far easier are *self-organizing lists*, where the order of the keys changes in response to the queries. The simplest and best self-organizing scheme is move-to-front; that is, we move the most recently searched-for key from its current position to the front of the list. Popular keys keep getting boosted to the front, while unsearched-for keys drift towards the back of the list. There is no need to keep track of the frequency of access; just move the keys on demand. Self-organizing lists also exploit *locality of reference*, since accesses to a given key are likely to occur in clusters. Any key will be maintained near the top of the list during a cluster of accesses, even if other keys have proven more popular in the past.

  Self-organization can extend the useful size range of sequential search. However, you should switch to binary search beyond 50-100 elements.

- *Is the key close by?* – Suppose we know that the target key is to the right of position p, and we think it is close by. Sequential search is fast if we are correct, but we will be punished severely whenever we guess wrong. A better idea is to test repeatedly at larger intervals ($p + 1, p + 2, p + 4, p + 8, p + 16, \ldots$) to the right until we find a key to the right of our target. After this, we have a window containing the target and we can proceed with binary search.

  Such a *one-sided binary search* finds the target at position $p + l$ using at most $2\lceil \lg l \rceil$ comparisons, so it is faster than binary search when $l << n$, yet it can never be much worse. One-sided binary search is particularly useful in unbounded search problems, such as in numerical root finding.

- *Is my data structure sitting on external memory?* – Once the number of keys grows *too* large, as in a CD-ROM telephone directory of all the people in the United States, binary search loses its status as the best search technique. Binary search jumps wildly around the set of keys looking for midpoints to compare, and it becomes very expensive to read in a new page from a secondary storage device for each comparison. Much better are data structures such as B-trees (see Section 8.1.1), which cluster the keys into pages so as to minimize the number of disk accesses per search.

- *Can I guess where the key should be?* – In *interpolation search*, we exploit our understanding of the distribution of keys to guess where to look next. Interpolation search is probably a more accurate description of how we use a telephone book than binary search. For example, suppose we are searching for *Washington, George* in a sorted telephone book. We would certainly be safe making our first comparison three-fourths of the way down the list, essentially doing two comparisons for the price of one.

    Although interpolation search is an appealing idea, we caution against it for three reasons: First, you have to work very hard to optimize your search algorithm before you can hope for a speedup over binary search. Second, even if you get lucky and beat binary search, it is unlikely to be by enough to have justified the exercise. Third, your program will be much less robust and efficient when the distribution changes, such as when your application gets put to work on French words instead of English.

Implementations: The basic sequential and binary search algorithms are simple enough to implement that you should likely do them yourself. Still, the routines described below may be useful as models.

Gonnet and Baeza-Yates provides code fragments in C and Pascal for sequential, binary, and interpolation search, as well as for related dictionary structures. LEDA (see Section 9.1.1) provides a sorted array data type in C++ that supports binary search. Many textbooks include implementations of binary search, including [MS91]. See Section 9.1.6.4 for details.

Notes: Mehlhorn and Tsakalidis [MT90b] give a through survey of the state-of-the-art in modern data structures. Knuth [Knu73a] provides a detailed analysis and exposition on all fundamental search algorithms and dictionary data structures but omits such modern data structures as red-black and splay trees. Gonnet and Baeza-Yates [GBY91] provide detailed references and experimental results for a wide variety of search algorithms.

Manber [Man89] provides an interesting discussion of variations of binary search, including one-sided binary search and searching for an index in A where $a_i = i$.

In linear interpolation search on an array of sorted numbers, the next position probed is given by

$$next = (low - 1) + \lceil \frac{q - S[low - 1]}{S[high + 1] - S[low - 1]} \times (high - low + 1) \rceil$$

where q is the query numerical key and S the sorted numerical array. If the keys are drawn independently from a uniform distribution, the expected search time is $O(\lg \lg n)$ [YY76]. Expositions on interpolation search include [Raw92].

Nonuniform access patterns can be exploited in binary search trees by structuring them so that popular keys are located near the root, thus minimizing search time. Dynamic programming can be used to construct such optimal search trees in $O(n \lg n)$ time [Knu73b]. Expositions include [AHU74]. Splay trees are self-organizing tree structures, as discussed in Section 8.1.1.

Related Problems: Dictionaries (see page 175), sorting (see page 236).

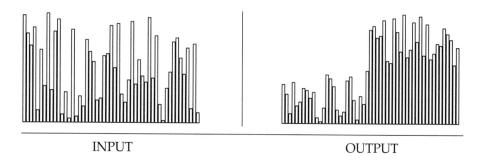

INPUT   OUTPUT

8.3.3   Median and Selection

Input description: A set of n numbers or keys, and an integer k.

Problem description: Find the key that is smaller than exactly k of the n keys.

Discussion: Median finding is an essential problem in statistics, where it provides a more robust notion of average than the *mean*. The mean wealth of people who have published research papers on sorting is significantly inflated by the presence of one William Gates [GP79], although his effect on the *median* wealth is merely to cancel out one starving graduate student.

Median finding is a special case of the more general *selection* problem, which asks for the ith element in sorted order. Selection arises in several applications:

- *Filtering outlying elements* – In dealing with noisy data samples, it is usually a good idea to throw out the largest and smallest 10% or so of them. Selection can be used to identify the items defining the tenth and ninetieth percentiles, and the outliers are then filtered out by comparing each item to the two selected elements.

- *Identifying the most promising candidates* – In a computer chess program, we might quickly evaluate all possible next moves, and then decide to study the top 25% more carefully. Selection followed by filtering is the way to go.

- *Order statistics* – Particularly interesting special cases of selection include finding the smallest element ($i = 1$), the largest element ($i = n$), and the median element ($i = n/2$).

The mean of n numbers can be easily computed in linear time by summing the elements and dividing by n. However, finding the median is a more difficult problem. Algorithms that compute the median can easily be generalized to arbitrary selection.

The most elementary median-finding algorithm sorts the items in $O(n \lg n)$ time and then returns the item sitting the $(n/2)$nd position. The good thing is that this gives much more information than just the median, enabling you to select the ith element (for all $1 \le i \le n$) in constant time after the sort. However, there are faster algorithms if all you want is the median.

In particular, there is an $O(n)$ *expected*-time algorithm based on quicksort. Select a random element in the data set as a pivot, and use it to partition the data into sets of elements less than and greater than the pivot. From the sizes of these sets, we know the position of the pivot in the total order, and hence whether the median lies to the left or right of the pivot. Now we recur on the appropriate subset until it converges on the median. This takes (on average) $O(\lg n)$ iterations, with the cost of each iteration being roughly half that of the previous one. This defines a geometric series that converges to a linear-time algorithm, although if you are very unlucky it takes the same time as quicksort, $O(n^2)$.

More complicated algorithms are known that find the median in worst-case linear time. However, the expected-time algorithm will likely win in practice. Just make sure to select random pivots in order to avoid the worst case.

Beyond mean and median, a third notion of average is the *mode*, defined to be the element that occurs the greatest number of times in the data set. The best way to compute the mode sorts the set in $O(n \lg n)$ time, which places all identical elements next to each other. By doing a linear sweep from left to right on this sorted set, we can count the length of the longest run of identical elements and hence compute the mode in a total of $O(n \lg n)$ time.

In fact, there is no faster worst-case algorithm possible to compute the mode, since the problem of testing whether there exist two identical elements in a set (called element uniqueness) can be shown to have an $\Omega(n \log n)$ lower bound. Element uniqueness is equivalent to asking if the mode is ≥ 2. Possibilities exist, at least theoretically, for improvements when the mode is large by using fast median computations.

Implementations: A bare bones implementation in C of the recursive k-selection algorithm appears in [GBY91]. See Section 9.1.6.2 for further details.

XTango (see Section 9.1.5) is an algorithm animation system for UNIX and X-windows, which includes an animation of the linear-time selection algorithm. This animation is a good one.

Notes: The linear expected-time algorithm for median and selection is due to Hoare [Hoa61]. Floyd and Rivest [FR75] provide an algorithm that uses fewer comparisons on average. Good expositions on linear-time selection include [AHU74, Baa88, CLR90, Raw92], with [Raw92] being particularly enlightening.

A sport of considerable theoretical interest is determining *exactly* how many comparisons are sufficient to find the median of n items. The linear-time algorithm of Blum et. al. [BFP+72] proves that $c \cdot n$ suffice, but we want to know what c is. A lower bower bound of $2n$ comparisons for median finding was given by Bent and John [BJ85]. In 1976, Schönhage, Paterson, and Pippenger [SPP76] presented an algorithm using $3n$ comparisons. Recently, Dor and Zwick [DZ95] proved that $2.95n$ comparisons suffice. These algorithms attempt to minimize the number of element comparisons but not the total number of operations, and hence do not lead to faster algorithms in practice.

Tight combinatorial bounds for selection problems are presented in [Aig88]. An optimal algorithm for computing the mode is presented in [DM80].

Related Problems: Priority queues (see page 180), sorting (see page 236).

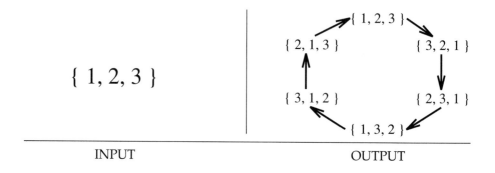

| INPUT | OUTPUT |

8.3.4   Generating Permutations

Input description: An integer n.

Problem description: Generate (1) all or (2) a random or (3) the next permutation of length n.

Discussion: A permutation describes an arrangement, or ordering, of things. Many algorithmic problems in this catalog seek the best way to order a set of objects, including traveling salesman (the least-cost order to visit n cities), bandwidth (order the vertices of a graph on a line so as to minimize the length of the longest edge), and graph isomorphism (order the vertices of one graph so that it is identical to another). Any algorithm for solving such problems exactly must construct a series of permutations along the way.

There are $n!$ permutations of n items, which grows so quickly that you can't expect to generate all permutations for $n > 11$, since $11! = 39,916,800$. Numbers like these should cool the ardor of anyone interested in exhaustive search and help explain the importance of generating random permutations.

Fundamental to any permutation-generation algorithm is a notion of order, the sequence in which the permutations are constructed, from first to last. The most natural generation order is *lexicographic*, the order they would appear if they were sorted numerically. Lexicographic order for $n = 3$ is $\{1,2,3\}$, $\{1,3,2\}$, $\{2,1,3\}$, $\{2,3,1\}$, $\{3,1,2\}$, and finally $\{3,2,1\}$. Although lexicographic order is aesthetically pleasing, there is often no particular reason to use it. For example, if you are searching through a collection of files, it does not matter whether the filenames are encountered in sorted order, so long as you search through all of them. Indeed, nonlexicographic orders lead to faster and simpler permutation generation algorithms.

There are two different paradigms for constructing permutations: ranking/unranking and incremental change methods. Although the latter are more efficient, ranking and unranking can be applied to solve a much wider class of problems, including the other combinatorial generation problems in this book. The key is to define functions *rank* and *unrank* on all permutations p and integers n, m, where $|p| = n$ and $0 \le m < n!$.

- *Rank(p)* – What is the position of p in the given generation order? A typical ranking function is recursive, such as $Rank(p) = (p_1 - 1) \cdot (|p| - 1)! + Rank(p_2, \ldots, p_{|p|})$, with $Rank(\{1\}) = 0$. Getting this right means relabeling the elements of the smaller permutation to reflect the deleted first element. Thus

$$Rank(\{2,1,3\}) = 1 \cdot 2! + Rank(\{1,2\}) = 2 + 0 \cdot 1! + Rank(\{1\}) = 2$$

- *Unrank(m,n)* – Which permutation is in position m of the $n!$ permutations of n items? A typical unranking function finds the number of times $(n - 1)!$ goes into m and proceeds recursively. $Unrank(2,3)$ tells us that the first element of the permutation must be '2', since $(2 - 1) \cdot (3 - 1)! \leq 2$ but $(3 - 1) \cdot (3 - 1)! > 2$. Deleting $(2 - 1) \cdot (3 - 1)!$ from m leaves the smaller problem $Unrank(0,2)$. The ranking of 0 corresponds to the total order, and the total order on the two remaining elements (since '2' has been used) is $\{1,3\}$, so $Unrank(2,3) = \{3,1,2\}$.

What the actual rank and unrank functions are does not matter as much as the fact that they must be inverses. In other words, $p = Unrank(Rank(p), n)$ for all permutations p. Once you define ranking and unranking functions for permutations, you can solve a host of related problems:

- *Sequencing permutations* – To determine the *next* permutation that occurs in order after p, we can $Rank(p)$, add 1, and then $Unrank(p)$. Similarly, the permutation right before p in order is $Unrank(Rank(p) - 1, |p|)$. Counting through the integers from 0 to $n! - 1$ and unranking them is equivalent to generating all permutations.
- *Generating random permutations* – Select a random integer from 0 to $n! - 1$ and unrank it.
- *Keep track of a set of permutations* – Suppose we want to construct random permutations and act only when we encounter one we have not seen before. We can set up a bit vector (see Section 8.1.5) with $n!$ bits, and set bit i to 1 if permutation $Unrank(i,n)$ has been seen. A similar technique was employed with k-subsets in the Lotto application of Section 1.9.

The rank/unrank method is best suited for small values of n, since $n!$ quickly exceeds the capacity of machine integers, unless arbitrary-precision arithmetic is available (see Section 8.2.9). The incremental change methods work by defining the *next* and *previous* operations to transform one permutation into another, typically by swapping two elements. The tricky part is to schedule the swaps so that permutations do not repeat until all of them have been generated. See the output picture above for an ordering of the six permutations of $\{1,2,3\}$ with a single swap between successive permutations.

Incremental change algorithms for sequencing permutations are tricky, but they are concise enough that they can be expressed in a dozen lines of code. See the implementation section for pointers to code. Because the incremental change is a single swap, these algorithms can be extremely fast – on average, constant time – which is independent of the size of the permutation! The secret is to represent the permutation using an n-element array to facilitate the swap. In certain applications, only the change between permutations is important. For example, in a brute-force program to search for the optimal tour, the cost of the tour associated with the new permutation will be that of the previous permutation, with the addition and deletion of four edges.

Throughout this discussion, we have assumed that the items we are permuting are all distinguishable. However, if there are duplicates (meaning our set is a *multiset*), you can save considerable time and effort by avoiding identical permutations. For example, there are only ten permutations of $\{1, 1, 2, 2, 2\}$, instead of 120. To avoid duplicates use backtracking and generate the permutations in lexicographic order.

Generating random permutations is an important little problem that people stumble across often, and often botch up. The right way is the following two-line, linear-time algorithm. We assume that $Random[i,n]$ generates a random integer between i and n, inclusive.

> for $i = 1$ to n do $a[i] = i$;
> for $i = 1$ to $n - 1$ do $swap[a[i], a[Random[i, n]]]$;

That this algorithm generates all permutations uniformly at random is not obvious. If you think so, explain convincingly why the following algorithm *does not* generate permutations uniformly:

> for $i = 1$ to n do $a[i] = i$;
> for $i = 1$ to $n - 1$ do $swap[a[i], a[Random[1, n]]]$;

Such subtleties demonstrate why you must be very careful with random generation algorithms. Indeed, we recommend that you try some reasonably extensive experiments with any random generator before really believing it. For example, generate 10,000 random permutations of length 4 and see whether all 24 of them occur approximately the same number of times. If you understand how to measure statistical significance, you are in even better shape.

Implementations: The best source on generating combinatorial objects is Nijenhuis and Wilf [NW78], who provide efficient Fortran implementations of algorithms to construct random permutations and to sequence permutations in minimum-change order. Also included are routines to extract the cycle structure of a permutation. See Section 9.1.6.3 for details.

An exciting WWW site developed by Frank Ruskey of the University of Victoria contains a wealth of material on generating combinatorial objects of different types, including permutations, subsets, partitions, and certain graphs. Specifically, there is an interactive interface that lets you specify which type of objects you would like to construct and quickly returns the objects to you. It is well worth checking this out at http://sue.csc.uvic.ca/~cos/.

Combinatorica [Ski90] provides Mathematica implementations of algorithms that construct random permutations and sequence permutations in minimum change and lexicographic orders. It also provides a backracking routine to construct all distinct permutations of a multiset, and it supports various permutation group operations. See Section 9.1.4.

The Stanford GraphBase (see Section 9.1.3) contains routines to generate all permutations of a multiset.

Notes: The primary reference on permutation generation is the survey paper by Sedgewick [Sed77]. Good expositions include [NW78, RND77, Rus97].

The fast permutation generation methods make only a single swap between successive permutations. The Johnson-Trotter algorithm [Joh63, Tro62] satisfies an even stronger condition, namely that the two elements being swapped are always adjacent.

In the days before ready access to computers, books with tables of random permutations [MO63] were used instead of algorithms. The swap-based random permutation algorithm presented above was first described in [MO63].

Related Problems: Random number generation (see page 217), generating subsets (see page 250), generating partitions (see page 253).

$$\{ 1, 2, 3 \}$$

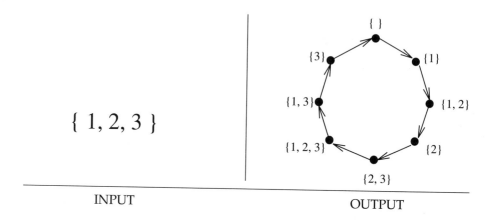

INPUT   OUTPUT

8.3.5 Generating Subsets

Input description: An integer n.

Problem description: Generate (1) all or (2) a random or (3) the next subset of the integers $\{1, \ldots, n\}$.

Discussion: A subset describes a selection of objects, where the order among them does not matter. Many of the algorithmic problems in this catalog seek the best subset of a group of things: vertex cover seeks the smallest subset of vertices to touch each edge in a graph; knapsack seeks the most profitable subset of items of bounded total size; and set packing seeks the smallest subset of subsets that together cover each item exactly once.

There are 2^n distinct subsets of an n-element set, including the empty set as well as the set itself. This grows exponentially, but at a considerably smaller rate than the $n!$ permutations of n items. Indeed, since $2^{20} = 1,048,576$, a brute-force search through all subsets of 20 elements is easily manageable, although by $n = 30$, $2^{30} = 1,073,741,824$, so you will certainly be pushing things.

By definition, the relative order among the elements does not distinguish different subsets. Thus $\{1, 2, 5\}$ is the same as $\{2, 1, 5\}$. However, it is a very good idea to maintain your subsets in a sorted or *canonical* order, in order to speed up such operations as testing whether two subsets are identical or making them look right when printed.

As with permutations (see Section 8.3.4), the key to subset generation problems is establishing a numerical sequence among all 2^n subsets. There are three primary alternatives:

- *Lexicographic order* – Lexicographic order is sorted order, and often the most natural order for generating combinatorial objects. The eight subsets of $\{1, 2, 3\}$ in lexicographic order are $\{\}, \{1\}, \{1, 2\}, \{1, 2, 3\}, \{1, 3\}, \{2\}, \{2, 3\}$, and $\{3\}$. Unfortunately, it is surprisingly difficult to generate subsets in lexicographic order. Unless you have a compelling reason to do so, forget about it.

- *Gray Code* – A particularly interesting and useful subset sequence is the minimum change order, wherein adjacent subsets differ by the insertion or deletion of exactly one element. Such an ordering, called a *Gray code*, appears in the output picture above.

  Subset generation in Gray code order can be very fast, because there is a nice recursive construction to sequence subsets. Further, since only one element changes between subsets, exhaustive search algorithms built on Gray codes can be quite efficient. A set cover program would only have to update the change in coverage by the addition or deletion of one subset. See the implementation section below for Gray code subset generation programs.

- *Binary counting* – The simplest approach to subset generation problems is based on the observation that any subset S' of S is defined by which of the $n = |S|$ items are in S'. We can represent S' by a binary string of n bits, where bit i is 1 iff the ith element of S is in S'. This defines a bijection between the 2^n binary strings of length n, and the 2^n subsets of n items. For $n = 3$, binary counting generates subsets in the following order: {}, {3}, {2}, {2,3}, {1}, {1,3}, {1,2}, {1,2,3}.

  This binary representation is the key to solving all subset generation problems. To generate all subsets in order, simply count from 0 to $2^n - 1$. For each integer, successively mask off each of the bits and compose a subset of exactly the items corresponding to '1' bits. To generate the *next* or *previous* subset, increment or decrement the integer by one. *Unranking* a subset is exactly the masking procedure, while *ranking* constructs a binary number with 1's corresponding to items in S and then converts this binary number to an integer.

  To generate a random subset, you could generate a random integer from 0 to $2^n - 1$ and unrank, although you are probably asking for trouble because any flakiness with how your random number generator rounds things off means that certain subsets can never occur. Therefore, a better approach is simply to flip a coin n times, with the ith flip deciding whether to include element i in the subset. A coin flip can be robustly simulated by generating a random real or large integer and testing whether it is bigger or smaller than half the range. A Boolean array of n items can thus be used to represent subsets as a sort of premasked integer. The only complication is that you must explicitly handle the carry if you seek to generate all subsets.

Generation problems for two closely related problems arise often in practice:

- *k-subsets* – Instead of constructing all subsets, we may only be interested in the subsets containing exactly k elements. There are $\binom{n}{k}$ such subsets, which is substantially less than 2^n, particularly for small values of k.

  The best way to construct all k-subsets is in lexicographic order. The ranking function is based on the observation that there are $\binom{n-f}{k-1}$ k-subsets whose smallest element is f. Using this, it is possible to determine the smallest element in the mth k-subset of n items. We then proceed recursively for subsequent elements of the subset. See the implementations below for details.

- *Strings* – Generating all subsets is equivalent to generating all 2^n strings of true and false. To generate all or random strings on alphabets of size α, the same basic techniques apply, except there will be α^n strings in total.

Implementations: Nijenhuis and Wilf [NW78] provide efficient Fortran implementations of algorithms to construct random subsets and to sequence subsets in Gray code and lexicographic order. They also provide routines to construct random k-subsets and sequence them in lexicographic order. See Section 9.1.6.3 for details on ftp-ing these programs. Algorithm 515 [BL77] of the *Collected Algorithms of the ACM* is another Fortran implementation of lexicographic k-subsets, available from Netlib (see Section 9.1.2).

An exciting WWW site developed by Frank Ruskey of the University of Victoria contains a wealth of material on generating combinatorial objects of different types, including permutations, subsets, partitions, and certain graphs. Specifically, it provides an interactive interface that lets you specify which type of objects you would like to construct and then returns the objects to you. Check this out at http://sue.csc.uvic.ca/~cos/.

Combinatorica [Ski90] provides Mathematica implementations of algorithms to construct random subsets and to sequence subsets in Gray code, binary, and lexicographic order. They also provide routines to construct random k-subsets and strings, and sequence them lexicographically. See Section 9.1.4 for further information on Combinatorica.

Notes: The primary expositions on subset generation include [NW78, RND77, Rus97]. Wilf [Wil89] provides an update of [NW78], including a thorough discussion of modern Gray code generation problems.

Gray codes were first developed [Gra53] to transmit digital information in a robust manner over an analog channel. By assigning the code words in Gray code order, the ith word differs only slightly from the $(i + 1)$st, so minor fluctuations in analog signal strength corrupts only a few bits. Gray codes have a particularly nice correspondence to Hamiltonian cycles on the hypercube. See any of the references above for details. An exposition on the more general problem of constructing Gray codes for k items (instead of 2^n subsets) appears in [Man89].

The popular puzzle *Spinout*, manufactured by Binary Arts Corporation, can be solved using Gray codes.

Related Problems: Generating permutations (see page 246), generating partitions (see page 253).

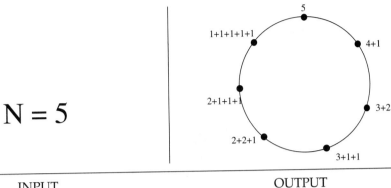

$$N = 5$$

INPUT   OUTPUT

8.3.6 Generating Partitions

Input description: An integer n.

Problem description: Generate (1) all or (2) a random or (3) the next integer or set partitions of length n.

Discussion: There are two different types of combinatorial objects denoted by the term "partition", namely integer partitions and set partitions. Although they are quite different beasts, it is a good idea to make both a part of your vocabulary:

- *Integer partitions* of n are sets of nonzero integers that add up to exactly n. For example, the seven distinct integer partitions of 5 are {5}, {4,1}, {3,2}, {3,1,1}, {2,2,1}, {2,1,1,1}, and {1,1,1,1,1}. An interesting application I encountered that required the generation of integer partitions was in a simulation of nuclear fission. When an atom is smashed, the nucleus of protons and neutrons is broken into a set of smaller clusters. The sum of the particles in the set of clusters must equal the original size of the nucleus. As such, the integer partitions of this original size represent all the possible ways to smash the atom.
- *Set partitions* divide the elements $1, \ldots, n$ into nonempty subsets. For example, there are fifteen distinct set partitions of $n = 4$: {1234}, {123,4}, {124,3}, {12,34}, {12,3,4}, {134,2}, {13,24}, {13,2,4}, {14,23}, {1,234}, {1,23,4}, {14,2,3}, {1,24,3}, {1,2,34}, and {1,2,3,4}. Several of the problems in this catalog return set partitions as results, such as vertex coloring and connected components.

Although the number of integer partitions grows exponentially with n, they do so at a refreshingly slow rate. There are only 627 partitions of $n = 20$, and it is even possible to enumerate all partitions of $n = 100$, since there there are only 190,569,292 of them.

The best way to generate all partitions is to construct them in lexicographically decreasing order. The first partition is $\{n\}$ itself. The general rule is to subtract 1 from the smallest part that is > 1 and then collect all the 1's so as to match the new smallest part > 1. For example, the partition after $\{4, 3, 3, 3, 1, 1, 1, 1\}$ is $\{4, 3, 3, 2, 2, 2, 1\}$, since the five 1's left after $3 - 1 = 2$ becomes the smallest part are best packaged as 2,2,1. When the partition is all 1's, we have completed one trip through all the partitions.

This algorithm is not particularly complicated, but it is sufficiently intricate that you should consider using one of the implementations below. In either case, test it to make sure that you get exactly 627 distinct partitions for $n = 20$.

Generating integer partitions uniformly at random is a trickier matter than generating random permutations or subsets. This is because selecting the first (i.e. largest) element of the partition has a dramatic effect on the number of possible partitions that can be generated. Observe that no matter how large n is, there is only one partition of n whose largest part is 1. The number of partitions of n with largest part at most k is given by the recurrence

$$P_{n,k} = P_{n-k,k} + P_{n,k-1}$$

with the two boundary conditions $P_{x-y,x} = P_{x-y,x-y}$ and $P_{n,1} = 1$. This function can be used to select the largest part of your random partition with the correct probabilities and, by proceeding recursively, to eventually construct the entire random partition. Implementations are cited below.

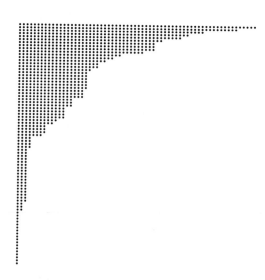

Figure 8–3. The Ferrers diagram of a random partition of $n = 1000$

Random partitions tend to have large numbers of fairly small parts, best visualized by a Ferrers diagram as in Figure 8-3. Each row of the diagram corresponds to one part of the partition, with the size of each part represented by that many dots.

Set partitions can be generated using techniques similar to integer partitions. Each set partition is encoded as a *restricted growth function*, $a_1, \ldots, a_n$, where $a_1 = 0$ and $a_i \leq 1 + \max(a_1, \ldots, a_i)$, for $i = 2, \ldots, n$. Each distinct digit identifies a subset, or *block*, of the partition, while the growth condition ensures that the blocks are sorted into a canonical order based on the smallest element in each block. For example, the restricted growth function $0, 1, 1, 2, 0, 3, 1$ defines the set partition $\{\{1, 5\}, \{2, 3, 7\}, \{4\}, \{6\}\}$.

Since there is a one-to-one equivalence between set partitions and restricted growth functions, we can use lexicographic order on the restricted growth functions to order the set partitions. Indeed, the fifteen partitions of $\{1, 2, 3, 4\}$ listed above are sequenced according to the lexicographic order of their restricted growth function (check it out).

To randomly generate set partitions, we use a similar counting strategy as with integer partitions. The Stirling numbers of the second kind $\{^n_k\}$ count the number of partitions of $\{1, \ldots, n\}$ with exactly k blocks. They are computed using the recurrence

$$\{^n_k\} = \{^{n-1}_{k-1}\} + k\{^{n-1}_{k}\}$$

with the boundary conditions $\{^n_n\} = \{^n_1\} = 1$. The reader is referred to the references and implementations for more details and code.

Implementations: The best source on generating combinatorial objects is Nijenhuis and Wilf [NW78], who provide efficient Fortran implementations of algorithms to construct random and sequential integer partitions, set partitions, compositions, and Young tableaux. See Section 9.1.6.3 for details on ftp-ing these programs.

An exciting WWW site developed by Frank Ruskey of the University of Victoria contains a wealth of material on generating combinatorial objects of different types, including permutations, subsets, partitions, and certain graphs. It is well worth checking this out at http://sue.csc.uvic.ca/~cos/.

Combinatorica [Ski90] provides Mathematica implementations of algorithms to construct random and sequential integer partitions, compositions, strings, and Young tableaux, as well as to count and manipulate these objects. See Section 9.1.4.

Algorithm 403 [CT71] of the *Collected Algorithms of the ACM* is a Fortran code for constructing integer partitions with k parts. It is available from Netlib (see Section 9.1.2). The Stanford GraphBase (see Section 9.1.3) also contains generators for constructing all integer partitions.

Notes: The standard references on combinatorial generation [NW78, Rus97, Ski90, SW86] all present algorithms for generating integer and/or set partitions. Andrews [And76] is the primary reference on integer partitions and related topics.

Interestingly, the set of all 52 set partitions for $n = 5$ appears in the form of Murasaki diagrams in the oldest novel known, *The Tale of Genji*.

Two related combinatorial objects are Young tableaux and integer compositions, although they are less likely to emerge in applications. Generation algorithms for both are presented in [NW78, Rus97, Ski90].

Young tableaux are two-dimensional configurations of integers $\{1, \ldots, n\}$ where the number of elements in each row is defined by an integer partition of n. Further, the elements of each row and column are sorted in increasing order, and the rows are left-justified. This notion of shape captures a wide variety of structures as special cases. They have many interesting properties, including the existance of a bijection between pairs of tableaux and permutations.

Compositions represent the set of possible assignments of a set of n indistinguishable balls to k distinguishable boxes. For example, we can place three balls into two boxes as $\{3,0\}$, $\{2,1\}$, $\{1,2\}$, or $\{0,3\}$. Compositions are most easily constructed sequentially in lexicographic order. To construct them randomly, pick a random $(k-1)$-subset of $n+k-1$ items using the algorithm of Section 8.3.5, and count the number of unselected items between the selected ones. For example, if $k = 5$ and $n = 10$, the $(5-1)$-subset $\{1,3,7,14\}$ of $1, \ldots, (n+k-1) = 14$ defines the composition $\{0,1,3,6,0\}$, since there are no items to the left of element 1 nor right of element 14.

Related Problems: Generating permutations (see page 246), generating subsets (see page 250).

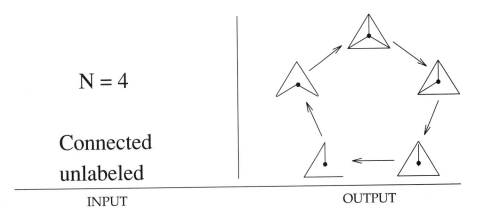

N = 4

Connected
unlabeled

INPUT   OUTPUT

8.3.7 Generating Graphs

Input description: Parameters describing the desired graph, such as the number of vertices n, the number of edges m, or the edge probability p.

Problem description: Generate (1) all or (2) a random or (3) the next graph satisfying the parameters.

Discussion: Graph generation typically arises in constructing test data for programs. Perhaps you have two different programs that solve the same problem, and you want to see which one is faster or make sure that they always give the same answer. Another application is experimental graph theory, verifying whether a particular property is true for all graphs or how often it is true. It is much easier to conjecture the four-color theorem once you have demonstrated 4-colorings for all planar graphs on 15 vertices.

A different application of graph generation arises in network design. Suppose you need to design a network linking ten machines using as few cables as possible, such that the network can survive up to two vertex failures. One approach is to test all the networks with a given number of edges until you find one that will work. For larger graphs, more heuristic approaches, like simulated annealing, will likely be necessary.

Many factors complicate the problem of generating graphs. First, make sure you know what you want to generate:

- *Do I want labeled or unlabeled graphs?* – The issue here is whether the names of the vertices matter in deciding whether two graphs are the same. In generating *labeled graphs*, we seek to construct all possible labelings of all possible graph topologies. In generating *unlabeled graphs*, we seek only one representative for each topology and ignore labelings. For example, there are only two connected unlabeled graphs on three vertices – a triangle and a simple path. However, there are four connected labeled graphs on three vertices – one triangle and three 3-vertex paths, each distinguished by their central vertex. In general, labeled graphs are much easier to generate. However, there are so many more of them that you quickly get swamped with isomorphic copies of the same few graphs.

- *What do I mean by random?* – There are two primary models of random graphs, both of which generate graphs according to different probability distributions. The first model is parameterized by a given edge probability p. Typically, $p = 0.5$, although smaller values can be used to construct sparser random graphs. In this model a coin is flipped for each pair of vertices x and y to decide whether to add an edge (x, y). All *labeled* graphs will be generated with equal probability when $p = 1/2$.

  The second model is parameterized by the desired number of edges m. It selects m distinct edges uniformly at random. One way to do this is by drawing random (x, y)-pairs and creating an edge if that pair is not already in the graph. An alternative approach to computing the same things constructs the set of $\binom{n}{2}$ possible edges and selects a random m-subset of them, as discussed in Section 8.3.5.

Which of these options best models your application? Probably none of them. Random graphs, by definition, have very little structure. In most applications, graphs are used to model relationships, which are often highly structured. Experiments conducted on random graphs, although interesting and easy to perform, often fail to capture what you are looking for.

An alternative to random graphs is to use "organic" graphs, graphs that reflect the relationships among real-world objects. The Stanford GraphBase, discussed below, is an outstanding source of organic graphs. Further, there are many raw sources of relationships electronically available via the Internet that can be turned into interesting organic graphs with a little programming and imagination. Consider the graph defined by a set of WWW pages, with any hyperlink between two pages defining an edge. Or what about the graph implicit in railroad, subway, or airline networks, with vertices being stations and edges between two stations connected by direct service? As a final example, every large computer program defines a call graph, where the vertices represent subroutines, and there is an edge (x, y) if x calls y.

Two special classes of graphs have generation algorithms that have proven particularly useful in practice:

- *Trees* – Prüfer codes provide a simple way to rank and unrank *labeled* trees and thus solve all the standard generation problems discussed in Section 8.3.4. There are exactly n^{n-2} labeled trees on n vertices, and exactly that many strings of length $n - 2$ on the alphabet $\{1, 2, \dots, n\}$.

  The key to Prüfer's bijection is the observation that every tree has at least two vertices of degree 1. Thus in any labeled tree, the vertex v incident on the leaf with lowest label is well-defined. We take v to be S_1, the first character in the code. We then delete the associated leaf and repeat the procedure until only two vertices are left. This defines a unique code S for any given labeled tree that can be used to rank the tree. To go from code to tree, observe that the degree of vertex v in the tree is one more than the number of times v occurs in S. The lowest-labeled leaf will be the smallest integer missing from S, which when paired with S_1 determines the first edge of the tree. The entire tree follows by induction.

Algorithms for efficiently generating unlabeled rooted trees are presented in the implementation section below.

- *Fixed degree sequence graphs* – The *degree sequence* of a graph G is an integer partition $p = (p_1, \ldots, p_n)$ where p_i is the degree of the ith highest-degree vertex of G. Since each edge contributes to the degree of two vertices, p is a partition of $2m$, where m is the number of edges in G.

  Not all partitions correspond to degree sequences of graphs. However, there is a recursive construction that constructs a graph with a given degree sequence if one exists. If a partition is realizable, the highest-degree vertex v_1 can be connected to the next p_1 highest-degree vertices in G, or the vertices corresponding to parts $p_2, \ldots, p_{p_1+1}$. Deleting p_1 and decrementing $p_2, \ldots, p_{p_1+1}$ yields a smaller partition, which we recur on. If we terminate without ever creating negative numbers, the partition was realizable. Since we always connect the highest-degree vertex to other high-degree vertices, it is important to reorder the parts of the partition by size after each iteration.

  Although this construction is deterministic, a semirandom collection of graphs realizing this degree sequence can be generated from G using *edge-flipping* operations. Suppose edges (x, y) and (w, z) are in G, but (x, w) and (y, z) are not. Exchanging these pairs of edges creates a different (not necessarily connected) graph without changing the degrees of any vertex.

Implementations: The Stanford GraphBase [Knu94] is perhaps most useful as an instance generator for constructing a wide variety of graphs to serve as test data for other programs. It incorporates graphs derived from interactions of characters in famous novels, Roget's Thesaurus, the Mona Lisa, expander graphs, and the economy of the United States. It also contains routines for generating binary trees, graph products, line graphs, and other operations on basic graphs. Finally, because of its machine-independent random number generators, it provides a way to construct random graphs such that they can be reconstructed elsewhere, thus making them perfect for experimental comparisons of algorithms. See Section 9.1.3 for additional information.

Combinatorica [Ski90] provides Mathematica generators for basic graphs such as stars, wheels, complete graphs, random graphs and trees, and graphs with a given degree sequence. Further, it includes operations to construct more interesting graphs from these, including join, product, and line graph. Graffiti [Faj87], a collection of almost 200 graphs of graph-theoretic interest, are available in Combinatorica format. See Section 9.1.4.

The graph isomorphism testing program nauty (see Section 8.5.9), by Brendan D. McKay of the Australian National University, has been used to generate catalogs of all nonisomorphic graphs with up to 11 vertices. This extension to nauty, named makeg, can be obtained by anonymous ftp from bellatrix.anu.edu.au (150.203.23.14) in the directory pub/nauty19.

Nijenhuis and Wilf [NW78] provide efficient Fortran routines to enumerate all labeled trees via Prüfer codes and to construct random unlabeled rooted trees. See Section 9.1.6.3. A collection of generators for standard families of graphs is included with LEDA (see Section 9.1.1).

Notes: An extensive literature exists on generating graphs uniformly at random. Surveys include [Gol93, Tin90]. Closely related to the problem of generating classes of graphs is counting them. Harary and Palmer [HP73] survey results in graphical enumeration.

Random graph theory is concerned with the properties of random graphs. Threshold laws in random graph theory define the edge density at which properties such as connectedness become highly likely to occur. Expositions on random graph theory include [ES74, Pal85].

An integer partition is *graphic* if there exists a simple graph with that degree sequence. Erdős and Gallai [EG60] proved that a degree sequence is graphic if and only if the sequence observes the following condition for each integer $r < n$:

$$\sum_{i=1}^{r} d_i \leq r(r-1) + \sum_{i=r+1}^{n} \min(r, d_i)$$

The bijection between $n - 2$ strings and labeled trees is due to Prüfer [Prü18]. Good expositions on this result include [Eve79a, NW78].

Related Problems: Generating permutations (see page 246), graph isomorphism (see page 335).

September 27

1752 ?

September, 1752

S	M	Tu	W	Th	F	S
		1	2	14	15	16
17	18	19	20	21	22	23
24	25	26	27	28	29	30

INPUT   OUTPUT

8.3.8 Calendrical Calculations

Input description: A particular calendar date d, specified by month, day, and year.

Problem description: Which day of the week did d fall on according to the given calendar system?

Discussion: Many business applications need to perform calendrical calculations. Perhaps we want to display a calendar of a specified month and year. Maybe we need to compute what day of the week or year some event occurs, as in figuring out the date on which a 180-day futures contract comes due. The importance of correct calendrical calculations is perhaps best revealed by the furor over the "millennium bug," the crisis in legacy programs that allocate only two digits for storing the year.

More complicated questions arise in international applications, because different nations and ethnic groups around the world use different calendar systems. Some of these, like the Gregorian calendar used in most of the world, are based on the sun, while others, like the Hebrew calendar, are lunar calendars. How would you tell today's date according to the Chinese or Arabic calendar?

The algorithms associated with calendrical calculations are different from the other problems in this book, because calendars are historical objects, not mathematical ones. The issues revolve around specifying the rules of the calendrical system and simply implementing them correctly, rather than designing efficient shortcuts for the computation.

The basic approach behind calendar systems is to start with a particular reference date and count from there. The particular rules for wrapping the count around into months and years is what distinguishes a given calendar system from another. To implement a calendar, we need two functions, one that given a date returns the integer number of days that have elapsed since the reference start date, the other of which takes an integer n and returns the calendar date exactly n days from the reference date. This is analogous to the ranking and unranking rules for combinatorial objects, such as permutations (see Section 8.3.4).

The major source of complications in calendar systems is that the solar year is not an integer number of days long. Thus if a calendar seeks to keep its annual dates in sync with the seasons, leap days must be added at both regular and irregular intervals. Since a solar year is 365 days and 5:49:12 hours long, an extra 10:48 minutes would have to be accounted for at the end of each year if we were simply to add a leap day every four years.

The original Julian calendar (from Julius Caesar) did not account for these extra minutes, which had accumulated to ten days by 1582 when Pope Gregory XIII proposed the Gregorian calendar used today. Gregory deleted the ten days and eliminated leap days in years that are multiples of 100 but not 400. Supposedly, riots ensued because the masses feared their lives were being shortened by ten days. Outside the Catholic church, resistance to change slowed the reforms. The deletion of days did not occur in England and America until September 1752, and not until 1927 in Turkey.

The rules for most calendrical systems are sufficiently complicated and pointless that you should lift code from a reliable place rather than attempt to write your own. We identify suitable implementations below.

There are a variety of "impress your friends" algorithms that enable you to compute in your head on which day of the week a particular date occurred. Such algorithms often fail to work reliably outside the given century and should be avoided for computer implementation.

Implementations: Dershowitz and Reingold provide a uniform algorithmic presentation [DR90, RDC93] for a variety of different calendar systems, including the Gregorian, ISO, Julian, Islamic, and Hebrew calendars, as well as other calendars of historical interest. Further, they provide Common Lisp and C++ routines to convert dates between calendars, day of the week computations, and the determination of secular and religious holidays. These are likely to be the most comprehensive and reliable calendrical routines you will be able to get your hands on, and are available at http://emr.cs.uiuc.edu:80/~reingold/calendars.html.

A nice package for calendrical computations in Mathematica by Ilan Vardi is available in the Packages/Miscellaneous directory of the standard Mathematica distribution, and also from MathSource. Vardi's book [Var91] discusses the theory behind the implementation, which provides support for the Gregorian, Julian, and Islamic calendars.

Gregorian calendar computations implemented in C appear in [BR95]. This code uses 1582 as the date of the calendar reform, instead of the standard UNIX date of 1752. The code for these algorithms is printed in the text and are available on disk for a modest fee.

Notes: The most comprehensive discussion of calendrical computation algorithms are the papers by Dershowitz and Reingold [DR90, RDC93]. These papers are superseded by their book [DR97]. Histories of the Gregorian calendar appear in [BR95].

Related Problems: Arbitrary-precision arithmetic (see page 224), generating permutations (see page 246).

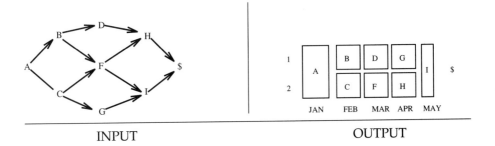

INPUT   OUTPUT

8.3.9 Job Scheduling

Input description: A directed acyclic graph $G = (V, E)$, where the vertices represent jobs and the edge (u, v) implies that task u must be completed before task v.

Problem description: What schedule of tasks completes the job using the minimum amount of time or processors?

Discussion: Devising a proper schedule to satisfy a set of constraints is fundamental to many applications. A critical aspect of any parallel processing system is the algorithm mapping tasks to processors. Poor scheduling can leave most of the expensive machine sitting idle while one bottleneck task is performed. Assigning people to jobs, meetings to rooms, or courses to final exam periods are all different examples of scheduling problems.

Scheduling problems differ widely in the nature of the constraints that must be satisfied and the type of schedule desired. For this reason, several other catalog problems have a direct application to various kinds of scheduling:

- To construct a schedule consistent with the precedence constraints, see topological sorting in Section 8.4.2.
- To assign a set of jobs to people who have the appropriate skills for them, see bipartite matching in Section 8.4.6.
- To assign a set of jobs to time slots such that no two jobs that interfere are assigned the same time slot, see vertex and edge coloring in Sections 8.5.7 and 8.5.8.
- To construct the most efficient route for a delivery person to visit a given set of locations, see the traveling salesman problem in Section 8.5.4. To construct the most efficient route for a snowplow or mailman to completely traverse a given set of edges, see the Eulerian cycle problem in Section 8.4.7.

In this section, we focus on precedence-constrained scheduling problems for directed acyclic graphs. These problems are often called PERT/CPM, for *Program Evaluation and Review Technique/Critical Path Method*. Suppose you have broken a big job into a large number of smaller tasks. For each task you know how long it should take (or perhaps an upper bound on how long it might take). Further, for each pair of tasks you know whether it is essential that one task be performed before another. The fewer constraints we have to enforce, the better our schedule can be. These constraints must define a directed acyclic graph, acyclic because a cycle in the precedence constraints represents a Catch-22 situation that can never be resolved.

We are interested in several problems on these networks:

- *Minimum completion time* – assuming that we have an unlimited number of workers, what is the fastest we can get this job completed while respecting precedence constraints. If there were *no* precedence constraints, each task could be assigned its own worker, and the total time would be that of the longest single task. If there were such strict precedence constraints that each task had to follow the completion of its immediate predecessor, the minimum completion time would be obtained by summing up the times for each task.

  The minimum completion time for a DAG can be easily computed in $O(n + m)$ time. Initialize the completion time for all the vertices to 0, except the start vertex, which is initialized to the length of the start task. Perform a topological sort to order the vertices such that all precedences will have been considered by the time we evaluate each vertex. For each vertex u, consider all the edges (u, v) leaving u. The completion time of these vertices is the maximum of the current completion time for v plus the completion time of u plus the task time of v.

- *Critical path* – The longest path from the start vertex to the completion vertex defines the *critical path*. This can be important to know, for the only way to shorten the minimum total completion time is to reduce the time of one of the tasks on each critical path. The tasks on the critical paths can be determined in $O(n + m)$ time using the simple dynamic programming presented in Section 8.4.4.

- *What is the tradeoff between number of workers and completion time?* – What we would really be interested in knowing is how best to complete the schedule with a given number of workers. Unfortunately, this and most similar problems are NP-complete.

An even more general formulation seeks critical paths in networks where certain jobs are restricted to certain people. Such networks are known as disjunctive networks. Finding critical paths on such networks is more complicated than CPM/PERT, but implementations are described below.

In any sufficiently large and sufficiently real scheduling application, there will be combinations of constraints that are difficult or impossible to model using these techniques. There are two reasonable ways to deal with such problems. First, we can ignore enough constraints that the problem reduces to one of the types that we have described here, solve it, and then see how bad it is using the other constraints. Perhaps the schedule can be easily modified by hand to satisfy constraints like keeping Joe and Bob apart so they can't kill each other. Another approach is to formulate your scheduling problem via linear-integer programming (see Section 8.2.6) in all its complexity. This method can be better only if you really know enough about your desired schedule to formulate the right linear program, and if you have the time to wait for the program to give you the solution. I would start out with something simpler and see what happens first.

Another fundamental scheduling problem takes a set of jobs without precedence constraints and assign them to identical machines so as to minimize the total elapsed time. Consider a copy shop with k Xerox machines and a stack of jobs to finish by the end of the day. Such tasks are called *job-shop scheduling*. They can be modeled as bin-packing problems (see Section 8.6.9), where each job is assigned a number equal to the number of hours it will take and each machine is represented by a bin with space equal to the number of hours in a day.

More sophisticated variations of job-shop scheduling provide each task with allowable start and required finishing times. Effective heuristics are known, based on sorting the tasks by size and finishing time. We refer the reader to the references for more information. Note that these scheduling problems become hard only when the tasks cannot be broken up onto multiple machines or interrupted (preempted) and then rescheduled. If your application has these degrees of freedom, you should exploit them.

Implementations: Pascal implementations of Balas's algorithm for disjunctive network scheduling and Hu's algorithm for assigning jobs to processors with precedence constraints appear in [SDK83]. See Section 9.1.6.1.

Algorithm 520 [WBCS77] of the *Collected Algorithms of the ACM* is a Fortran code for multiple-resource network scheduling. It is available from Netlib (see Section 9.1.2).

Notes: The literature on scheduling algorithms is a vast one. For a more detailed survey of the field, we refer the reader to [Cof76, LLK83].

Good expositions on CPM/PERT include [Eve79a, Law76, PGD82, SDK83]. Good expositions on job-shop scheduling include [AC91, SDK83].

Related Problems: Topological sorting (see page 273), matching (see page 287), vertex coloring (see page 329), edge coloring (see page 333), bin packing (see page 374).

(X1 or X2 or $\overline{X3}$)	(X1 or X2 or $\overline{X3}$)
($\overline{X1}$ or $\overline{X2}$ or X3)	($\overline{X1}$ or X2 or X3)
($\overline{X1}$ or $\overline{X2}$ or $\overline{X3}$)	($\overline{X1}$ or $\overline{X2}$ or $\overline{X3}$)
($\overline{X1}$ or X2 or X3)	($\overline{X1}$ or X2 or X3)

INPUT   OUTPUT

8.3.10 Satisfiability

Input description: A set of clauses in conjunctive normal form.

Problem description: Is there a truth assignment to the Boolean variables such that every clause is simultaneously satisfied?

Discussion: Satisfiability arises whenever we seek a configuration or object that must be consistent with (i.e. satisfy) a given set of constraints. For example, consider the problem of drawing name labels for cities on a map. For the labels to be legible, we do not want the labels to overlap, but in a densely populated region many labels need to be drawn in a small space. How can we avoid collisions?

For each of the n cities, suppose we identify two possible places to position its label, say right above or right below each city. We can represent this choice by a Boolean variable v_i, which will be true if city c_i's label is above c_i, otherwise v_i = false. Certain pairs of labels may be forbidden, such as when c_i's above label would obscure c_j's below label. This pairing can be forbidden by the two-element clause ($\overline{v}_i$ or v_j), where $\overline{v}$ means "not v". Finding a satisfying truth assignment for the resulting set of clauses yields a mutually legible map labeling if one exists.

Satisfiability is *the* original NP-complete problem. Despite its applications to constraint satisfaction, logic, and automatic theorem proving, it is perhaps most important theoretically as the root problem from which all other NP-completeness proofs originate.

- *Is your formula in CNF or DNF?* – In satisfiability, the constraints are specified as a logical formula. There are two primary ways of expressing logical formulas, conjunctive normal form (CNF) and disjunctive normal form (DNF). In CNF formulas, we must satisfy all clauses, where each clause is constructed by and-ing or's of literals together, such as

$$(v_1 \text{ or } \overline{v}_2) \text{ and } (v_2 \text{ or } v_3)$$

In DNF formulas, we must satisfy any one clause, where each clause is constructed by or-ing ands of literals together. The formula above can be written in DNF as

$$(\bar{v}_1 \text{ and } \bar{v}_2 \text{ and } v_3) \text{ or } (\bar{v}_1 \text{ and } v_2 \text{ and } \bar{v}_3)$$

$$\text{or } (\bar{v}_1 \text{ and } v_2 \text{ and } v_3) \text{ or } (v_1 \text{ and } \bar{v}_2 \text{ and } v_3)$$

Solving DNF-satisfiability is trivial, since any DNF formula can be satisfied unless *every* clause contains both a literal and its complement (negation). However, CNF-satisfiability is NP-complete. This seems paradoxical, since we can use De Morgan's laws to convert CNF-formulae into equivalent DNF-formulae and vice versa. The catch is that an exponential number of terms might be constructed in the course of translation, so that the translation itself might not run in polynomial time.

- *How big are your clauses?* – k-SAT is a special case of satisfiability when each clause contains at most k literals. The problem of 1-SAT is trivial, since we must set true any literal appearing in any clause. The problem of 2-SAT is not trivial, but it can still be solved in linear time. This is very interesting, because many problems can be modeled as 2-SAT using a little cleverness. Observe that the map labeling problem described above is an instance of 2-SAT and hence can be solved in time linear in the number of clauses, which might be quadratic in the number of variables.

  The good times end as soon as clauses contain three literals each, i.e. 3-SAT, for 3-SAT is NP-complete. Thus in general it will not be helpful to model a problem as satisfiability unless we can do it with two-element clauses.

- *Does it suffice to satisfy* most *of the clauses?* – Given an instance of general satisfiability, there is not much you can do to solve it except by backtracking algorithms such as the Davis-Putnam procedure. In the worst case, there are 2^m truth assignments to be tested, but fortunately, there are lots of ways to prune the search. Although satisfiability is NP-complete, how hard it is in practice depends on how the instances are generated. Naturally defined "random" instances are often surprisingly easy to solve, and in fact it is nontrivial to generate instances of the problem that are truly hard.

  Still, we would likely benefit by relaxing the problem so that the goal is to satisfy as many clauses as possible. Here optimization techniques such as simulated annealing can be put to work to refine random or heuristic solutions. Indeed, any random truth assignment to the variables will satisfy any particular k-SAT clause with probability $1 - (1/2)^k$, so our first attempt is likely to satisfy most of the clauses. Finishing off the job is the hard part. Finding an assignment that satisfies the maximum number of clauses is NP-complete even for nonsatisfiable instances.

When faced with a problem of unknown complexity, proving the problem NP-complete can be an important first step. If you think your problem might be hard, the first thing to do is skim through Garey and Johnson [GJ79] looking for your problem. If you don't find it, my recommendation is to put the book away and try to prove hardness from first principles, using one of the basic problems in this catalog, particularly 3-SAT, vertex cover, independent set, integer partition, clique, and Hamiltonian cycle. I find it much easier to start from these than some complicated problem I stumble over in the book, and more insightful too, since the reason for the hardness is not obscured by the hidden hardness proof for the complicated problem. Chapter 6 focuses on strategies for proving hardness.

Implementations: Programs for solving satisfiability problems were sought for the Second DIMACS Implementation Challenge, held in October 1993. Programs and data from the challenge are available by anonymous ftp from dimacs.rutgers.edu in the directory /pub/challenge/sat. In particular, *sato* is a decision procedure for propositional logic written in C by Hantao Zhang. There is also a random formula generator named mwff.c for constructing hard satisfiability instances in C by Bart Selman. Several other solvers and instance generators are also available from this site.

The propositional satisfiability tester POSIT, by Jon W. Freeman, is based on a highly optimized version of the Davis-Putnum procedure. It is available by anonymous ftp from ftp.cis.upenn.edu in /pub/freeman/posit-1.0.tar.Z.

Notes: The primary reference on NP-completeness is [GJ79], featuring a list of roughly four hundred NP-complete problems. Although the list is over fifteen years old, it remains an extremely useful reference; it is perhaps the book I reach for most often. An occasional column by David Johnson in the *Journal of Algorithms* has helped to update the book. In [Joh90], Johnson gives a thorough and readable survey of the relationship between different complexity classes.

Good expositions of Cook's theorem [Coo71], where satisfiability is proven hard, include [CLR90, GJ79, PS82]. The importance of Cook's result became clear in Karp's paper [Kar72] showing that it implied the hardness of over twenty different combinatorial problems.

A linear-time algorithm for 2-satisfiability appears in [APT79]. The application of 2-satisfiability to map labeling is taken from [WW95].

Related Problems: Constrained optimization (see page 209), traveling salesman problem (see page 319).

8.4 Graph Problems: Polynomial-Time

Algorithmic graph problems constitute approximately one third of all the problems in this catalog. Further, several problems from other sections can be formulated strictly in terms of graphs. Identifying the name of a graph-theoretic invariant or problem is one of the primary skills of a good algorist. Indeed, this catalog will tell you exactly how to proceed as soon as you figure out your particular problem's name.

We have partitioned the bulk of the algorithmic graph problems in this book between this and the subsequent section. Here, we deal only with problems for which there exist efficient algorithms to solve them. As there is often more than one way to model a given application, it makes sense to look here before proceeding on to the harder formulations.

The algorithms presented here have running times that grow slowly with the size of the graph. We adopt throughout the standard convention that n refers to the number of vertices in a graph, while m is the number of edges.

Although graphs are combinatorial objects, describing a binary relation on a set of objects, graphs are usually best understood as drawings. Beyond just the problems of visualization, many interesting graph properties follow from the nature of a particular type of drawing, such as planar graphs. In this chapter, we also present a variety of different problems and algorithms associated with graph drawing.

Most advanced graph algorithms are difficult to program. However, good implementations are often available if you know where to look. The best single source is LEDA, discussed in Section 9.1.1, although faster special-purpose codes exist for many problems.

Books specializing in graph algorithms include:

- *Even* [Eve79a] – This is a good book on graph algorithms, fairly advanced, with a particularly thorough treatment of planarity-testing algorithms.
- *Ahuja, Magnanti, and Orlin* [AMO93] – While purporting to be a book on network flows, it covers the gamut of graph algorithms with emphasis on operations research. Strongly recommended.
- *van Leeuwen* [vL90a] – A 100+ page survey on research results in graph algorithms, this is the best source to determine what is known in algorithmic graph theory.
- *McHugh* [McH90] – A more elementary but comprehensive treatment of basic graph algorithms, including parallel algorithms.

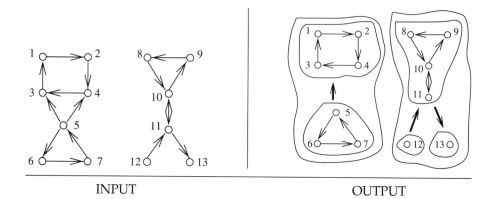

INPUT   OUTPUT

8.4.1 Connected Components

Input description: A directed or undirected graph G.

Problem description: Traverse each edge and vertex of all connected components of G.

Discussion: The connected components of a graph represent, in grossest terms, the pieces of the graph. Two vertices are in the same component of G if and only if there is some path between them.

Finding connected components is at the heart of many graph applications. For example, consider the problem of identifying clusters in a set of items. We can represent each item by a vertex and add an edge between each pair of items that are deemed "similar." The connected components of this graph correspond to different classes of items.

Testing whether a graph is connected is an essential preprocessing step for every graph algorithm. Such tests can be performed so quickly and easily that you should always verify that your input graph is connected, even when you *know* it has to be. Subtle, difficult-to-detect bugs often result when your algorithm is run only on one component of a disconnected graph.

Testing the connectivity of any undirected graph is a job for either depth-first or breadth-first search, as discussed in Section 4.4. Which one you choose doesn't really matter. Both traversals initialize a *component-number* field for each vertex to 0, and then start the search for component 1 from vertex v_1. As each vertex is visited, the value of this field is set to the current component number. When the initial traversal ends, the component number is incremented, and the search begins again from the first vertex with *component-number* still 0. Properly implemented using adjacency lists, this runs in $O(n + m)$, or time linear in the number of edges and vertices.

Other notions of connectivity also arise in practice:

- *What if my graph is directed?* – There are two distinct definitions of connected components for directed graphs. A directed graph is *weakly connected* if it would be connected by ignoring the direction of edges. Thus a weakly connected graph consists of a single piece. A directed graph is *strongly connected* if there is a directed path between every pair of vertices. This distinction is best made clear by considering a network of one- and two-way streets in a city. The network is strongly connected if it is possible to drive legally between every two positions. The network is weakly connected if it is possible to drive legally or *illegally* between every two positions. The network is disconnected if there is no possible way to drive from a to b.

The weakly and strongly connected components define unique partitions on the vertices. The output figure above illustrates a directed graph consisting of two weakly connected or five strongly connected components (also called *blocks* of G).

Testing whether a directed graph is weakly connected can be done easily in linear time. Simply turn all edges into undirected edges and use the DFS-based connected components algorithm described above. Tests for strong connectivity are somewhat more complicated. The simplest algorithm performs a breadth-first search from each vertex and verifies that all vertices have been visited on each search. Thus in $O(mn)$ time, it can be confirmed whether the graph is strongly connected. Further, this algorithm can be modified to extract all strongly connected components if it is not.

In fact, strongly connected components can be found in linear time using one of two more sophisticated DFS-based algorithms. See the references below for details. It is probably easier to start from an existing implementation below than a textbook description.

- *How reliable is my network; i.e. how well connected is it?* – A chain is only as strong as its weakest link. When it is missing one or more links, it is disconnected. The notion of *connectivity* of graphs measures the strength of the graph – how many edges or vertices must be removed in order to break it, or disconnect it. Connectivity is an essential invariant for network design and other structural problems.

  Algorithmic connectivity problems are discussed in Section 8.4.8. In particular, *biconnected components* are pieces of the graph that result by cutting the edges incident on a single vertex. All biconnected components can be found in linear time using a DFS-based algorithm. Vertices whose deletion disconnects the graph belong to more than one biconnected component, although edges are uniquely partitioned among them.

- *Is the graph a tree? How can I find a cycle if one exists?* – The problem of cycle identification often arises, particularly with respect to directed graphs. For example, testing if a sequence of conditions can deadlock often reduces to cycle detection. If I am waiting for Fred, and Fred is waiting for Mary, and Mary is waiting for me, we are all deadlocked.

  For undirected graphs, the analogous problem is tree identification. A tree is, by definition, an undirected, connected graph without any cycles. As described above, a depth-first search can be used to test whether it is connected. If the graph is connected and has $n - 1$ edges for n vertices, it is a tree.

Depth-first search can be used to find cycles in both directed and undirected graphs. Whenever we encounter a back edge in our DFS, i.e. an edge to an ancestor vertex in the DFS tree, the back edge and the tree together define a directed cycle. No other such cycle can exist in a directed graph. Directed graphs without cycles are called DAGs (directed acyclic graphs). Topological sorting (see Section 8.4.2) is the fundamental operation on DAGs.

Implementations: LEDA (see Section 9.1.1) provides good implementations of breadth-first and depth-first search, connected components and strongly connected components, all in C++. XTango (see Section 9.1.5) is an algorithm animation system for UNIX and X-windows, which includes an animation of depth-first search.

Pascal implementations of BFS, DFS, and biconnected and strongly connected components appear in [MS91]. See Section 9.1.6.4 for details. Combinatorica [Ski90] provides Mathematica implementations of the same routines. See Section 9.1.4.

The Stanford GraphBase (see Section 9.1.3) contains routines to compute biconnected and strongly connected components. An expository implementation of BFS and DFS in Fortran appears in [NW78] (see Section 9.1.6.3).

Notes: Depth-first search was first used in algorithms for finding paths out of mazes, and dates back to the nineteenth century [Luc91, Tar95]. Breadth-first search was first reported to find the shortest path out of mazes by Moore in 1957 [Moo59].

Hopcroft and Tarjan [HT73b, Tar72] first established depth-first search as a fundamental technique for efficient graph algorithms. Expositions on depth-first and breadth-first search appear in every book discussing graph algorithms, with [CLR90] perhaps the most thorough description available.

The first linear-time algorithm for strongly connected components is due to Tarjan [Tar72], with expositions including [Baa88, Eve79a, Man89]. Another algorithm, simpler to program and slicker, to find strongly connected components is due to Sharir and Kosaraju. Good expositions of this algorithm appear in [AHU83, CLR90].

Related Problems: Edge-vertex connectivity (see page 294), shortest path (see page 279).

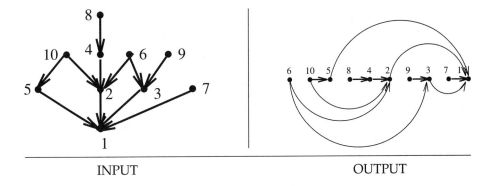

INPUT   OUTPUT

8.4.2 Topological Sorting

Input description: A directed acyclic graph $G = (V, E)$, also known as a *partial order* or *poset*.

Problem description: Find a linear ordering of the vertices of V such that for each edge $(i, j) \in E$, vertex i is to the left of vertex j.

Discussion: Topological sorting arises as a natural subproblem in most algorithms on directed acyclic graphs. Topological sorting orders the vertices and edges of a DAG in a simple and consistent way and hence plays the same role for DAGs that depth-first search does for general graphs.

Topological sorting can be used to schedule tasks under precedence constraints. Suppose we have a set of tasks to do, but certain tasks have to be performed before other tasks. These precedence constraints form a directed acyclic graph, and any topological sort (also known as a *linear extension*) defines an order to do these tasks such that each is performed only after all of its constraints are satisfied.

Three important facts about topological sorting are:

- Only directed acyclic graphs can have linear extensions, since any directed cycle is an inherent contradiction to a linear order of tasks.

- Every DAG can be topologically sorted, so there must always be at least one schedule for any reasonable precedence constraints among jobs.

- DAGs typically allow many such schedules, especially when there are few constraints. Consider n jobs without any constraints. Any of the $n!$ permutations of the jobs constitutes a valid linear extension.

A linear extension of a given DAG is easily found in linear time. The basic algorithm performs a depth-first search of the DAG to identify the complete set of *source vertices*, where source vertices are vertices without incoming edges. At least one such source must exist in any DAG. Note that source vertices can appear at the start of any schedule without violating any constraints. After deleting all the outgoing edges of the source vertices, we will create new source vertices, which can sit comfortably to the immediate right of the first set. We repeat until all vertices have been accounted for. Only a modest amount of care with data structures (adjacency lists and queues) is needed to make this run in $O(n + m)$ time.

This algorithm is simple enough that you should be able to code up your own implementation and expect good performance, although implementations are described below. Two special considerations are:

- *What if I need all the linear extensions, instead of just one of them?* – In certain applications, it is important to construct *all* linear extensions of a DAG. Beware, because the number of linear extensions can grow exponentially in the size of the graph. Even the problem of counting the number of linear extensions is NP-hard.

  Algorithms for listing all linear extensions in a DAG are based on backtracking. They build all possible orderings from left to right, where each of the in-degree zero vertices are candidates for the next vertex. The outgoing edges from the selected vertex are deleted before moving on. Constructing truly random linear extensions is a hard problem, but pseudorandom orders can be constructed from left to right by selecting randomly among the in-degree zero vertices.

- *What if your graph is not acyclic?* – When the set of constraints is not a DAG, but it contains some inherent contradictions in the form of cycles, the natural problem becomes to find the smallest set of jobs or constraints that if eliminated leaves a DAG. These smallest sets of offending jobs (vertices) or constraints (edges) are known as the *feedback vertex set* and the *feedback arc set*, respectively, and are discussed in Section 8.5.11. Unfortunately, both of them are NP-complete problems.

  Since the basic topological sorting algorithm will get stuck as soon as it identifies a vertex on a directed cycle, we can delete the offending edge or vertex and continue. This quick-and-dirty heuristic will eventually leave a DAG.

Implementations: Many textbooks contain implementations of topological sorting, including [MS91] (see Section 9.1.6.4) and [Sed92] (see Section 9.1.6.6). LEDA (see Section 9.1.1) includes a linear-time implementation of topological sorting in C++.

XTango (see Section 9.1.5) is an algorithm animation system for UNIX and X-Windows, which includes an animation of topological sorting.

Combinatorica [Ski90] provides Mathematica implementations of topological sorting and other operations on directed acyclic graphs. See Section 9.1.4.

Notes: Good expositions on topological sorting include [CLR90, Man89]. Brightwell and Winkler [BW91] proved that it is #P-complete to count the number of linear extensions of a partial order. The complexity class #P includes NP, so any #P-complete problem is at least NP-hard.

Related Problems: Sorting (see page 236), feedback edge/vertex set (see page 343).

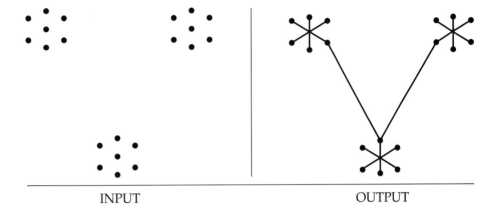

INPUT   OUTPUT

8.4.3 Minimum Spanning Tree

Input description: A graph $G = (V, E)$ with weighted edges.

Problem description: The subset of $E' \subset E$ of minimum weight forming a tree on V.

Discussion: The minimum spanning tree (MST) of a graph defines the cheapest subset of edges that keeps the graph in one connected component. Telephone companies are particularly interested in minimum spanning trees, because the minimum spanning tree of a set of sites defines the wiring scheme that connects the sites using as little wire as possible. It is the mother of all network design problems.

Minimum spanning trees prove important for several reasons:

- They can be computed quickly and easily, and they create a sparse subgraph that reflects a lot about the original graph.

- They provide a way to identify clusters in sets of points. Deleting the long edges from a minimum spanning tree leaves connected components that define natural clusters in the data set, as shown in the output figure above.

- They can be used to give approximate solutions to hard problems such as Steiner tree and traveling salesman.

- As an educational tool, minimum spanning tree algorithms provide graphic evidence that greedy algorithms can yield provably optimal solutions.

Two classical algorithms efficiently construct minimum spanning trees, namely Prim's and Kruskal's. Brief overviews of both algorithms are given below, with correctness arguments in Section 4.7. We refer the reader to the codes below for implementation details.

Prim's algorithm starts with an arbitrary vertex v and "grows" a tree from it, repeatedly finding the lowest-cost edge that will link some new vertex into this tree. During execution we will label each vertex as either in the tree, *fringe -*

meaning there exists an edge from a tree vertex, or *unseen* - meaning the vertex is more than one edge away from the tree.

> Prim(G)
>> Select an arbitrary vertex to start
>> While (there are fringe vertices)
>>> select minimum-weight edge between tree and fringe
>>> add the selected edge and vertex to the tree

This creates a spanning tree for any connected graph, since no cycle can be introduced via edges between tree and fringe vertices. That it is in fact a tree of minimum weight can be proven by contradiction, and the proof is in Section 4.7.1. With simple data structures, Prim's algorithm can be implemented in $O(n^2)$ time.

Kruskal's algorithm is also greedy. It starts with each vertex as a separate tree and merges these trees together by repeatedly adding the lowest cost edge that merges two distinct subtrees (i.e. does not create a cycle).

> Kruskal(G)
>> Sort the edges in order of increasing weight
>> *count* = 0
>> while (*count* < n − 1) do
>>> get next edge (v, w)
>>> if (component (v) ≠ component(w))
>>>> add to T
>>>> component(v) = component(w)

The "which component?" tests are efficiently implemented using the union-find data structure of Section 8.1.5, to yield an $O(m \lg m)$ algorithm.

Minimum spanning tree is only one of several spanning tree problems that arise in practice. The following questions will help you sort your way through them:

- *Are the weights of all edges of your graph identical?* – Every spanning tree on n points contains exactly $n − 1$ edges. Thus if your graph is unweighted, *any* spanning tree will be a minimum spanning tree. Either breadth-first or depth-first search can be used to find a rooted spanning tree in linear time. Depth-first search trees tend to be long and thin, while breadth-first search trees better reflect the distance structure of the graph, as discussed in Section 4.4.

- *Should I use Prim's or Kruskal's algorithm?* – As described, Prim's algorithm runs in $O(n^2)$, while Kruskal's algorithm takes $O(m \log m)$ time. Thus Prim's algorithm is faster on dense graphs, while Kruskal's is faster on sparse graphs. Although Prim's algorithm can be implemented in $O(m + n \lg n)$ time using more advanced data structures (in particular, Fibonacci heaps), this will not be worth the trouble unless you have extremely large, fairly sparse graphs.

  I personally find Kruskal's algorithm easier to understand and implement than Prim's, but that is just a matter of taste.

- *What if my input is points in the plane, instead of a graph?* – Geometric instances, comprising n points in d-dimensions, can be solved by constructing the complete distance graph in $O(n^2)$ and then finding the MST of this complete graph. However, for points in two or even three dimensions, it can be more efficient to solve the geometric version of the problem directly. To find the minimum spanning tree of n points, first construct the Delaunay triangulation of these points (see Sections 8.6.3 and 8.6.4). In two dimensions, this gives a graph with $O(n)$ edges that contains all the edges of the minimum spanning tree of the point set. Running Kruskal's algorithm on this sparse graphs finishes the job in $O(n \lg n)$ time.

- *How do I find a spanning tree that avoids vertices of high degree?* – Another common goal of spanning tree problems is to minimize the maximum degree, typically to minimize the fan out in an interconnection network. Unfortunately, finding a spanning tree of maximum degree 2 is clearly NP-complete, since this is identical to the Hamiltonian path problem. Efficient algorithms are known, however, that construct spanning trees whose maximum degree at most one more than required. This is likely to suffice in practice. See the references below.

Implementations: Pascal implementations of Prim's, Kruskal's, and the Cheriton-Tarjan algorithm are provided in [MS91], along with extensive empirical analysis that shows that the implementation of Prim's algorithm with the appropriate priority queue is fastest on most graphs. See Section 9.1.6.4.

The Stanford GraphBase (see Section 9.1.3) contains implementations of four different minimum spanning tree algorithms, and the result of timing experiments suggesting that Kruskal's algorithm is best. The results are reported in terms of memory accesses (mems) instead of seconds, to make them independent of processor speed.

A C++ implementation of Kruskal's algorithm is provided in LEDA (see Section 9.1.1). Alternative implementations of Prim's and Kruskal's algorithms are provided in Pascal [SDK83] and C++ [Sed92]. See Section Section 9.1.6.6. XTango (see Section 9.1.5) includes an animation of both Prim's and Kruskal's algorithms.

Algorithm 479 [Pag74] and Algorithm 613 [HJS84] of the *Collected Algorithms of the ACM* are Fortran codes for minimum spanning tree, the former in an implementation of a point clustering algorithm. They are available from Netlib (see Section 9.1.2). A bare bones Fortran implementation is provided in [NW78], including the enumeration of all spanning trees. See Section 9.1.6.3.

Combinatorica [Ski90] provides Mathematica implementations of Kruskal's minimum spanning tree algorithm and quickly counting the number of spanning trees of a graph. See Section 9.1.4.

Notes: Good expositions on Prim's [Pri57] and Kruskal's [Kru56] algorithms will appear in any textbook on algorithms, but include [Baa88, CLR90, Man89, Tar83]. The fastest implementations of Prim's and Kruskal's algorithms use Fibonacci heaps [FT87]. Expositions of faster algorithms for geometric instances include [PS85].

A recent breakthrough on the minimum spanning tree problem is the linear-time randomized algorithm of Karger, Klein, and Tarjan [KKT95]. Simplifications will be needed before this becomes the algorithm of choice. The history of the minimum spanning tree problem dates back at least to Boruvka, in 1926, and is presented in [GH85]. Interestingly, it is Boruvka's algorithm that serves as the foundation to the new randomized one.

Fürer and Raghavachari [FR94] give an algorithm that constructs a spanning tree whose maximum degree is almost minimized, indeed is at most one more than the lowest-degree spanning tree. The situation is analogous to Vizing's theorem for edge coloring, which also gives an approximation algorithm to within additive factor one.

Minimum spanning tree algorithms have an interpretation in terms of *matroids*, which are systems of subsets closed under inclusion, for which the maximum weighted independent set can be found using a greedy algorithm. The connection between greedy algorithms and matroids was established by Edmonds [Edm71]. Expositions on the theory of matroids include [Law76, PS82].

Algorithms for generating spanning trees in order from minimum to maximum weight are presented in [Gab77]. Good expositions on the matrix-tree theorem, which counts the number of spanning trees of a graph, include [Eve79a].

Related Problems: Steiner tree (see page 339), traveling salesman (see page 319).

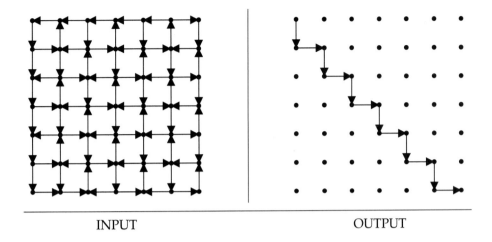

INPUT   OUTPUT

8.4.4 Shortest Path

Input description: An edge-weighted graph G, with start vertex s and end vertex t.

Problem description: Find the shortest path from s to t in G.

Discussion: The problem of finding shortest paths in a graph has a surprising range of applications:

- The most obvious applications arise in transportation or communications, such as finding the best route to drive between Chicago and Phoenix or figuring how to direct packets to a destination across a network.

- Consider the problem of image segmentation, that is, separating two characters in a scanned, bit-mapped image of printed text. We need to find the separating line between two points that cuts through the fewest number of black pixels. This grid of pixels can be modeled as a graph, with any edge across a black pixel given a high cost. The shortest path from top to bottom defines the best separation between left and right.

- A major problem in speech recognition is distinguishing between words that sound alike (homophones), such as *to*, *two*, and *too*. We can construct a graph whose vertices correspond to possible words, with an edge between possible neighboring words. If the weight of each edge measures the likelihood of transition, the shortest path across the graph defines the best interpretation of a sentence. For a more detailed account of such an application, see Section 4.10.

- Suppose we want to draw an informative picture of a graph. The center of the page should correspond to the "center" of the graph, whatever that means. A good definition of the center is the vertex that minimizes the maximum distance to any other vertex in the graph. Finding this center point requires knowing the distance (i.e. shortest path) between all pairs of vertices.

The primary algorithm for finding shortest paths is *Dijkstra's algorithm*, which efficiently finds the shortest paths from a given vertex x to all $n-1$ other vertices. Dijkstra's algorithm starts from x. In each iteration, it identifies a new vertex v for which the shortest path from x to v is known. We maintain a set of vertices S to which we currently know the shortest path from v, and this set grows by one vertex in each iteration. In each iteration, we identify the edge (u, v) where $u \in S$ and $v \in V - S$ such that

$$dist(x, u) + weight(u, v) = \min_{(u',v') \in E} dist(x, u') + weight(u', v')$$

This edge (u, v) gets added to a *shortest path tree*, whose root is x and which describes all the shortest paths from x. See the discussion in Section 4.8.1 for more details.

The straightforward implementation of this algorithm is $O(mn)$. However, with simple data structures it can be reduced to $O(n^2)$ or $O(m \lg n)$ time. Theoretically faster times can be achieved using significantly more complicated data structures, as described below. If we are just interested in knowing the shortest path from x to y, simply stop as soon as y enters S.

Dijkstra's algorithm is the right choice for single-source shortest path on positively weighted graphs. However, special circumstances dictate different choices:

- *Is your graph weighted or unweighted?* – If your graph is unweighted, a simple breadth-first search starting from the source vertex will find the shortest path in linear time. It is only when edges have different weights that you need more sophisticated algorithms. Breadth-first search is both simpler and faster than Dijkstra's algorithm.

- *Does your graph have negative cost weights?* – Dijkstra's algorithm assumes that all edges have positive cost. If your graph has edges with negative weights, you must use the more general but less efficient Bellman-Ford algorithm. If your graph has a cycle of negative cost, then the shortest path between any two vertices in the graph is not defined, since we can detour to the negative cost cycle and loop around it an arbitrary number of times, making the total cost as small as desired. Note that adding the same amount of weight to each edge to make it positive and running Dijkstra's algorithm *does not* find the shortest path in the original graph, since paths that use more edges will be rejected in favor of longer paths using fewer edges.

    Why might one ever need to find shortest paths in graphs with negative cost edges? An interesting application comes in currency speculation. Construct a graph where each vertex is a nation and there is an edge weighted $\lg(w(x, y))$ from x to y if the exchange rate from currency x to currency y is $w(x, y)$. In *arbitrage*, we seek a cycle to convert currencies so that we end up with more money than we started with. For example, if the exchange rates are 12 pesos per dollar, 5 pesos per franc, and 2 francs per dollar, by simply moving money around we can convert \$1 to \$1.20. In fact, there will be a profit opportunity whenever there exists a negative cost cycle in this weighted graph.

- *Is your input a set of geometric obstacles instead of a graph?* – If you seek the shortest path between two points in a geometric setting, like an obstacle-filled room, you may either convert your problem into a graph of distances and feed it to Dijkstra's algorithm or use a more efficient geometric algorithm to compute the shortest path directly from the arrangement of obstacles. For such geometric algorithms, see Section 8.6.14 on motion planning.

- *Does your graph have cycles in it, or is it a DAG?* – If your graph is a directed acyclic graph (DAG), than the shortest path can be found in linear time. Perform a topological sort to order the vertices such that all edges go from left to right, then do dynamic programming on the left-to-right ordered vertices. Indeed, most dynamic programming problems can be easily formulated as shortest paths on specific DAGs. The algorithm is discussed in [Man89] if you cannot figure it out from this description. Note that the same algorithm (replacing min with max) also suffices for finding the *longest path* in a DAG. which is useful in many applications like scheduling (see Section 8.3.9).

- *Do you need the shortest path between all pairs of points?* – If you are interested in the shortest path between all pairs of vertices, one solution is to run Dijkstra n times, once with each vertex as the source. However, the Floyd-Warshall algorithm is a slick $O(n^3)$ dynamic programming algorithm for all-pairs shortest path, which is faster and easier to program than Dijkstra and which works with negative cost edges (but not cycles). It is discussed more thoroughly in Section 4.8.2. Let M denote the distance matrix, where $M_{ij} = \infty$ if there is no edge (i, j).

$$D^0 = M$$
for $k = 1$ to n do
    for $i = 1$ to n do
        for $j = 1$ to n do
$$D_{ij}^k = \min(D_{ij}^{k-1}, D_{ik}^{k-1} + D_{kj}^{k-1})$$
Return D^n

The key to understanding Floyd's algorithm is that D_{ij}^k denotes "the length of the shortest path from i to j that goes through possible intermediate vertices $v_1, \ldots, v_k$." Note that $O(n^2)$ space suffices, since we need keep only D^k and D^{k-1} around at time k.

- *How do I find the shortest cycle in a graph?* – One application of all-pairs shortest path is to find the shortest cycle in a graph, called its *girth*. Floyd's algorithm can be used to compute d_{ii} for $1 \leq i \leq n$, which is the length of the shortest way to get from vertex i to i, in other words the shortest cycle through i.

    This *might* be exactly what you want. However, the shortest cycle through x is likely to go from x to y back to x, using the same edge twice. A *simple* cycle is one that visits no edge or vertex twice. To find the shortest simple cycle, the easiest approach is to compute the lengths of the shortest paths from i to all other vertices, and then explicitly check whether there is an acceptable edge from each vertex back to i.

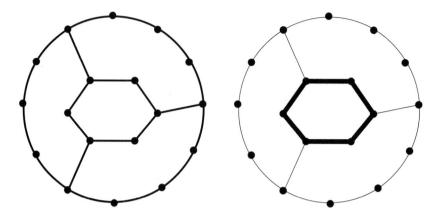

Figure 8–4. The girth, or shortest cycle, of a graph

Finding the *longest* cycle in a graph includes the special case of Hamiltonian cycle (see 8.5.5), so it is NP-complete.

The all-pairs shortest path matrix can be used to compute several useful invariants of any graph G, that are related to the center of G. The *eccentricity* of a vertex v in a graph is the shortest-path distance to the farthest vertex from v. From the eccentricity come other graph invariants. The *radius* of a graph is the smallest eccentricity of any vertex, while the *center* is the set of vertices whose eccentricity is the radius. The *diameter* of a graph is the maximum eccentricity of any vertex.

Implementations: The highest performance code (for both Dijkstra and Bellman-Ford) available for finding shortest paths in graphs is SPLIB [CGR93], developed in C language by Cherkassky, Goldberg, and Radzik. They report solving instances with over one million vertices in under two minutes on a Sun Sparc-10 workstation. Their codes are available from http://www.neci.nj.nec.com/homepages/avg.html for noncommercial use.

LEDA (see Section 9.1.1) provides good implementations in C++ for all of the shortest-path algorithms we have discussed, including Dijkstra, Bellman-Ford, and Floyd's algorithms.

Pascal implementations of Dijkstra, Bellman-Ford, and Floyd's algorithms are given in [SDK83]. See Section 9.1.6.1.

XTango (see Section 9.1.5) includes animations of both Dijkstra's and Floyd's shortest-path algorithms.

Combinatorica [Ski90] provides Mathematica implementations of Dijkstra's and Floyd's algorithms for shortest paths, acyclicity testing, and girth computation for directed/undirected and weighted/unweighted graphs. See Section 9.1.4.

The Stanford GraphBase (see Section 9.1.3) contains an implementation of Dijkstra's algorithm, used for computing word ladders in a graph defined by five-letter words, as well as an implementation of a program to bound the girth of graphs. Algorithm 562 [Pap80] of the *Collected Algorithms of the ACM* is a Fortran program to find shortest paths in graphs (see Section 9.1.2).

Notes: Good expositions on Dijkstra's algorithm [Dij59] and Floyd's all-pairs-shortest-path algorithm [Flo62] include [Baa88, CLR90, Man89]. Good expositions of the Bellman-Ford algorithm [Bel58, FF62] are slightly rarer, but include [CLR90, Eve79a, Law76]. Expositions on finding the shortest path in a DAG include [Law76].

A survey on shortest-path algorithms with 222 references appears in [DP84]. Included are citations to algorithms for related path problems, like finding the kth-shortest path and shortest paths when edge costs vary with time. Expositions on finding the kth-shortest path include [Law76].

The theoretically fastest algorithms known for single-source shortest path for positive edge weight graphs are variations of Dijkstra's algorithm with Fibonacci heaps [FT87]. Experimental studies of shortest-path algorithms include [DF79, DGKK79]. However, these experiments were done before Fibonacci heaps were developed. See [CGR93] for a more recent study.

Theoretically faster algorithms exist when the weights of the edges are small; i.e. their absolute values are each bounded by W. For positive edge weights, the single-source-shortest-path can be found in $O(m + n\sqrt{\lg W})$ [AMOT88], while $O(\sqrt{nm}\lg(nW))$ suffices for graphs with negative edge weights [GT89]

Related Problems: Network flow (see page 297), motion planning (see page 389).

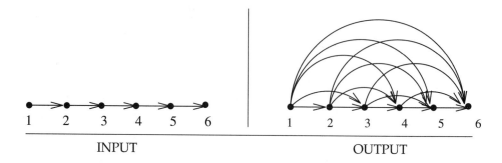

INPUT   OUTPUT

8.4.5   Transitive Closure and Reduction

Input description: A directed graph $G = (V, E)$.

Problem description: For *transitive closure*, construct a graph $G' = (V, E')$ with edge $(i, j) \in E'$ iff there is a directed path from i to j in G. For *transitive reduction*, construct a small graph $G' = (V, E')$ with a directed path from i to j in G' iff $(i, j) \in E$.

Discussion: Transitive closure can be thought of as establishing a data structure that makes it possible to solve reachability questions (can I get to x from y?) efficiently. After the preprocessing of constructing the transitive closure, all reachability queries can be answered in constant time by simply reporting a matrix entry. Transitive closure is fundamental in propagating the consequences of modified attributes of a graph G. For example, consider the graph underlying any spreadsheet model, where the vertices are cells and there is an edge from cell i to cell j if the result of cell j depends on cell i. When the value of a given cell is modified, the values of all reachable cells must also be updated. The identity of these cells is revealed by the transitive closure of G. Many database problems reduce to computing transitive closures, for analogous reasons.

There are three basic algorithms for computing transitive closure:

- The simplest algorithm just performs a breadth-first or depth-first search from each vertex and keeps track of all vertices encountered. Doing n such traversals gives an $O(n(n + m))$ algorithm, which degenerates to cubic time if the graph is dense. This algorithm is easily implemented, runs well on sparse graphs, and is likely the right answer for your application.

- If the transitive closure of G will be dense, a better algorithm exploits the fact that the strongly connected components of G can be computed in linear time (see Section 8.4.1). All pairs of vertices in each strongly connected component are mutually reachable. Further, if (x, y) is an edge between two vertices in different strongly connected components, every vertex in y's component is reachable from each vertex in x's component. Thus the problem reduces to finding the transitive closure on a graph of strongly connected components, which should have considerably fewer edges and vertices than G.

- Warshall's algorithm constructs transitive closures in $O(n^3)$ with a simple, slick algorithm that is identical to Floyd's all-pairs-shortest-path algorithm of Section 8.4.4. If we are interested only in the transitive closure, and not the length of the resulting paths, we can reduce storage by retaining only one bit for each matrix element. Thus $D_{ij}^k = 1$ iff j is reachable from i using only vertices $1, \ldots, k$ as intermediates.

  Another related algorithm, discussed in the references, runs in the same time as matrix multiplication. You might conceivably win for large n by using Strassen's fast matrix multiplication algorithm, although I for one wouldn't bother trying. Since transitive closure is provably as hard as matrix multiplication, there is little hope for a significantly faster algorithm.

Transitive reduction (also known as *minimum equivalent digraph*) is essentially the inverse operation of transitive closure, namely reducing the number of edges while maintaining identical reachability properties. The transitive closure of G is identical to the transitive closure of the transitive reduction of G. The primary application of transitive reduction is space minimization, by eliminating redundant edges from G that do not effect reachability. Transitive reduction also arises in graph drawing, where it is important to eliminate as many unnecessary edges as possible in order to reduce the visual clutter.

Although the transitive closure of G is uniquely defined, a graph may have many different transitive reductions, including G itself. We want the smallest such reduction, but there can be multiple formulations of the problem:

- A linear-time, quick-and-dirty transitive reduction algorithm identifies the strongly connected components of G, replaces each by a simple directed cycle, and adds these edges to those bridging the different components. Although this reduction is not provably minimal, it is likely to be pretty close on typical graphs.

  One catch with this heuristic is that it can add edges to the transitive reduction of G that are not in G. This may or may not be a problem for your application.

- If, in fact, all edges of the transitive reduction of G must be in G, we must abandon hope of finding the minimum size reduction. To see why, consider a directed graph consisting of one strongly connected component, so that every vertex can reach every other vertex. The smallest possible transitive reduction will be a simple directed cycle, consisting of exactly n edges. This is possible as a subset of edges only if G is Hamiltonian, thus proving that finding the smallest subset of edges is NP-complete.

  A heuristic for finding such a transitive reduction is to consider each edge successively and delete it if its removal does not change the transitive reduction. Implementing this efficiently means minimizing the time spent on reachability tests. Observe that directed edge (i, j) can be eliminated whenever there is another path from i to j avoiding this edge.

- If we are allowed to have arbitrary pairs of vertices as edges in the reduction and need the minimum size reduction, it can be found in $O(n^3)$ time. See the references below for details. However, the quick-and-dirty algorithm above will likely suffice for most applications and will be easier to program as well as more efficient.

Implementations: LEDA (see Section 9.1.1) provides an implementation of transitive closure in C++ using $O(nm)$ time [GK79].

Combinatorica [Ski90] provides Mathematica implementations of transitive closure and reduction, as well as the display of partial orders requiring transitive reduction. See Section 9.1.4.

Notes: Van Leeuwen [vL90a] provides an excellent survey on transitive closure and reduction, including 33 references. Good expositions of Warshall's algorithm [War62] include [Baa88, CLR90, Man89]. The equivalence between matrix multiplication and transitive closure was proven by Fischer and Meyer [FM71], with good expositions including [AHU74].

The equivalence between transitive closure and reduction, as well as the $O(n^3)$ reduction algorithm, was established in [AGU72]. Empirical studies of transitive closure algorithms include [SD75].

Estimating the size of the transitive closure is important in database query optimization. A linear-time algorithm for estimating the size of the closure is given by Cohen [Coh94].

Related Problems: Connected components (see page 270), shortest path (see page 279).

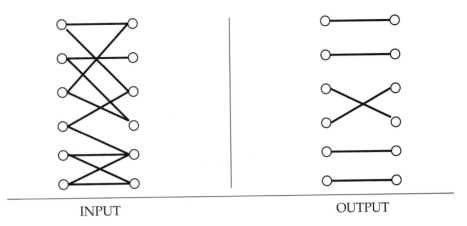

INPUT   OUTPUT

8.4.6 Matching

Input description: A (weighted) graph $G = (V, E)$.

Problem description: Find the largest-size set of edges S from E such that each vertex in V is incident to at most one edge of S.

Discussion: Consider a set of employees, each of whom is capable of doing some subset of the tasks that must be performed. We seek to find an assignment of employees to tasks such that each task is assigned to a unique employee. Each mapping between an employee and a task they can handle defines an edge, so what we need is a set of edges with no employee or job in common, i.e. a matching.

Efficient algorithms for constructing matchings are based on constructing *augmenting paths* in graphs. Given a (partial) matching M in a graph G, an augmenting path P is a path of edges where every odd-numbered edge (including the first and last edge) is not in M, while every even-numbered edge is. Further, the first and last vertices must not be already in M. By deleting the even-numbered edges of P from M and replacing them with the odd-numbered edges of P, we enlarge the size of the matching by one edge. Berge's theorem states that a matching is maximum if and only if it does not contain any augmenting path. Therefore, we can construct maximum-cardinality matchings by searching for augmenting paths and stopping when none exist.

This basic matching framework can be enhanced in several ways, while remaining essentially the same *assignment* problem:

- *Is your graph weighted or unweighted?* – Many matching applications are based on unweighted graphs. Perhaps we seek to maximize the number of tasks performed, where each task is as good as another. Such a problem seeks a maximum *cardinality* matching. We say that a matching is *perfect* if every vertex is involved in the matching.

  For certain applications, we need to augment each edge with a weight, perhaps reflecting the salary of the employee or their effectiveness at a given task. The problem now becomes constructing a maximum weighted matching, i.e. the set of independent edges of maximum total cost. By setting all edge weights to be 1, any algorithm for finding weighted matchings can be used to solve maximum cardinality matching.

- *What if certain employees can be given multiple jobs?* – In a natural generalization of the assignment problem, certain employees can given more than one task to do. We do not seek a matching so much as a covering with small "stars". Such multiple jobs can be modeled by simply replicating the employee vertex as many times as the number of jobs she can handle. By adding sufficiently complicated constraints on the solution, we will eventually require the use of full integer programming.

- *Is your graph bipartite?* – Many common matching problems involve bipartite graphs, as in the classical assignment problem of jobs to workers. This is fortunate because faster and simpler algorithms exist for bipartite matching. General graphs prove trickier because it is possible to have augmenting paths that are odd-length cycles, i.e. the first and last vertices are the same. Such cycles (or blossoms) are impossible in bipartite graphs, which by definition do not contain odd-length cycles.

  The standard algorithms for bipartite matching are based on network flow, using a simple transformation to convert a bipartite graph into an equivalent flow graph.

Another common "application" of bipartite matching is in marrying off a set of boys to a set of girls such that each boy gets a girl he likes. This can be modeled as a bipartite matching problem, with an edge between any compatible boy and girl. This is possible only for graphs with perfect matchings. An interesting related problem seeks a matching such that no parties can be unhappy enough to seek to break the matching. That is, once each of the boys has ranked each of the girls in terms of desirability, and the girls do the same to the boys, we seek a matching with the property that there are no marriages of the form (B_1, G_1) and (B_2, G_2), where B_1 and G_2 in fact prefer each other to their own spouses. In real life, these two would run off with each other, breaking the marriages. A marriage without any such couples is said to be *stable*.

It is a surprising fact that no matter how the boys and girls rate each other, there is always at least one stable marriage. Further, such a marriage can be found in $O(n^2)$ time. An important application of stable marriage occurs in the annual matching of medical residents to hospitals.

Implementations: The highest performance code available for constructing a maximum-cardinality bipartite matching of maximum weight in graphs is CSA [GK93], developed in the C language by Goldberg and Kennedy. This code is based on a cost-scaling network flow algorithm. They report solving instances with over 30,000 vertices in a few minutes on a Sun Sparc-2 workstation. Their codes are available for noncommercial use from http://www.neci.nj.nec.com/homepages/avg.html

The First DIMACS Implementation Challenge [JM93] focused on network flows and matching. Several instance generators and implementations for maximum weight and maximum cardinality matching were collected, which can be obtained by anonymous ftp from dimacs.rutgers.edu in the directory pub/netflow/matching. These include:

- A maximum-cardinality matching solver in Fortran 77 by R. Bruce Mattingly and Nathan P. Ritchey that seems capable of solving instances of 5,000 nodes and 60,000 edges in under 30 seconds.
- A maximum-cardinality matching solver in C by Edward Rothberg, that implements Gabow's $O(n^3)$ algorithm.
- A maximum-weighted matching solver in C by Edward Rothberg. This is slower but more general than his unweighted solver described above. For example, it took over 30 seconds on a weighted graph with 500 nodes and 4,000 edges.

LEDA (see Section 9.1.1) provides efficient implementations in C++ for both maximum cardinality and maximum weighted matching, on both bipartite and general graphs. Sedgewick [Sed92] provides a simple implementation of the stable marriage theorem in C++. See Section 9.1.6.6 for details.

Pascal implementations of maximum cardinality matching appears in [SDK83]. Alternative Pascal maximum-cardinality and bipartite matching codes appear in [MS91]. All are discussed in Section 9.1.6.

The Stanford GraphBase (see Section 9.1.3) contains an implementation of the Hungarian algorithm for bipartite matching. To provide readily visualized weighted bipartite graphs, Knuth uses a digitized version of the Mona Lisa and seeks row/column disjoint pixels of maximum brightness. Matching is also used to construct clever, resampled "domino portraits."

Algorithm 548 [CT80] presents a Fortran code for the assignment problem. Algorithm 575 [Duf81] permutes a matrix so as to minimize the number of zeros along the diagonal, which involves solving a matching problem. Both codes are available from Netlib (see Section 9.1.2).

Combinatorica [Ski90] provides a (slow) Mathematica implementations of bipartite and maximal matching, as well as the stable marriage theorem. See Section 9.1.4.

Notes: Lovász and Plummer [LP86] is the definitive reference on matching theory and algorithms. Survey articles on matching algorithms include [Gal86]. Good expositions on network flow algorithms for bipartite matching include [CLR90, Eve79a, Man89], and those on the Hungarian method include [Law76, PS82]. The best algorithm for maximum bipartite matching, due to Hopcroft and Karp [HK73], repeatedly finds the shortest augmenting paths instead of using network flow, and runs in $O(\sqrt{n}m)$. Expositions on the augmenting path method include [Man89, PS82, SDK83].

Edmond's algorithm [Edm65] for maximum-cardinality matching is of great historical interest for provoking questions about what problems can be solved in polynomial time. Expositions on Edmond's algorithm include [Law76, PS82, Tar83]. Gabow's [Gab76] implementation of Edmond's algorithm runs in $O(n^3)$ time. The best algorithm known for general matching runs in $O(\sqrt{n}m)$ [MV80]. A faster algorithm for matching in geometic graphs appears in [Vai88].

The theory of stable matching is thoroughly treated in [GI89]. The original algorithm for finding stable marriages is due to Gale and Shapely [GS62] with expositions including [Law76].

Related Problems: Eulerian cycle (see page 291), network flow (see page 297).

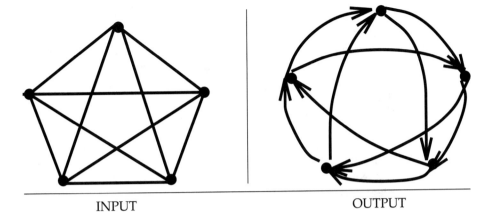

INPUT   OUTPUT

8.4.7   Eulerian Cycle / Chinese Postman

Input description: A graph $G = (V, E)$.

Problem description: Find the shortest tour of G visiting each edge at least once.

Discussion: Suppose you are given the map of a city and charged with designing the routes for garbage trucks, snow plows, or postmen. In all of these applications, every road in the city must be completely traversed at least once in order to ensure that all deliveries or pickups are made. For efficiency, you seek to minimize total drive time, or equivalently, the total distance or number of edges traversed.

Such applications are variants of the *Eulerian cycle* problem, best characterized by the children's puzzle that asks them to draw a given figure completely without lifting their pencil off the paper and without repeating any edges. We seek a path or cycle through a graph that visits each edge exactly once.

There are well-known conditions for determining whether a graph contains an Eulerian cycle, or path:

- An *undirected* graph contains an Eulerian *cycle* iff (1) it is connected and (2) each vertex is of even degree.
- An *undirected* graph contains an Eulerian *path* iff (1) it is connected and (2) all but two vertices are of even degree. These two vertices will be the start and end points of the path.
- A *directed* graph contains an Eulerian *cycle* iff (1) it is connected and (2) each vertex has the same in-degree as out-degree.
- Finally, a *directed* graph contains an Eulerian *path* iff (1) it is connected and (2) all but two vertices have the same in-degree as out-degree, and these two vertices have their in-degree and out-degree differ by one.

Given this characterization of Eulerian graphs, it is easy to test in linear time whether such a cycle exists: test whether the graph is connected using DFS or

BFS, and then count the number of odd-degree vertices. Actually constructing such a cycle also takes linear time. Use DFS to find a cycle in the graph. Delete this cycle and repeat until the entire set of edges has been partitioned into a set of edge-disjoint cycles. Since deleting a cycle reduces each vertex degree by an even number, the remaining graph will continue to satisfy the same Eulerian degree-bound conditions. For any connected graph, these cycles will have common vertices, and so by splicing these cycles in a "figure eight" at a shared vertex, we can construct a single circuit containing all of the edges.

An Eulerian cycle, if one exists, solves the motivating snowplow problem, since any tour that visits each edge only once must have minimum length. However, it is unlikely that any real road network would happen to satisfy the degree conditions that make it Eulerian. We need to solve the more general *Chinese postman problem*, which minimizes the length of a cycle that traverses every edge at least once. In fact, it can be shown that this minimum cycle never visits any edge more than twice, so good tours exist for any road network.

The optimal postman tour can be constructed by adding the appropriate edges to the graph G so as to make it Eulerian. Specifically, we find the shortest path between each pair of odd-degree vertices in G. Adding a path between two odd-degree vertices in G turns both of them to even-degree, thus moving us closer to an Eulerian graph. Finding the best set of shortest paths to add to G reduces to identifying a minimum-weight perfect matching in a graph on the odd-degree vertices, where the weight of edge (i, j) is the length of the shortest path from i to j. For directed graphs, this can be solved using bipartite matching, where the vertices are partitioned depending on whether they have more ingoing or outgoing edges. Once the graph is Eulerian, the actual cycle can be extracted in linear time using the procedure described above.

Implementations: Unfortunately, we have not able to identify a suitable Chinese postman implementation. However, it should not be difficult for you to roll your own by using a matching code from Section 8.4.6 and all-pairs shortest path from Section 8.4.4. Matching is by far the hardest part of the algorithm.

Combinatorica [Ski90] provides Mathematica implementations of Eulerian cycles and de Bruijn sequences. See Section 9.1.4.

Nijenhuis and Wilf [NW78] provide an efficient Fortran routine to enumerate all Eulerian cycles of a graph by backtracking. See Section 9.1.6.3.

Notes: The history of graph theory began in 1736, when Euler [Eul36] first solved the seven bridges of Königsberg problem. Königsberg (now Kaliningrad) is a city on the banks of the Pregel river. In Euler's day there were seven bridges linking the banks and two islands, which can be modeled as a multigraph with seven edges and four vertices. Euler sought a way to walk over each of the bridges exactly once and return home, i.e. an Eulerian cycle. Since all four of the vertices had odd degree, Euler proved that such a tour is impossible. The bridges were destroyed in World War II. See [BLW76] for a translation of Euler's original paper and a history of the problem.

Expositions on linear algorithms for constructing Eulerian cycles [Ebe88] include [Eve79a, Man89]. Fleury's algorithm [Luc91] is a direct and elegant approach to constructing Eulerian cycles. Start walking from any vertex, and erase any edge that has been traversed. The only criterion in picking the next edge is that we avoid using a bridge (edges whose deletion) unless there is no other alternative. No Eulerian graph contains a bridge, but what remains at some point on the walk ceases to be biconnected.

The Euler's tour technique is an important paradigm in parallel graph algorithms. See [Man89] for an exposition. Efficient algorithms exist to count the number of Eulerian cycles in a graph [HP73].

The problem of finding the shortest tour traversing all edges in a graph was introduced by Kwan [Kwa62], hence the name *Chinese* postman. The bipartite matching algorithm for solving Chinese postman is due to Edmonds and Johnson [EJ73]. This algorithm works for both directed and undirected graphs, although the problem is NP-complete for mixed graphs [Pap76a]. Mixed graphs contain both directed and undirected edges. Expositions of the Chinese postman algorithm include [Law76].

A *de Bruijn* sequence S of span n on an alphabet σ of size α is a circular string of length α^n containing all strings of length n as substrings of S, each exactly once. For example, for $n = 3$ and $\sigma = \{0, 1\}$, the circular string 00011101 contains the following substrings in order: 000, 001, 011, 111, 110, 101, 010, 100. De Bruijn sequences can be thought of as "safe cracker" sequences, describing the shortest sequence of dial turns with α positions sufficient to try out all combinations of length n.

De Bruijn sequences can be constructed by building a graph whose vertices are all α^{n-1} strings of length $n - 1$, where there is an edge (u, v) iff $u = s_1 s_2 \ldots s_{n-1}$ and $v = s_2 \ldots s_{n-1} s_n$. Any Eulerian cycle on this graph describes a de Bruijn sequence. For expositions on de Bruijn sequences and their construction, see [Eve79a, Ski90].

Related Problems: Matching (see page 287), Hamiltonian cycle (see page 287).

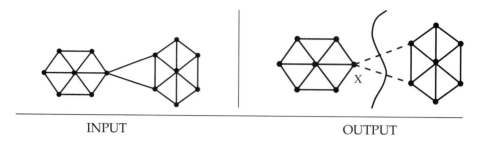

INPUT   OUTPUT

8.4.8   Edge and Vertex Connectivity

Input description: A graph G. Optionally, a pair of vertices s and t.

Problem description: What is the smallest subset of vertices (edges) whose deletion will disconnect G? Alternatively, what is the smallest subset of vertices (edges) that will separate s from t?

Discussion: Graph connectivity often arises in problems related to network reliability. In the context of telephone networks, the vertex connectivity is the smallest number of switching stations that a terrorist must bomb in order to separate the network, i.e. prevent two unbombed stations from talking to each other. The edge connectivity is the smallest number of wires that need to be cut to accomplish the same thing. One well-placed bomb or snipping the right pair of cables suffices to disconnect the network above.

The edge (vertex) connectivity of a graph G is the smallest number of edge (vertex) deletions sufficient to disconnect G. There is a close relationship between the two quantities. The vertex connectivity is always no smaller than the edge connectivity, since deleting one vertex incident on each edge in a cut set succeeds in disconnecting the graph. Of course, smaller vertex subsets may be possible. The minimum vertex degree is an upper bound on both the edge and vertex connectivity, since deleting all its neighbors (or the edges to all its neighbors) disconnects the graph into one big and one single-vertex component.

Several connectivity problems prove to be of interest:

- *Is the graph already disconnected?* – The simplest connectivity problem is testing whether the graph is in fact connected. A simple depth-first or breadth-first search suffices to identify all components in linear time, as discussed in Section 8.4.1. For directed graphs, the issue is whether the graph is *strongly connected*, meaning there is a directed path between any pair of vertices. In a *weakly connected* graph, there may exist paths to nodes from which there is no way to return.

- *What if I want to split the graph into equal-sized pieces?* – Often, what is sought is not the smallest set of edges or vertices whose deletion will disconnect the graph, but a small set that breaks the graph into roughly equal-sized pieces. For example, suppose we want to break a computer program spread across several files into two maintainable units. Construct a graph where the vertices are subroutines, with an edge between any two subroutines that interact, say by one calling the other. We seek to partition the routines into equal-sized sets so that the fewest pairs of interacting routines are spread across the set.

This is the *graph partition* problem, which is further discussed in Section 8.5.6. Although the problem is NP-complete, reasonable heuristics exist to solve it.

- *Is there one weak link in my graph?* – We say that G is *biconnected* if no single vertex deletion is sufficient to disconnect G. Any vertex that is such a weak point is called an *articulation vertex*. A *bridge* is the analogous concept for edges, meaning a single edge whose deletion disconnects the graph.

  The simplest algorithms for identifying articulation vertices (or bridges) would try deleting vertices (or edges) one by one, and then using DFS or BFS to test whether the resulting graph is still connected. More complicated but linear-time algorithms exist for both problems, based on depth-first search. Implementations are described below and in Section 8.4.1.

- *Are arbitrary cuts OK, or must I separate a given pair of vertices?* – There are two flavors of the general connectivity problem. One asks for the smallest cutset for the graph, the other for the smallest set to separate s from t. Any algorithm for $(s - t)$-connectivity can be used with each of the $n(n - 1)/2$ possible pairs of vertices to give an algorithm for general connectivity. Less obviously, $n - 1$ runs will suffice, since we know that vertex v_1 must end up in a different component from at least one of the other $n - 1$ vertices in any cut set.

Both edge and vertex connectivity can be found using network flow techniques. Network flow, discussed in Section 8.4.9, interprets a weighted graph as a network of pipes where the maximum capacity of an edge is its weight, and seeks to maximize the flow between two given vertices of the graph. The maximum flow between v_i, v_j in G is exactly the weight of the smallest set of edges to disconnect G with v_i and v_j in different components. Thus the edge connectivity can be found by minimizing the flow between v_i and each of the $n - 1$ other vertices in an unweighted graph G. Why? After deleting the minimum-edge cut set, v_i must be separated from some other vertex.

Vertex connectivity is characterized by *Menger's theorem*, which states that a graph is k-connected if and only if every pair of vertices is joined by at least k vertex-disjoint paths. Network flow can again be used to perform this calculation, since in an unweighted graph G a flow of k between a pair of vertices implies k edge-disjoint paths. We must construct a graph G' with the property that any set of edge-disjoint paths in G' corresponds to vertex-disjoint paths in G. This can be done by replacing each vertex v_i of G with two vertices $v_{i,1}$ and $v_{i,2}$, such that edge $(v_{i,1}, v_{i,2}) \in G'$ for all $v_i \in G$, and by replacing every edge $(v_i, x) \in G$ by edges $(v_{i,j}, x_k)$, $j \neq k \in \{0, 1\}$ in G'. Thus two edge-disjoint paths in G' correspond to vertex-disjoint paths in G, and as such, the minimum maximum-flow in G' gives the vertex connectivity of G.

Implementations: Combinatorica [Ski90] provides Mathematica implementations of edge and vertex connectivity, as well as connected, biconnected, and strongly connected components with bridges and articulation vertices. See Section 9.1.4.

LEDA *does not* currently seem to have biconnected components and bridges, but it does contain all the tools to implement connectivity algorithms, including network flow.

Pascal implementations of biconnected and strongly connected components appear in [MS91]. See Section 9.1.6.4 for details.

The Stanford GraphBase (see Section 9.1.3) contains routines to compute biconnected and strongly connected components.

Notes: Good expositions on the network-flow approach to edge and vertex connectivity include [Eve79a, Ski90]. The correctness of these algorithms is based on Menger's theorem [Men27] that the connectivity is determined by the number of edge and vertex disjoint paths separating a pair of vertices. The maximum-flow, minimum-cut theorem is due to Ford and Fulkerson [FF62].

Efficient randomized algorithms for computing graph connectivity have recently been developed by Karger. See Motwani and Raghavan [MR95] for an excellent treatment of randomized algorithms.

A nonflow-based algorithm for edge k-connectivity in $O(kn^2)$ is due to Matula [Mat87]. Faster k-connectivity algorithms are known for certain small values of k. All three-connected components of a graph can be generated in linear time [HT73a], while $O(n^2)$ suffices to test 4-connectivity [KR91].

Related Problems: Connected components (see page 270), network flow (see page 297), graph partition (see page 326).

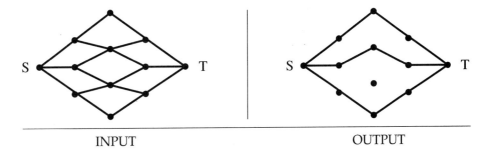

| INPUT | OUTPUT |

8.4.9 Network Flow

Input description: A graph G, where each edge $e = (i, j)$ has a capacity c_e. A source node s and sink node t.

Problem description: What is the maximum flow you can route from s to t while respecting the capacity constraint of each edge?

Discussion: Applications of network flow go far beyond plumbing. Finding the most cost-effective way to ship goods between a set of factories and a set of stores defines a network flow problem, as do resource-allocation problems in communications networks and a variety of scheduling problems.

The real power of network flow is that a surprising variety of linear programming problems that arise in practice can be modeled as network flow problems, and that special-purpose network flow algorithms can solve such problems much faster than general-purpose linear programming methods. Several of the graph problems we have discussed in this book can be modeled as network flow, including bipartite matching, shortest path, and edge/vertex connectivity.

The key to exploiting this power is recognizing that your problem can be modeled as network flow. This is not easy, and it requires experience and study. My recommendation is to first construct a linear programming model for your problem and then compare it with the linear program for minimum-cost flow on a directed network $G = (V, E)$. Let x_{ij} be a variable accounting for the flow from vertex i through edge j. The flow through edge j is constrained by its capacity, so

$$0 \le x_{ij} \le c_j \text{ for } 1 \le j \le m$$

Further, at each nonsource or sink vertex, as much flow comes in as goes out, so

$$\sum_{j=1}^{m} x_{ij} - \sum_{j=1}^{m} x_{ji} = 0$$

We seek the assignment that minimizes

$$\sum_{j=1}^{m} d_{ij} \cdot x_{ij}$$

where d_{ij} is the cost of a unit of flow from i through j.
Special considerations include:

- *What if all my costs are identical?* – Simpler and faster algorithms exist for solving the simple (as opposed to min-cost) maximum flow problem. This problem arises in many applications, including connectivity and bipartite matching.

- *What if all arc capacities are identical, either 0 or 1?* – Faster algorithms exist for 0-1 network flows. See the references for details.

- *What if I have multiple sources and/or sinks?* – Such problems can be handled by modifying the network to create a vertex to serve as a super-source that feeds all the sources and a super-sink that drains all the sinks.

- *What if I have multiple types of material moving through the network?* – In modeling a telecommunications network, every message has a given source and destination. Each destination needs to receive exactly those calls sent to it, not a given quantity of communication from arbitrary places. This can be modeled as a *multicommodity flow* problem, where each call defines a different commodity and we seek to satisfy all demands without exceeding the total capacity of any edge.

    Linear programming will suffice for multicommodity flow if fractional flows are permitted. Unfortunately, multicommodity integral flow is NP-complete, even with only two commodities.

Network flow algorithms can be complicated, and significant engineering is required to optimize performance. Thus we strongly recommend that you use an existing code if possible, rather than implement your own. Excellent codes are available and are described below. The two primary classes of algorithms are:

- *Augmenting path methods* – These algorithms repeatedly find a path of positive capacity from source to sink and add it to the flow. It can be shown that the flow through a network of rational capacities is optimal if and only if it contains no augmenting path, and since each augmentation adds to the flow, we will eventually find the maximum. The difference between network flow algorithms is in *how* they select the augmenting path. If we are not careful, each augmenting path will add but a little to the total flow, and so the algorithm might take a long time to converge.

- *Preflow-push methods* – These algorithms push flows from one vertex to another, ignoring until the end the constraint that the in-flow must equal the out-flow at each vertex. Preflow-push methods prove faster than augmenting path methods, essentially because multiple paths can be augmented simultaneously. These algorithms are the method of choice and are implemented in the best codes described below.

Implementations: The highest-performance code available for solving maximum-flow in graphs is PRF [CG94], developed in the C language by Cherkassky and Goldberg. They report solving instances with over 250,000 vertices in under two minutes on a Sun Sparc-10 workstation. For minimum-cost max-flow, the highest-performance code available is CS [Gol92], capable of solving instances of over 30,000 vertices in a few minutes on Sun Sparc-2 workstations. Both of their codes are available by ftp for noncommercial use from http://www.neci.nj.nec.com/homepages/avg.html.

The First DIMACS Implementation Challenge on Network Flows and Matching [JM93] collected several implementations and generators for network flow, which can be obtained by anonymous ftp from dimacs.rutgers.edu in the directory pub/netflow/maxflow. These include:

- A preflow-push network flow implementation in C by Edward Rothberg. It took under a second on a test graph of 500 nodes and 4,000 edges, but over an hour with 5,000 nodes and 40,000 edges.

- An implementation of eleven network flow variants in C, including the older Dinic and Karzanov algorithms by Richard Anderson and Joao Setubal. On an instance of 8,000 vertices and 12,000 edges, all options finished within two seconds.

Nijenhuis and Wilf [NW78] provide a Fortran implementation of Karzanov's algorithm for network flow. See Section 9.1.6.3. Fortran minimum-cost flow codes are given in [PGD82] and [KH80].

LEDA (see Section 9.1.1) provides C++ implementations of maximum-flow and minimum-cost max-flow algorithms. It also provides an implementation of minimum cut.

Pascal implementations of max-flow and minimum-cost flow algorithms are provided in [SDK83]. Alternative Pascal max-flow implementations appear in [MS91]. For both codes, see Section 9.1.6.4.

Combinatorica [Ski90] provides a (slow) Mathematica implementation of network flow, with applications to connectivity testing and matching. See Section 9.1.4.

Notes: The definitive book on network flows and its applications is [AMO93]. Good expositions on preflow-push algorithms [GT88] include [CLR90]. Older augmenting path algorithms are discussed in [Eve79a, Man89, PS82]. Expositions on min-cost flow include [Law76, PS82, SDK83]. Expositions on the hardness of multicommodity flow [Ita78] include [Eve79a].

Conventional wisdom holds that network flow should be computable in $O(nm)$ time, and there has been steady progress in lowering the time complexity. See [AMO93] for a history of algorithms for the problem. The fastest known general network flow algorithm runs in $O(nm \lg(n^2/m))$ time [GT88]. Empirical studies of minimum-cost flow algorithms include [GKK74, Gol92].

Although network flow can be used to find minimum cut sets in graphs, faster algorithms are available, including [SW94] and [MR95].

Related Problems: Linear programming (see page 213), matching (see page 287), connectivity (see page 294).

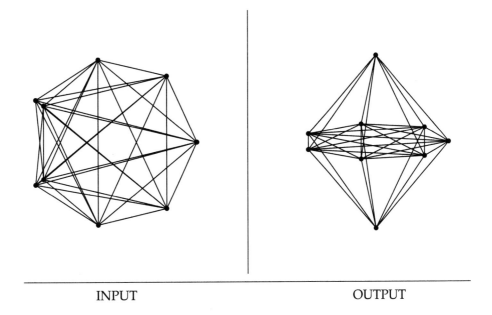

INPUT   OUTPUT

8.4.10   Drawing Graphs Nicely

Input description: A graph G.

Problem description: Draw a graph G so as to accurately reflect its structure.

Discussion: Drawing graphs nicely is a problem that constantly arises in applications, such as displaying file directory trees or circuit schematic diagrams. Yet it is inherently ill-defined. What exactly does *nicely* mean? We seek an algorithm that shows off the structure of the graph so the viewer can best understand it. We also seek a drawing that looks aesthetically pleasing. Unfortunately, these are "soft" criteria for which it is impossible to design an optimization algorithm. Indeed, it is possible to come up with two or more radically different drawings of certain graphs and have each be most appropriate in certain contexts. For example, page 335 contains three different drawings of the Petersen graph. Which of these is the "right" one?

Several "hard" criteria can partially measure the quality of a drawing:

- *Crossings* – We seek a drawing with as few pairs of crossing edges as possible, since they are distracting.
- *Area* – We seek a drawing that uses as little paper as possible, while ensuring that no pair of vertices are placed too close to each other.
- *Edge length* – We seek a drawing that avoids long edges, since they tend to obscure other features of the drawing.
- *Aspect ratio* – We seek a drawing whose aspect ratio (width/height) reflects that of the desired output medium (typically a computer screen at 4/3) as close as possible.

Unfortunately, these goals are mutually contradictory, and the problem of finding the best drawing under any nonempty subset of them will likely be NP-complete.

Two final warnings before getting down to business. For graphs without inherent symmetries or structure to display, it is likely that no really nice drawing exists, especially for graphs with more than 10 to 15 vertices. Even when a large, dense graph has a natural drawing, the shear amount of ink needed to draw it can easily overwhelm any display. A drawing of the complete graph on 100 vertices, K_{100}, contains approximately 5,000 edges, which on a 1000×1000 pixel display works out to 200 pixels an edge. What can you hope to see except a black blob in the center of the screen?

Once all this is understood, it must be admitted that certain graph drawing algorithms can be quite effective and fun to play with. To choose the right one, first ask yourself the following questions:

- *Must the edges be straight, or can I have curves and/or bends?* – Straight-line drawing algorithms are simpler than those with polygonal lines, but to visualize complicated graphs such as circuit designs, orthogonal polyline drawings seem to work best. *Orthogonal* means that all lines must be drawn either horizontal or vertical, with no intermediate slopes. *Polyline* means that each graph edge is represented by a chain of straight-line segments, connected by vertices or bends.

- *Can you build a natural, application-specific drawing algorithm?* – If your graph represents a network of cities and roads, you are unlikely to find a better drawing than placing the vertices in the same position as the cities on a map. This same principle holds for many different applications.

- *Is your graph either planar or a tree?* – If so, use one of the special planar graph or tree drawing algorithms of Sections 8.4.11 and 8.4.12.

- *How fast must your algorithm be?* – If it is being used for interactive update and display, your graph drawing algorithm had better be very fast. You are presumably limited to using incremental algorithms, which change the positions of the vertices only in the immediate neighborhood of the edited vertex. If you need to print a pretty picture for extended study, you can afford to be a little more extravagant.

As a first, quick-and-dirty drawing for most applications, I recommend simply spacing the vertices evenly on a circle, and then drawing the edges as straight lines between vertices. Such drawings are easy to program and fast to construct, and have the substantial advantage that no two edges can obscure each other, since no three vertices will be nearly collinear. As soon as you allow internal vertices into your drawing, such artifacts can be hard to avoid. An unexpected pleasure with circular drawings is the symmetry that is sometimes revealed because consecutive vertices appear in the order they were inserted into the graph. Simulated annealing can be used to permute the circular vertex order so as to minimize crossings or edge length, and thus significantly improve the drawing.

A good, general-purpose heuristic for drawing graphs models the graph as a system of springs and then uses energy minimization to space the vertices. Let adjacent vertices attract each other with a force proportional to the logarithm of their separation, while all nonadjacent vertices repel each other with a force proportional to their separation distance. These weights provide incentive for all edges to be as short as possible, while spreading the vertices apart. The behavior of such a system can be approximated by determining the force acting on each vertex at a particular time and then moving each vertex a small amount in the appropriate direction. After several such iterations, the system should stabilize on a reasonable drawing. The input and output figures above demonstrate the effectiveness of the spring embedding on a particular small graph.

If you need a polyline graph drawing algorithm, my recommendation is that you study several of the implementations presented below, particularly graphEd and GraphViz, and see whether one of them can do the job. You will have to do a significant amount of work before you can hope to develop a better algorithm.

Once you have a graph drawn, this opens another can of worms, namely where to place the edge/vertex labels. We seek to position the labels very close to the edges or vertices they identify, and yet to place them such that they do not overlap each other or important graph features. Map labeling heuristics are described in [WW95]. Optimizing label placement can be shown to be an NP-complete problem, but heuristics related to bin packing (see Section 8.6.9) can be effectively used.

Implementations: Georg Sander maintains a comprehensive WWW page on graph drawing at http://www.cs.uni-sb.de/RW/users/sander/html/gstools.html. This is well worth checking out and probably should be your first stop in hunting down programs for graph drawing.

The best ftp-able package of graph drawing algorithms is GraphEd, by Michael Himsolt. GraphEd [Him94] is a powerful interactive editor that enables the user to construct and manipulate both directed and undirected graphs. It contains a variety of graph and tree drawing algorithms, including planar drawings, polyline drawings, upward drawings of directed acyclic graphs (DAGs), and spring embeddings, and allows variations in node, edge, and label styles. Sgraph is an interface to GraphEd to support user-specific extensions written in C. It includes a modest library of algorithms for planarity testing, maximum flow, matching, and connectivity testing. GraphEd can be obtained by anonymous ftp from forwiss.uni-passau.de (132.231.20.10) in directory /pub/local/graphed. GraphEd is free for noncommercial use. Graphlet is a more recent project by the same group, available at http://www.fmi.uni-passau.de/Graphlet.

GraphViz is a popular graph drawing program developed by Stephen North of Bell Laboratories. It represents edges as splines and can construct useful drawings of quite large and complicated graphs. I recommend it, even though licensing considerations make it impossible to include on the Algorithm Repository or CD-ROM. A noncommercial license is available from http://portal.research.bell-labs.com/orgs/ssr/book/reuse/.

Combinatorica [Ski90] provides Mathematica implementations of several graph drawing algorithms, including circular, spring, and ranked embeddings. See Section 9.1.4 for further information on Combinatorica.

daVinci is a graph drawing and editing system whose layout algorithm seeks to minimize edge crossings and line bends, from Michael Froehlich at the University of Bremen. Information about daVinci is available from http://www.informatik.uni-bremen.de/~davinci. Binaries are available for a variety of UNIX workstations, although source code is not available.

Notes: A significant community of researchers in graph drawing has emerged in recent years, fueled by or fueling an annual conference on graph drawing, the proceedings of which are published by Springer-Verlag's Lecture Notes in Computer Science series. Perusing a volume of the proceedings will provide a good view of the state of the art and of what kinds of ideas people are thinking about.

The best reference available on graph drawing is the annotated bibliography on graph drawing algorithms by Giuseppe Di Battista, Peter Eades, and Roberto Tamassia [BETT94], which is also available from http://www.cs.brown.edu/~rt. See [BGL$^+$95] for an experimental study of graph drawing algorithms.

Although it is trivial to space n points evenly along the boundary of a circle, the problem is considerably more difficult on the surface of a sphere. Extensive tables of such spherical codes for $n \leq 130$ in up to five dimensions have been construction by Sloane, Hardin, and Smith, and are available from netlib (see Section 9.1.2) in att/math/sloane.

Related Problems: Drawing trees (see page 305), planarity testing (see page 308).

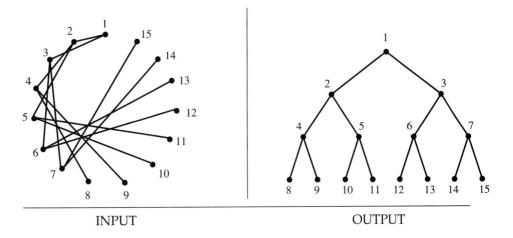

INPUT   OUTPUT

8.4.11 Drawing Trees

Input description: A tree *T*, which is a graph without any cycles.

Problem description: A nice drawing of the tree *T*.

Discussion: There are as many reasons to want to draw trees as there are types of structures that trees represent. Consider software and debugging tools that illustrate the hierarchical structure of file system directories, or that trace the execution of a program through its subroutine calls.

The primary issue in drawing trees is establishing whether you are drawing free or rooted trees:

- *Rooted trees* define a hierarchical order, emanating from a single source node identified as the root. Any drawing must reflect this hierarchical structure, as well as any additional application-dependent constraints on the order in which children must appear. For example, family trees are rooted, with sibling nodes typically drawn from left to right in the order of birth.
- *Free trees* do not encode any structure beyond their connection topology. For example, there is no root associated with the minimum spanning tree of any graph, so a hierarchical drawing can be misleading. Such free trees might well inherit their drawing from that of the full underlying graph, such as the map of the cities whose distances define the minimum spanning tree.

Since trees are always planar graphs, they can and should be drawn such that no two edges cross. Any of the planar drawing algorithms of Section 8.4.12 could be used to do so. However, such algorithms are overkill, because much simpler algorithms can be used to construct planar drawings of trees. The spring-embedding heuristics of Section 8.4.10 also work well on free trees, although they may be too slow for many applications.

The most natural tree-drawing algorithms work with rooted trees. However, they can be used equally well with free trees by selecting one vertex to serve as the root of the drawing. This faux-root can be selected arbitrarily, or, even better, by using a *center* vertex of the tree. A center vertex minimizes the maximum distance to other vertices. For trees, the center always consists of either one vertex or two adjacent vertices, so pick either one of them. Further, the center of a tree can be identified in linear time by repeatedly trimming all the leaves until only the center remains.

Your two primary options for drawing rooted trees are ranked and radial embeddings:

- *Ranked embeddings* – Place the root in the top center of your page, and then partition the page into the root-degree number of top-down strips. Deleting the root creates the root-degree number of subtrees, each of which is assigned to its own strip. Draw each subtree recursively, by placing its new root (the vertex adjacent to the old root) in the center of its strip a fixed distance from the top, with a line from old root to new root. The output figure above is a nicely ranked embedding of a balanced binary tree.

  Such ranked embeddings are particularly effective for rooted trees used to represent a hierarchy, be it a family tree, a data structure, or a corporate ladder. The top-down distance illustrates how far each node is from the root. Unfortunately, such a repeated subdivision eventually produces very narrow strips, until most of the vertices are crammed into a small region of the page. You should adjust the width of each strip to reflect the total number of nodes it will contain, or even better, the maximum number of nodes on a single level.

- *Radial embeddings* – A better way to draw free trees is with a radial embedding, where the root/center of the tree is placed in the center of the drawing. The space around this center vertex is divided into angular sectors for each subtree. Although the same problem of cramping will eventually occur, radial embeddings make better use of space than ranked embeddings and appear considerably more natural for free trees. The rankings of vertices in terms of distance from the center is illustrated by the concentric circles of vertices.

Implementations: Georg Sander maintains a comprehensive WWW page on graph drawing at http://www.cs.uni-sb.de/RW/users/sander/html/gstools.html. This should probably be your first stop in hunting down programs for tree drawing.

The best FTP-able package of graph drawing algorithms is GraphEd, by Michael Himsolt (himsolt@fmi.uni-passau.de). It contains a variety of graph and tree drawing algorithms and an interface to support user-specific extensions written in C. See Section 8.4.10 for more details on GraphEd and other graph drawing systems.

Combinatorica [Ski90] provides Mathematica implementations of several tree drawing algorithms, including radial and rooted embeddings. See Section 9.1.4 for further information on Combinatorica.

Notes: The best reference available on graph drawing is the annotated bibliography on graph drawing algorithms by Giuseppe Di Battista, Peter Eades, and Roberto Tamassia [BETT94], also available via http://www.cs.brown.edu/~rt.

Heuristics for tree layout have been studied by several researchers [RT81, Vau80, WS79], although under certain aesthetic criteria the problem is NP-complete [SR83].

Related Problems: Drawing graphs (see page 301), planar drawings (see page 308).

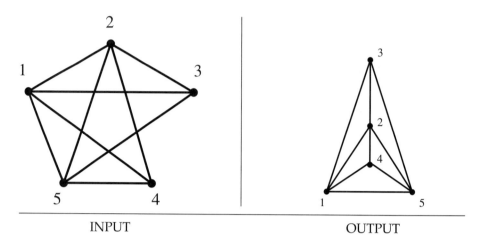

| INPUT | OUTPUT |

8.4.12   Planarity Detection and Embedding

Input description: A graph G.

Problem description: Can G be drawn in the plane such that no two edges cross? If so, produce such a drawing.

Discussion: Planar drawings (or *embeddings*) make it easy to understand the structure of a given graph by eliminating crossing edges, which are often confused as additional vertices. Graphs arising in many applications, such as road networks or printed circuit boards, are naturally planar because they are defined by surface structures.

Planar graphs have a variety of nice properties, which can be exploited to yield faster algorithms for many problems on planar graphs. Perhaps the most important property is that every planar graph is *sparse*. Euler's formula shows that for planar graph $G = (V, E)$, $|E| \leq 3|V| - 6$, so every planar graph contains a linear number of edges, and further, every planar graph must contain a vertex of degree at most 5. Since every subgraph of a planar graph is planar, this means that there is always a sequence of low-degree vertices whose deletion from G eventually leaves the empty graph.

The study of planarity has motivated much of the development of graph theory. To get a better appreciation of the subtleties of planar drawings, the reader is urged to construct a planar (noncrossing) embedding for the graph $K_5 - e$, the complete graph on five vertices with any single edge deleted. Then construct such an embedding where all the edges are straight. Finally, attempt to do the same for K_5 itself.

It is useful to distinguish the problem of planarity testing (does a graph have a planar drawing) from constructing planar embeddings (actually finding the drawing), although both can be done in linear time. Surprisingly, many efficient algorithms for planar graphs do not make use of the drawing but use the low-degree deletion sequence described above.

Algorithms for planarity testing begin by embedding an arbitrary cycle from the graph in the plane and then considering additional paths in G between vertices on this cycle. Whenever two such paths cross, one must be drawn outside the cycle and one inside. When three such paths mutually cross, there is no way to resolve the problem, and so the graph cannot be planar. Linear-time algorithms for planarity detection are based on depth-first search, but they are subtle and complicated enough that you would be wise to use an existing implementation if you can.

Such path-crossing algorithms can be used to construct a planar embedding by inserting the paths into the drawing one by one. Unfortunately, because they work in an incremental manner, nothing prevents them from inserting many vertices and edges into a relatively small area of the drawing. Such cramping makes the drawing ugly and hard to understand, and is a major problem with planar-embedding algorithms. More recently, algorithms have been devised that construct *planar-grid embeddings*, where each vertex lies on a $(2n - 4) \times (n - 2)$ grid. Thus no region can get too cramped and no edge can get too long. Still, the resulting drawings tend not to look as natural as one might hope.

For nonplanar graphs, what is often sought is a drawing that minimizes the number of crossings. Unfortunately, computing the crossing number is NP-complete. A useful heuristic extracts a large planar subgraph of G, embeds this subgraph, and then inserts the remaining edges one by one so as to minimize the number of crossings. This won't do much for dense graphs, which are doomed to have a large number of crossings, but it will work well for graphs that are almost planar, such as road networks with overpasses or printed circuit boards with multiple layers. Large planar subgraphs can be found by modifying planarity-testing algorithms to delete troublemaking edges.

Implementations: LEDA (see Section 9.1.1) provides a nice set of data structures and algorithms to support working on planar subdivisions. Included are both linear-time planarity testing and constructing straight-line planar-grid embeddings.

GraphEd includes an implementation of both planarity testing and planar graph layout. See Section 8.4.10 for more details on GraphEd. Combinatorica [Ski90] provides a (slow) Mathematica implementation of planarity testing. See Section 9.1.4.

Notes: Kuratowski [Kur30] gave the first characterization of planar graphs, namely that they do not contain a subgraph homeomorphic to $K_{3,3}$ or K_5. Thus if you are still working on the exercise to embed K_5, now is an appropriate time to give it up. Fary's theorem [F48] states that every planar graph can be drawn such that each edge is straight.

Hopcroft and Tarjan [HT74] gave the first linear-time algorithm for drawing graphs. Expositions on linear-time planarity testing include [Eve79a]. Nishizeki and Chiba [NC88] provide a good reference to the algorithmic theory of planar graphs. Efficient algorithms for planar grid embeddings were first developed by [dFPP88]. See [CHT90] for an algorithm to find the maximum planar subgraph of a nonplanar graph. Outer-planar graphs are those that can be drawn such that all vertices lie on the outer face of the drawing. Such graphs can be characterized as having no subgraph homeomorphic to $K_{2,3}$ and can be recognized and embedded in linear time.

Related Problems: Graph partition (see page 326), drawing trees (see page 305).

8.5 Graph Problems: Hard Problems

A cynical view of graph algorithms is that "everything we want to do is hard." Indeed, no polynomial-time algorithms are known for any of the problems in this section. Further, with the exception of graph isomorphism, all of them are provably NP-complete.

The theory of NP-completeness demonstrates that if any NP-complete problem has a polynomial-time algorithm, then polynomial-time algorithms must exist for all NP-complete problems. This seems sufficiently preposterous that NP-completeness suffices as a de facto proof that no efficient worst-case algorithm exists for the given problem.

Still, do not abandon hope if your problem resides in this chapter. For each of these problems, we provide a recommended line of attack, be it through combinatorial search, heuristics, or approximation algorithms. For every problem, there exist restricted input instances that are polynomial-time solvable, and if you are lucky, perhaps your data happens to fall into one of these classes. Hard problems require a different methodology to work with than polynomial-time problems, but with care they can usually be dealt with successfully.

The following books will help you deal with NP-complete problems:

- *Garey and Johnson* [GJ79] – This is the classic reference on the theory of NP-completeness. Most notably, it contains a concise catalog of over four hundred NP-complete problems, with associated references and comments. As soon as you begin to doubt the existence of an efficient algorithm for a given problem, browse through the catalog. Indeed, this is the book that I reach for most often.

- *Hochbaum* [Hoc96] – This book surveys the state of the art in approximation algorithms for NP-complete problems. Approximation algorithms efficiently produce solutions to problems that are always provably close to optimal.

- *Crescenzi and Kann* – This compendium of approximation algorithms for optimization problems is available at http://www.nada.kth.se/nada/theory /problemlist.html and on the enclosed CD-ROM and is the place to look first for a provably good heuristic for any given problem.

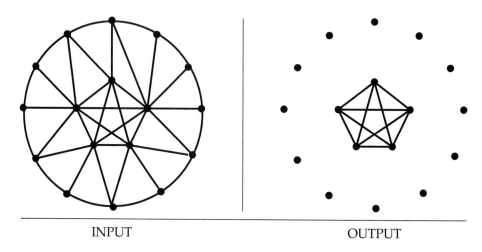

| INPUT | OUTPUT |

8.5.1 Clique

Input description: A graph $G = (V, E)$.

Problem description: What is the largest $S \subset V$ such that for all $x, y \in S$, $(x, y) \in E$?

Discussion: When I went to high school, everybody complained about the "clique", a group of friends who all hung around together and seemed to dominate everything social. Consider a graph whose vertices represent a set of people, with edges between any pair of people who are friends. Thus the clique in high school was in fact a clique in this friendship graph.

Identifying "clusters" of related objects often reduces to finding large cliques in graphs. One example is in a program recently developed by the Internal Revenue Service (IRS) to detect organized tax fraud, where groups of phony tax returns are submitted in the hopes of getting undeserved refunds. The IRS constructs graphs with vertices corresponding to submitted tax forms and with edges between any two tax-forms that appear suspiciously similar. A large clique in this graph points to fraud.

Since any edge in a graph represents a clique of two vertices, the challenge lies not in finding a clique, but in finding a large one. And it is indeed a challenge, for finding a maximum clique is NP-complete. To make matters worse, not only is no good approximation algorithm known, it is provably hard to approximate even to within a factor of $n^{1/3}$. Theoretically, clique is about as hard as a problem in this book can get. So what can we hope to do about it?

- *Will a maximal clique suffice?* – A *maximal* clique is a subset of vertices, each pair of which defines an edge, that cannot be enlarged by adding any additional vertex. This doesn't mean that it has to be large relative to the largest possible clique, but it might be. To find a nice maximal clique, sort the vertices from highest degree to lowest degree, put the first vertex in the clique, and then test each of the other vertices in order to see whether it is adjacent to all the clique vertices thus far. If so, add it; if not, continue down the list. In $O(nm)$ time you will have a maximal, and hopefully large, clique. An alternative approach would be to incorporate some randomness into your vertex ordering and accept the largest maximal clique you find after a certain number of trials.

- *What if I am looking for a large, dense subgraph instead of a perfect clique?* – Insisting on perfect cliques to define clusters in a graph can be risky, since the loss of a single edge due to error will eliminate that vertex from consideration. Instead, we should seek large *dense* subgraphs, i.e. subsets of vertices that contain a large number of edges between them. Cliques are, by definition, the densest subgraphs possible.

    A simple linear-time algorithm can be used to find the largest set of vertices whose induced (defined) subgraph has minimum vertex degree $\geq k$, beginning by repeatedly deleting all the vertices whose degree is less than k. This may reduce the degree of other vertices below k if they were adjacent to any deleted low-degree vertices. By repeating this process until all remaining vertices have degree $\geq k$, we eventually construct the largest dense subgraph. This algorithm can be implemented in $O(n + m)$ time by using adjacency lists and the constant-width priority queue of Section 8.1.2. By continuing to delete the lowest-degree vertices, we will eventually end up with a clique, which may or may not be large depending upon the graph.

- *What if the graph is planar?* – Planar graphs cannot have cliques of size larger than four, or else they cease to be planar. Since any edge defines a clique of size two, the only interesting cases are cliques of 3 and 4 vertices. Efficient algorithms to find such small cliques consider the vertices from lowest to highest degree. Any planar graph must contain a vertex of degree at most 5 (see Section 8.4.12), so there is only a constant-sized neighborhood to check exhaustively for a clique. Once we finish with this vertex, we delete it to leave a smaller planar graph with at least one low-degree vertex. Repeat until the graph is empty.

If you really need to find the largest clique in a graph, an exhaustive search via backtracking provides the only real solution. We search through all k-subsets of the vertices, pruning a subset as soon as it contains a vertex that is not connected to all the rest. We never need consider a subset of size larger than the highest vertex degree in the graph, since the maximum degree gives an upper bound on the size of the largest clique in the graph. Similarly, as soon as we discover a clique of size k, no vertex of degree $\leq k$ can help find a larger clique. To speed our search, we should delete all such useless vertices from G.

Heuristics for finding large cliques based on randomized techniques such as simulated annealing are likely to work reasonably well.

Implementations: Programs for the closely related problems of finding cliques and independent sets were sought for the Second DIMACS Implementation Challenge, held in October 1993. Programs and data from the challenge are available by anonymous ftp from dimacs.rutgers.edu. Source codes are available under pub/challenge/graph and test data under pub/djs.

314 Chapter 8 A Catalog of Algorithmic Problems

In particular, two C language programs by David S. Johnson and David L. Applegate are available. The dfmax.c implements a simple-minded branch-and-bound algorithm similar to that of [CP90]. The dmclique.c uses a "semi-exhaustive greedy" scheme for finding large independent sets described in [JAMS91]. Performance data for both programs is available in files results.dfmax and results.dmclique in directories /pub/challenge/graph/benchmarks/clique and /pub/challenge/graph/benchmarks/volume.

Combinatorica [Ski90] provides (slow) Mathematica implementations of cliques, independent sets, and vertex covers. See Section 9.1.4.

Notes: Good expositions of the proof that clique is NP-complete [Kar72] include [CLR90, GJ79, Man89]. It is also given in Section 6.2.3. This reduction established that clique, vertex cover, and independent set are very closely related problems, so heuristics and programs that solve one of them may also produce reasonable solutions for the other two.

The linear-time algorithm for constructing maximal induced subgraphs is discussed in [Man89]. That clique cannot be approximated to within a factor of $n^{1/3}$ is shown in [BGS95].

Related Problems: Independent set (see page 315), vertex cover (see page 317).

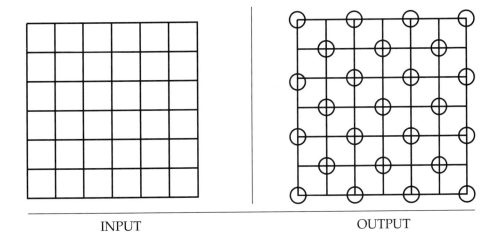

INPUT

OUTPUT

8.5.2   Independent Set

Input description: A graph $G = (V, E)$.

Problem description: What is the largest subset S of vertices of V such that no pair of vertices in S defines an edge of E between them?

Discussion: The need to find large independent sets typically arises in dispersion problems, where we seek a set of mutually separated points. For example, suppose you are trying to identify locations for a new franchise service such that no two locations are close enough to compete with each other. Construct a graph where the vertices are possible locations, and add edges between any two locations deemed close enough to interfere. The maximum independent set gives you the maximum number of franchises you can sell without cannibalizing sales.

Independent sets avoid conflicts between elements and hence arise often in coding theory and scheduling problems. Define a graph whose vertices represent the set of possible code words, and add edges between any two code words sufficiently similar to be confused due to noise. The maximum independent set of this graph defines the highest capacity code for the given communication channel.

Independent set is closely related to two other NP-complete problems:

- *Vertex coloring* – A coloring of a graph G is in fact a partitioning of the vertices of G into a small number of independent sets, since any two vertices of the same color cannot have an edge between them. In fact, most scheduling applications of independent set are really coloring problems, since all tasks eventually must be completed.

- *Clique* – The *complement* of a graph $G = (V, E)$ is a graph $G' = (V, E')$, where $(i, j) \in E'$ iff (i, j) is not in E. In other words, we replace each edge by a nonedge and vica versa. The maximum independent set in G is exactly the maximum clique in G', so these problems are essentially identical. Algorithms and implementations in Section 8.5.1 can thus be easily used for independent set.

The simplest reasonable heuristic is to find the lowest-degree vertex, add it to the independent set, and delete it and all vertices adjacent to it. Repeating this process until the graph is empty gives a *maximal* independent set, in that it can't be made larger just by adding vertices. Using randomization or perhaps some exhaustive search to distinguish among the low-degree vertices might result in somewhat larger independent sets.

The independent set problem is in some sense dual to the graph matching problem. The former asks for a large set of vertices with no edge in common, while the latter asks for a large set of edges with no vertex in common. This suggests trying to rephrase your problem as a matching problem, which can be computed in polynomial time, while the maximum independent set problem is NP-complete.

The maximum independent set of a tree can be found in linear time by (1) stripping off the leaf nodes, (2) adding them to the independent set, (3) deleting the newly formed leaves, and then (4) repeating from the first step on the resulting tree until it is empty.

Implementations: Programs for the closely related problems of finding cliques and independent sets were sought for the Second DIMACS Implementation Challenge, held in October 1993. Programs and data from the challenge are available by anonymous ftp from dimacs.rutgers.edu. Source codes are available under pub/challenge/graph and test data under pub/djs. See Section 8.5.1.

Combinatorica [Ski90] provides (slow) Mathematica implementations of cliques, independent sets, and vertex covers. See Section 9.1.4 for further information on Combinatorica.

Neural-network heuristics for vertex cover and related problems such as clique and vertex coloring have been implemented in C and Mathematica by Laura Sanchis and Arun Jagota, and are available in the algorithm repository http://www.cs.sunysb.edu/~algorith.

Notes: Independent set remains NP-complete for planar cubic graphs [GJ79]. However, it can be solved efficiently for bipartite graphs [Law76].

Related Problems: Clique (see page 312), vertex coloring (see page 329), vertex cover (see page 317).

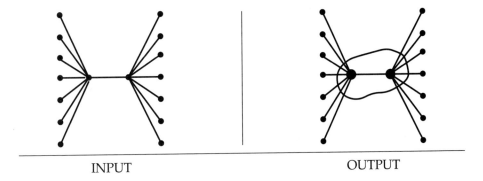

| INPUT | OUTPUT |

8.5.3 Vertex Cover

Input description: A graph $G = (V, E)$.

Problem description: What is the smallest subset of $S \subset V$ such that each $e \in E$ contains at least one vertex of S?

Discussion: Vertex cover is a special case of the more general *set cover* problem, which takes as input an arbitrary collection of subsets $S = (S_1, \ldots, S_n)$ of the universal set $U = \{1, \ldots, m\}$. We seek the smallest subset of subsets from S whose union is U. Set cover arises in many applications, including Boolean logic minimization. See Section 8.7.1 for a discussion of set cover.

To turn vertex cover into a set cover problem, let U be the complete set of edges, and create S_i to be the set of edges incident on vertex i. A set of vertices defines a vertex cover in graph G iff the corresponding subsets define a set cover in the given instance. However, since each edge can be in only two different subsets, vertex cover instances are simpler than general set cover. The primary reason for distinguishing between the two problems is that vertex cover is a relative lightweight among NP-complete problems, and so can be effectively solved.

Vertex cover and independent set are very closely related graph problems. Since every edge in E is (by definition) incident on a vertex in a cover S, there can be no edge for which both endpoints are not in S. Thus $V - S$ must be an independent set. Further, since minimizing S is the same as maximizing $V - S$, a minimum vertex cover defines a maximum independent set, and vice versa. This equivalence means that if you have a program that solves independent set, you can use it on your vertex cover problem. Having two ways of looking at it can be helpful if you expect that either the cover or independent set is likely to contain only a few vertices, for it might pay to search all possible pairs or triples of vertices if you think that it will pay off.

The simplest heuristic for vertex cover selects the vertex with highest degree, adds it to the cover, deletes all adjacent edges, and then repeats until the graph is empty. With the right data structures, this can be done in linear time, and the cover you get "usually" should be "pretty good". However, for certain input graphs the resulting cover can be $O(\lg n)$ times worse than the optimal cover.

Much better is the following approximation algorithm, discussed in Section 6.8.1, which always finds a vertex cover whose size is at most twice as large as optimal. Find a *maximal* matching in the graph, i.e. a set of edges no two of which share a vertex in common and that cannot be made larger by adding additional edges. Such a maximal matching can be built incrementally, by picking an arbitrary edge e in the graph, deleting any edge sharing a vertex with e, and repeating until the graph is out of edges. Taking *both* of the vertices for each edge in the matching gives us a vertex cover, since we only delete edges incident to one of these vertices and eventually delete all the edges. Because *any* vertex cover must contain *at least* one of the two vertices in each matching edge just to cover the matching, this cover must be at most twice as large as that of the minimum cover.

This heuristic can be tweaked to make it perform somewhat better in practice, without losing the performance guarantee or costing too much extra time. We can select the matching edges so as to "kill off" as many edges as possible, which should reduce the size of the maximal matching and hence the number of pairs of vertices in the vertex cover. Also, some vertices selected for our cover may in fact not be necessary, since all of their incident edges could also have been covered using other selected vertices. By making a second pass through our cover, we can identify and delete these losers. If we are really lucky, we might halve the size of our cover using these techniques.

A problem that might seem closely related to vertex cover is *edge cover*, which seeks the smallest set of edges such that each vertex is included in one of the edges. In fact, edge cover can be efficiently solved by finding a maximum cardinality matching in G (see Section 8.4.6) and then selecting arbitrary edges to account for the unmatched vertices.

Implementations: Programs for the closely related problems of finding cliques and independent sets were sought for the Second DIMACS Implementation Challenge, held in October 1993. See Section 8.5.1 for details.

Combinatorica [Ski90] provides (slow) Mathematica implementations of cliques, independent sets, and vertex covers. See Section 9.1.4 for further information on Combinatorica.

Neural-network heuristics for vertex cover and related problems such as clique and vertex coloring have been implemented in C and Mathematica by Laura Sanchis and Arun Jagota, and are available in the algorithm repository http://www.cs.sunysb.edu/~algorith.

Notes: Good expositions of the proof that vertex-cover is NP-complete [Kar72] include [CLR90, GJ79, Man89]. Good expositions on the 2-approximation algorithm include [CLR90]. The example that the greedy algorithm can be as bad as $\lg n$ times optimal is due to [Joh74] and is presented in [PS82].

Related Problems: Independent set (see page 315), set cover (see page 398).

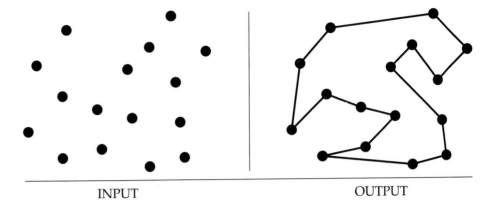

INPUT   OUTPUT

8.5.4 Traveling Salesman Problem

Input description: A weighted graph G.

Problem description: Find the cycle of minimum cost that visits each of the vertices of G exactly once.

Discussion: The traveling salesman problem is the most notorious NP-complete problem. This is a function of its general usefulness, and because it is easy to explain to the public at large. Imagine a traveling salesman who has to visit each of a given set of cities by car. What is the shortest route that will enable him to do so and return home, thus minimizing his total driving?

Although the problem arises in transportation applications, its most important applications come in optimizing the tool paths for manufacturing equipment. For example, consider a robot arm assigned to solder all the connections on a printed circuit board. The shortest tour that visits each solder point exactly once defines the most efficient path for the robot. A similar application arises in minimizing the amount of time taken by a graphics plotter to draw a given figure.

Several issues arise in solving TSPs:

- *Is the graph unweighted?* – If the graph is unweighted, or all the edges have one of two cost values, the problem reduces to finding a *Hamiltonian cycle*. See Section 8.5.5 for a discussion of this problem.

- *Are you given as input n points or a weighted graph?* – Geometric points are often easier to work with than a graph representation, for several reasons. First, they define a complete graph, so there is never an issue of finding a tour, just a good one. Second, although we always could construct the complete distance graph on the points and feed it to a graph solver, it might be more efficient to construct a sparse nearest-neighbor graph (see Section 8.6.5) and work primarily from that. Finally, geometric instances inherently satisfy the triangle inequality, discussed below.

- *How important is the restriction against visiting each vertex more than once?* – The restriction that the tour not revisit any vertex may be important in certain applications, but it often is irrelevant. For example, the cheapest way to visit all vertices might involve repeatedly visiting a hub site, as is common in modern air travel.

This issue does not arise whenever the graph observes the *triangle inequality*; that is, for all vertices $i, j, k \in V, d(i, j) \leq d(i, k) + d(k, j)$. In graphs that observe the triangle inequality, the shortest tour visits each vertex once. Heuristics work much better on graphs that do obey the triangle inequality.

- *How important is that that you find the optimal tour?* – If you insist on solving your TSP to optimality (and you probably shouldn't bother), there are two common approaches. *Cutting plane methods* model the problem as an integer program, then solve the linear programming relaxation of it. If the optimal solution is not at integer points, additional constraints designed to force integrality are added. *Branch-and-bound algorithms* perform a combinatorial search while maintaining careful upper and lower bounds on the cost of a tour or partial tour. In the hands of professionals, problems with thousands of vertices can be solved. In the hands of one gleaning their knowledge from this book, problems with 50 to maybe 100 vertices are potentially solvable, using the implementations discussed below.

Almost any flavor of TSP is going to be NP-complete, so the right way to proceed is with heuristics. These are often quite successful, typically coming within a few percent of the optimal solution, which is close enough for engineering work. Unfortunately, there have been literally dozens of heuristics proposed for TSPs, so the situation can be confusing. Empirical results in the literature are sometime contradictory. However, we recommend choosing from among the following heuristics:

- *Minimum spanning trees* – A simple and popular heuristic, especially when the sites represent points in the plane, is based on the minimum spanning tree of the points. By doing a depth-first search of this tree, we walk over each edge of the tree exactly twice, once going down when we discover the new vertex and once going up when we backtrack. We can then define a tour of the vertices according to the order in which they were discovered and use the shortest path between each neighboring pair of vertices in this order to connect them. This path must be a single edge if the graph is complete and obeys the triangle inequality, as with points in the plane. As discussed in Section 6.8.2, the resulting tour is always at most twice the length of the minimum TSP tour. In practice, it is usually better, typically 15% to 20% over optimal. Further, the time of the algorithm is bounded by that of computing the minimum spanning tree, only $O(n \lg n)$ in the case of points in the plane (see Section 8.4.3).

- *Incremental insertion methods* – A different class of heuristics inserts new points into a partial tour one at a time (starting from a single vertex) until the tour is complete. The version of this heuristic that seems to work best is *furthest point* insertion: of all remaining points, insert the point v into partial tour T

such that

$$\max_{v \in V} \min_{i=1}^{|T|}(d(v, v_i) + d(v, v_{i+1}))$$

The minimum ensures that we insert the vertex in the position that adds the smallest amount of distance to the tour, while the maximum ensures that we pick the worst such vertex first. This seems to work well because it first "roughs out" a partial tour before filling in details. Typically, such tours are only 5% to 10% longer than optimal.

- *k-optimal tours* – Substantially more powerful are the Kernighan-Lin, or *k-opt* class of heuristics. Starting from an arbitrary tour, the method applies local refinements to the tour in the hopes of improving it. In particular, subsets of $k \geq 2$ edges are deleted from the tour and the k remaining subchains rewired in a different way to see if the resulting tour is an improvement. A tour is k-optimal when no subset of k edges can be deleted and rewired so as to reduce the cost of the tour. Extensive experiments suggest that 3-optimal tours are usually within a few percent of the cost of optimal tours. For $k > 3$, the computation time increases considerably faster than solution quality. Two-opting a tour is a fast and effective way to improve any other heuristic. Simulated annealing provides an alternate mechanism to employ edge flips to improve heuristic tours.

Implementations: The world-record-setting traveling salesman program is by Applegate, Bixby, Chvatal, and Cook [ABCC95], which has solved instances as large as 7,397 vertices to optimality. At this time, the program is not being distributed. However, the authors seem willing to use it to solve TSPs sent to them. In their paper, they describe this work as neither theory nor practice, but sport – an almost recreational endeavor designed principally to break records. It is a very impressive piece of work, however.

The TSPLIB library of test instances for the traveling salesman problem is available from Netlib, and by anonymous ftp from softlib.cs.rice.edu. See Section 9.1.2.

Tsp_solve is a C++ code by Chad Hurwitz and Robert Craig that provides both heuristic and optimal solutions. Geometric problems of size up to 100 points are manageable. It is available from http://www.cs.sunysb.edu/~algorith or by e-mailing Chad Hurrwitz at churritz@cts.com. A heuristic Euclidean TSP solver in C due to Lionnel Maugis is available from http://www.cenaath.cena.dgac.fr/~maugis/tsp.shar.

Pascal implementations of branch-and-bound search and the insertion and Kerighan-Lin heuristics (for 2-opt and 3-opt) appear in [SDK83]. For details, see Section 9.1.6.1.

Algorithm 608 [Wes83] of the *Collected Algorithms of the ACM* is a Fortran implementation of a heuristic for the quadratic assignment problem, a more general problem that includes the traveling salesman as a special case. Algorithm

750 [CDT95] is a Fortran code for the exact solution of asymmetric TSP instances. See Section 9.1.2 for details.

XTango (see Section 9.1.5) includes animations of both the minimum spanning tree heuristic and a genetic algorithm for TSP. The latter converges sufficiently slowly to kill one's interest in genetic algorithms.

Combinatorica [Ski90] provides (slow) Mathematica implementations of exact and approximate TSP solutions. See Section 9.1.4.

Notes: The definitive reference on the traveling salesman problem is the book by Lawler et. al. [LLKS85]. Experimental results on heuristic methods for solving large TSPs include [Ben92a, GBDS80, Rei94]. Typically, it is possible to get within a few percent of optimal with such methods. TSPLIB [Rei91] provides the standard collection of hard instances of TSPs that arise in practice.

The Christofides heuristic is an improvement of the minimum-spanning tree heuristic and guarantees a tour whose cost is at most 3/2 times optimal on Euclidean graphs. It runs in $O(n^3)$, where the bottleneck is the time it takes to find a minimum-weight perfect matching (see Section 8.4.6). Good expositions of the Christofides heuristic [Chr76] include [Man89, PS85]. Expositions of the minimum spanning tree heuristic [RSL77] include [CLR90, O'R94, PS85].

Polynomial-time approximation schemes for Euclidean TSP have been recently developed by Arora [Aro96] and Mitchell [Mit96], which offer $1 + \epsilon$ factor approximations in polynomial time for any $\epsilon > 0$. They are of great theoretical interest, although any practical consequences remain to be determined.

The history of progress on optimal TSP solutions is somewhat inspiring. In 1954, Dantzig, Fulkerson, and Johnson solved a symmetric TSP instance of 42 United States cities [DFJ54]. In 1980, Padberg and Hong solved an instance on 318 vertices [PH80]. Applegate et. al. [ABCC95] have recently solved problems that are twenty times larger than this. Some of this increase is due to improved hardware, but most is due to better algorithms. The rate of growth demonstrates that exact solutions to NP-complete problems can be obtained for large instances if the stakes are high enough. Unfortunately, they seldom are. Good expositions on branch-and-bound methods include [PS82, SDK83]. Good expositions of the Kernighan-Lin heuristic [LK73, Lin65] include [MS91, PS82, SDK83].

Size is not the only criterion for hardness. One can easily construct an enormous graph consisting of one cheap cycle, for which it would be easy to find the optimal solution. For sets of points in convex position in the plane, the minimum TSP tour is described by its convex hull (see Section 8.6.2), which can be computed in $O(n \lg n)$ time. Other easy special cases are known.

Related Problems: Hamiltonian cycle (see page 323), minimum spanning tree (see page 275), convex hull (see page 351).

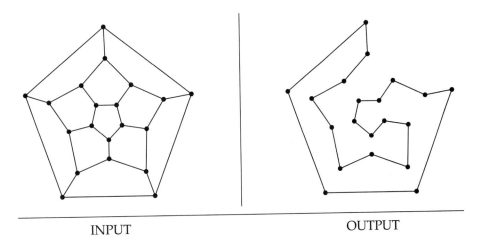

INPUT   OUTPUT

8.5.5 Hamiltonian Cycle

Input description: A graph $G = (V, E)$.

Problem description: Find an ordering of the vertices such that each vertex is visited exactly once.

Discussion: The problem of finding a Hamiltonian cycle or path in a graph is a special case of the traveling salesman problem, one where each pair of vertices with an edge between them has distance 1, while nonedge vertex pairs are separated by distance ∞.

Closely related is the problem of finding the longest path or cycle in a graph, which occasionally arises in pattern recognition problems. Let the vertices in the graph correspond to possible symbols, and let edges link symbols that can possibly be next to each other. The longest path through this graph is likely the correct interpretation.

The problems of finding longest cycles and paths are both NP-complete, even on very restrictive classes of unweighted graphs. There are several possible lines of attack, however:

- *Do you have to visit all the vertices or all the edges?* – First verify that you really have a Hamiltonian cycle problem. As discussed in Section 8.4.7, fast algorithms exist for edge-tour, or *Eulerian* cycle, problems, where you must visit all the edges without repetition. With a little cleverness, it is sometimes possible to reformulate a Hamiltonian cycle problem in terms of Eulerian cycles. Perhaps the most famous such instance is the problem of constructing de Bruijn sequences, discussed in Section 8.4.7.

- *Is there a serious penalty for visiting vertices more than once?* – By phrasing the problem as minimizing the total number of vertices visited on a complete tour, we have an optimization problem that now allows room for heuristics and approximation algorithms. For example, finding a spanning tree of the graph and doing a depth-first search, as discussed in Section 8.5.4, yields a tour with at most $2n$ vertices. Using randomization or simulated annealing might bring the size of this down considerably.

- *Am I seeking the longest path in a directed acyclic graph (DAG)?* – The problem of finding the longest path in a DAG can be solved in linear time using dynamic programming. This is about the only interesting case of longest path for which efficient algorithms exist.

- *Is my graph dense?* – For sufficiently dense graphs, there always exists at least one Hamiltonian cycle, and further, such a cycle can be found quickly. An efficient algorithm for finding a Hamiltonian cycle in a graph where all vertices have degree $\geq n/2$ is given in [Man89].

If you really must know whether your graph is Hamiltonian, backtracking with pruning is your only possible solution. Before you search, it pays to check whether your graph is biconnected (see Section 8.4.8). If not, this means that the graph has an articulation vertex whose deletion will disconnect the graph and so cannot be Hamiltonian.

Implementations: The football program of the Stanford GraphBase (see Section 9.1.3) uses a stratified greedy algorithm to solve the asymmetric longest path problem. The goal is to derive a chain of football scores in order to establish the superiority of one football team over another. After all, if Virginia beat Illinois by 30 points, and Illinois beat Stony Brook by 14 points, then by transitivity Virginia would beat Stony Brook by 44 points if they played, right? We seek the longest path in a graph where the weight of an edge (x, y) is the number of points x beat y by.

Nijenhuis and Wilf [NW78] provide an efficient routine to enumerate all Hamiltonian cycles of a graph by backtracking. See Section 9.1.6.3. Algorithm 595 [Mar83] of the *Collected Algorithms of the ACM* is a similar Fortran code that can be used as either an exact procedure or a heuristic by controlling the amount of backtracking. See Section 9.1.2.

XTango (see Section 9.1.5) is an algorithm animation system for UNIX and X-windows, which includes an animation of a backtracking solution to the knight's tour problem.

Combinatorica [Ski90] provides a Mathematica backtracking implementation of Hamiltonian cycle. See Section 9.1.4.

Notes: Hamiltonian cycles – circuits that visit each vertex of a graph exactly once – apparently first arose in Euler's study of the knight's tour problem, although they were popularized by Hamilton's "Around the World" game in 1839. Good expositions of the proof that Hamiltonian cycle is NP-complete [Kar72] include [Baa88, CLR90, GJ79].

Techniques for solving optimization problems in the laboratory using biological processes have recently attracted considerable attention. In the original application of these "biocomputing" techniques, Adleman [Adl94b] solved a seven-vertex instance of the directed Hamiltonian path problem. Unfortunately, this approach requires an exponential number of molecules, and Avogadro's number implies that such experiments are inconceivable for graphs beyond $n \approx 70$.

Related Problems: Eulerian cycle (see page 291), traveling salesman (see page 319).

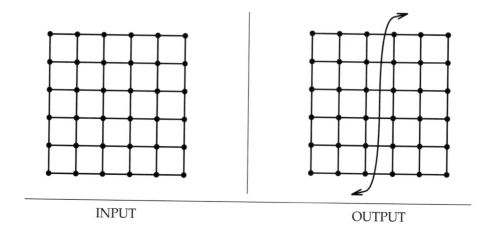

<table>
<tr><td>INPUT</td><td>OUTPUT</td></tr>
</table>

8.5.6 Graph Partition

Input description: A (weighted) graph $G = (V, E)$ and integers $j, k,$ and m.

Problem description: Partition the vertices into m subsets such that each subset has size at most j, while the cost of the edges spanning the subsets is bounded by k.

Discussion: Graph partitioning arises as a preprocessing step to divide-and-conquer algorithms, where it is often a good idea to break things into roughly equal-sized pieces. It also arises when dealing with extremely large graphs, when we need to cluster the vertices into logical components for storage (to improve virtual memory performance) or for drawing purposes (to collapse dense subgraphs into single nodes in order to reduce cluttering).

Several different flavors of graph partitioning arise depending on the desired objective function:

- *Minimum cut set* – The *smallest* set of edges to cut that will disconnect a graph can be efficiently found using network flow methods. See Section 8.4.8 for more on connectivity algorithms. Since the smallest cutset can split off only a single vertex, the resulting partition might be very unbalanced. Hence ...

- *Graph partition* – A better partition criterion seeks a small cut that partitions the vertices into roughly equal-sized pieces. Unfortunately, this problem is NP-complete. Fortunately, heuristics discussed below work well in practice.

    Certain special graphs always have small *separators*, which partition the vertices into balanced pieces. For any tree, there always exists a single vertex whose deletion partitions the tree so that no component contains more than $n/2$ of the original n vertices. These components need not always be connected. For example, consider the separating vertex of a star-shaped tree. However, the separating vertex can be found in linear time using depth first-search. Similarly, every planar graph has a set of $O(\sqrt{n})$ vertices whose

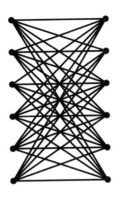 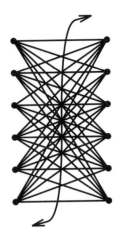

Figure 8–5. The maximum cut of a graph

deletion leaves no component with more than $2n/3$ vertices. Such separators provide a particularly useful way to decompose planar graphs.

- *Maximum cut* – Given an electronic circuit specified by a graph, the *maximum cut* defines the largest amount of data communication that can simultaneously take place in the circuit. The highest-speed communications channel should thus span the vertex partition defined by the maximum edge cut. Finding the maximum cut in a graph is NP-complete, despite the existence of algorithms for min-cut. However, heuristics similar to those of graph partitioning can work well.

The basic approach to dealing with graph partitioning or max-cut problems is to construct an initial partition of the vertices (either randomly or according to some problem-specific strategy) and then sweep through the vertices, deciding whether the size of the cut would increase or decrease if we moved this vertex over to the other side. The decision to move v can be made in time proportional to its degree by simply counting whether more of v's neighbors are on the same team as v or not. Of course, the desirable side for v will change if many of its neighbors jump, so multiple passes are likely to be needed before the process converges on a local optimum. Even such a local optimum can be arbitrarily far away from the global max-cut.

There are many variations of this basic procedure, by changing the order we test the vertices in or moving clusters of vertices simultaneously. Using some form of randomization, particularly simulated annealing, is almost certain to be a good idea.

Implementations: Jon Berry's implementations of several graph partitioning heuristics, including Kernighan-Lin, simulated annealing, and path optimization are available from http://www.elon.edu/users/f/berryj/www/.

A non-network-flow-based implementation of minimum cut is included with LEDA (see Section 9.1.1).

Notes: The fundamental heuristic for graph partitioning is due to Kernighan and Lin [KL70]. Empirical results on graph partitioning heuristics include [BG95, LR93].

The planar separator theorem and an efficient algorithm for finding such a separator are due to Lipton and Tarjan [LT79, LT80]. Although network flow can be used to find minimum cut sets in graphs, faster algorithms are available, including [SW94] and [Kar96a].

Expositions on the hardness of max-cut [Kar72] include [Eve79a]. Note that any random vertex partition will expect to cut half of the edges in the graph, since the probability that the two vertices defining an edge end up on different sides of the partition is 1/2. Goemans and Williamson [GW95] gave an 0.878-factor approximation algorithm for maximum-cut, based on semidefinite programming techniques. Tighter analysis of this algorithm followed by Karloff [Kar96b].

Related Problems: Edge/vertex connectivity (see page 294), network flow (see page 297).

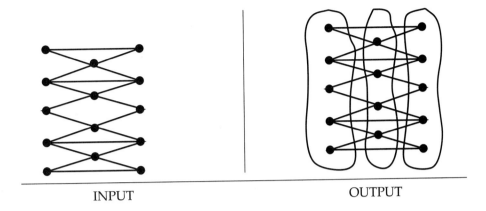

INPUT   OUTPUT

8.5.7   Vertex Coloring

Input description: A graph $G = (V, E)$.

Problem description: Color the vertices of V using the minimum number of colors such that for each edge $(i, j) \in E$, vertices i and j have different colors.

Discussion: Vertex coloring arises in a variety of scheduling and clustering applications. Compiler optimization is the canonical application for coloring, where we seek to schedule the use of a finite number of registers. In a program fragment to be optimized, each variable has a range of times during which its value must be kept intact, in particular, after it is initialized and before its final use. Any two variables whose life spans intersect cannot be placed in the same register. Construct a graph where there is a variable associated with each vertex and add an edge between any two vertices whose variable life spans intersect. A coloring of the vertices of this graph assigns the variables to classes such that two variables with the same color do not clash and so can be assigned to the same register.

No conflicts can occur if each vertex is colored with a distinct color. However, our goal is to find a coloring using the minimum number of colors, because computers have a limited number of registers. The smallest number of colors sufficient to vertex color a graph is known as its *chromatic number*.

Several special cases of interest arise in practice:

- *Can I color the graph using only two colors?* – An important special case is testing whether a graph is *bipartite*, meaning it can be colored using two different colors. Such a coloring of the vertices of a bipartite graph means that the graph can be drawn with the red vertices on the left and the blue vertices on the right such that all edges go from left to right. Bipartite graphs are fairly simple, yet they arise naturally in such applications as mapping workers to possible jobs. Fast, simple algorithms exist for problems such as matching (see Section 8.4.6) on bipartite graphs.

  Testing whether a graph is bipartite is easy. Color the first vertex blue, and then do a depth-first search of the graph. Whenever we discover a new, uncolored vertex, color it opposite that of its parent, since the same color would cause a clash. If we ever find an edge where both vertices have been colored identically, then the graph cannot be bipartite. Otherwise, this coloring will be a 2-coloring, and it is constructed in $O(n + m)$ time.

- *Is the graph planar, or are all vertices of low degree?* – The famous 4-color theorem states that every planar graph can be vertex colored using at most 4 distinct colors. Efficient algorithms for finding a 4-coloring are known, although it is NP-complete to decide whether a given planar graph is 3-colorable.

  There is a very simple algorithm that finds a vertex coloring of any planar graph using at most 6 colors. In any planar graph, there exists a vertex of degree at most five. Delete this vertex and recursively color the graph. This vertex has at most five neighbors, which means that it can always be colored using one of the six colors that does not appear as a neighbor. This works because deleting a vertex from a planar graph leaves a planar graph, so we always must have a low-degree vertex to delete. The same idea can be used to color any graph of maximum degree Δ using $\leq \Delta + 1$ colors in $O(n\Delta)$ time.

- *Is this an edge coloring problem?* – Certain vertex coloring problems can be modeled as *edge coloring*, where we seek to color the edges of a graph G such that no two edges with a vertex in common are colored the same. The payoff is that there is an efficient algorithm that always returns a near-optimal edge coloring. Algorithms for edge coloring are the focus of Section 8.5.8.

Computing the chromatic number of a graph is NP-complete, so if you need an exact solution you must resort to backtracking, which can be surprisingly effective in coloring certain random graphs. It remains hard to compute a provably good approximation to the optimal coloring, so expect no guarantees.

Incremental methods appear to be the heuristic of choice for vertex coloring. As in the previous algorithm for planar graphs, vertices are colored sequentially, with the colors chosen in response to colors already assigned in the vertex's neighborhood. These methods vary in how the next vertex is selected and how it is assigned a color. Experience suggests inserting the vertices in nonincreasing order of degree, since high-degree vertices have more color constraints and so are most likely to require an additional color if inserted late.

Incremental methods can be further improved by using *color interchange*. Observe that taking a properly colored graph and exchanging two of the colors (painting the red vertices blue and the blue vertices red) leaves a proper vertex coloring. Now suppose we take a properly colored graph and delete all but the red and blue vertices. If the remaining graph (the *induced subgraph*) consists of two or more connected components, we can repaint one or more of the components, again leaving a proper coloring. After such a recoloring, some vertex v previously adjacent to both red and blue vertices might now be only adjacent to blue vertices, thus freeing v to be colored red.

Color interchange is a win in terms of producing better colorings, at a cost of increased time and implementation complexity. Implementations are described below. Simulated annealing algorithms that incorporate color interchange to move from state to state are likely to be even more effective.

Implementations: Graph coloring has been blessed with two distinct and useful WWW resources. Michael Trick's page, http://mat.gsia.cmu.edu/COLOR/color.html, provides a nice overview of applications of graph coloring, an annotated bibliography, and a collection of over seventy graph coloring instances arising in applications such as register allocation and printed circuit board testing. Finally, it contains a C language implementation of an exact coloring algorithm, DSATUR. Joseph C. Culberson's WWW page on graph coloring, http://web.cs.ualberta.ca/~joe/Coloring/, provides an extensive bibliography and a collection of programs to generate hard graph coloring instances.

Programs for the closely related problems of finding cliques and vertex coloring graphs were sought for the Second DIMACS Implementation Challenge, held in October 1993. Programs and data from the challenge are available by anonymous ftp from dimacs.rutgers.edu. Source codes are available under pub/challenge/graph and test data under pub/djs, including a simple "semi-exhaustive greedy" scheme used in the graph coloring algorithm XRLF [JAMS91].

Pascal implementations of backtracking algorithms for vertex coloring and several heuristics, including largest-first and smallest-last incremental orderings and color interchange, appear in [SDK83]. See Section 9.1.6.1.

XTango (see Section 9.1.5) is an algorithm animation system for UNIX and X-windows, and includes an animation of vertex coloring via backtracking.

Nijenhuis and Wilf [NW78] provide an efficient Fortran implementation of chromatic polynomials and vertex coloring by backtracking. See Section 9.1.6.3.

Combinatorica [Ski90] provides Mathematica implementations of bipartite graph testing, heuristic colorings, chromatic polynomials, and vertex coloring by backtracking. See Section 9.1.4.

Notes: An excellent source on vertex coloring heuristics is Syslo, Deo, and Kowalik [SDK83], which includes experimental results. Heuristics for vertex coloring include Brèlaz [Brè79], Matula [MMI72], and Turner [Tur88]. Wilf [Wil84] proved that backtracking to test whether a random graph has chromatic number k runs in *constant time*, dependent on k but independent of n. This is not as interesting as it sounds, because only a vanishingly small fraction of such graphs are indeed k-colorable.

Expositions on algorithms to recognize bipartite graphs include [Man89]. Expositions on the hardness of 3-coloring graphs include [AHU74, Eve79a, Man89]. An interesting application of vertex coloring to scheduling traffic lights appears in [AHU83].

Baase [Baa88] gives a very good description of approximation algorithms for graph coloring, including Wigderson's [Wig83] factor of $n^{1-1/(\chi(G)-1)}$ approximation algorithm, where $\chi(G)$ is the chromatic number of G. Hardness of approximation results for vertex coloring include [BGS95].

Brook's theorem states that the chromatic number $\chi(G) \leq \Delta(G) + 1$, where $\Delta(G)$ is the maximum degree of a vertex of G. Equality holds only for odd-length cycles (which have chromatic number 2) and complete graphs.

The most famous problem in the history of graph theory is the four-color problem, first posed in 1852 and finally settled in 1976 by Appel and Haken using a proof involving extensive computation. Any planar graph can be 5-colored using a variation of the color interchange heuristic. Despite the four-color theorem, it is NP-complete to test whether a particular planar graph requires four colors or whether three suffice. See [SK86] for an exposition on the history of the four-color problem and the proof. An efficient algorithm to four-color a graph is presented in [RSST96].

Related Problems: Independent set (see page 315), edge coloring (see page 333).

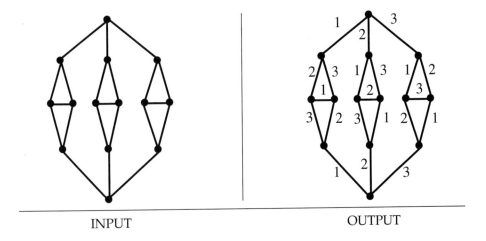

INPUT   OUTPUT

8.5.8   Edge Coloring

Input description: A graph $G = (V, E)$.

Problem description: What is the smallest set of colors needed to color the edges of E such that no two same-color edges share a vertex in common?

Discussion: The edge coloring of graphs arises in a variety of scheduling applications, typically associated with minimizing the number of noninterfering rounds needed to complete a given set of tasks. For example, consider a situation where we need to schedule a given set of two-person interviews, where each interview takes one hour. All meetings could be scheduled to occur at distinct times to avoid conflicts, but it is less wasteful to schedule nonconflicting events simultaneously. We can construct a graph whose vertices are the people and whose edges represent the pairs of people who want to meet. An edge coloring of this graph defines the schedule. The color classes represent the different time periods in the schedule, with all meetings of the same color happening simultaneously.

The National Football League solves such an edge coloring problem each season to make up its schedule. Each team's opponents are determined by the records of the previous season. Assigning the opponents to weeks of the season is the edge-coloring problem, presumably complicated by the constraints of spacing out rematches and making sure that there is a good game every Monday night.

The minimum number of colors needed to edge color a graph is called by some its *edge-chromatic number* and others its *chromatic index*. To gain insight into edge coloring, note that a graph consisting of an even-length cycle can be edge-colored with 2 colors, while odd-length cycles have an edge-chromatic number of 3.

Edge coloring has a better (if less famous) theorem associated with it than does vertex coloring. Vizing's theorem states that any graph with a maximum vertex degree of Δ can be edge colored using at most $\Delta + 1$ colors. To put this in perspective, note that *any* edge coloring must have at least Δ colors, since each of the edges incident on any vertex must be distinct colors.

Further, the proof of Vizing's theorem is constructive and can be turned into an $O(n^2)$ algorithm to find an edge-coloring with $\Delta + 1$ colors, which gives us an edge coloring using at most one extra color. Since it is NP-complete to decide whether we can save this one color, it hardly seems worth the effort to worry about it. An implementation of Vizing's theorem is described below.

Any edge coloring problem on G can be converted to the problem of finding a vertex coloring on the *line graph* $L(G)$, which has a vertex of $L(G)$ for each edge of G and an edge of $L(G)$ if and only if the two edges of G share a common vertex. Line graphs can be constructed in time linear to their size, and any vertex coloring code from Section 8.5.7 can be employed to color them.

Although any edge coloring problem can be so formulated as a vertex coloring problem, this is usually a bad idea, since the edge coloring problem is easier to solve. Vizing's theorem is our reward for the extra thought needed to see that we have an edge coloring problem.

Implementations: Yan Dong produced an implementation of Vizing's theorem in C++ as a course project for my algorithms course while a student at Stony Brook. It can be found on the algorithm repository WWW site http:// www.cs.sunysb.edu/~algorith, as can an alternative program by Mark Goldberg and Amir Sehic.

Combinatorica [Ski90] provides Mathematica implementations of edge coloring, via the line graph transformation and vertex coloring routines. See Section 9.1.4 for further information on Combinatorica.

Notes: Graph-theoretic results on edge coloring are surveyed in [FW77]. Vizing [Viz64] and Gupta [Gup66] both proved that any graph can be edge colored using at most $\Delta + 1$ colors. Despite these tight bounds, Holyer [Hol81] proved that computing the edge-chromatic number is NP-complete.

Whitney, in introducing line graphs [Whi32], showed that with the exception of K_3 and $K_{1,3}$, any two connected graphs with isomorphic line graphs are isomorphic. It is an interesting exercise to show that the line graph of an Eulerian graph is both Eulerian and Hamiltonian, while the line graph of a Hamiltonian graph is always Hamiltonian.

Related Problems: Vertex coloring (see page 329), scheduling (see page 263).

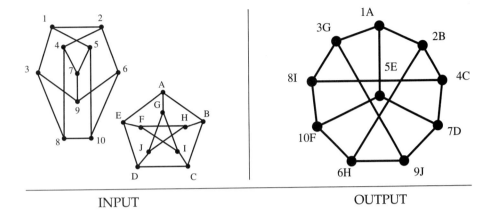

INPUT   OUTPUT

8.5.9 Graph Isomorphism

Input description: Two graphs, G and H.

Problem description: Find a (or all) mappings f of the vertices of G to the vertices of H such that G and H are identical; i.e. (x, y) is an edge of G iff $(f(x), f(y))$ is an edge of H.

Discussion: Isomorphism is the problem of testing whether two graphs are really the same. Suppose we are given a collection of graphs and must perform some operation on each of them. If we can identify which of the graphs are duplicates, they can be discarded so as to avoid redundant work.

We need some terminology to settle what is meant when we say two graphs are the same. Two labeled graphs $G = (V_g, E_g)$ and $H = (V_h, E_h)$ are *identical* when $(x, y) \in E_g$ iff $(x, y) \in E_h$. The isomorphism problem consists of finding a mapping from the vertices of G to H such that they are identical. Such a mapping is called an *isomorphism*.

Identifying symmetries is another important application of graph isomorphism. A mapping of a graph to itself is called an *automorphism*, and the collection of automorphisms (its automorphism *group*) provides a great deal of information about symmetries in the graph. For example, the complete graph K_n has $n!$ automorphisms (any mapping will do), while an arbitrary random graph is likely to have few or perhaps only one, since G is always identical to itself.

Several variations of graph isomorphism arise in practice:

- *Is graph G contained in (not identical to) graph H?* – Instead of testing equality, we are often interested in knowing whether a small pattern graph G is a *subgraph* of H. Such problems as clique, independent set, and Hamiltonian cycle are important special cases of subgraph isomorphism.

  There are two distinct notions of "contained in" with respect to graphs. *Subgraph isomorphism* asks whether there is a subset of edges and vertices of G that is isomorphic to a smaller graph H. *Induced* subgraph isomorphism asks whether there is a subset of vertices of G whose deletion leaves a subgraph isomorphic to a smaller graph H. For induced subgraph isomorphism, all edges of G must be present in H, but also all *nonedges* of G must be nonedges of H. Clique happens to be an instance of both subgraph problems, while Hamiltonian cycle is an example of vanilla subgraph isomorphism.

Be aware of this distinction in your application. Subgraph isomorphism problems tend to be harder than graph isomorphism, and induced subgraph problems tend to be even harder than subgraph isomorphism. Backtracking is your only viable approach.

- *Are your graphs labeled or unlabeled?* – In many applications, vertices or edges of the graphs are *labeled* with some attribute that must be respected in determining isomorphisms. For example, in comparing two bipartite graphs, each with "worker" vertices and "job" vertices, any isomorphism that equated a job with a worker would make no sense.

  Labels and related constraints can be factored into any backtracking algorithm. Further, such constraints can be used to significantly speed up the search by creating more opportunities for pruning whenever two vertex labels do not match up.

- *Are you testing whether two trees are isomorphic?* – There are faster algorithms for certain special cases of graph isomorphism, such as trees and planar graphs. Perhaps the most important case is detecting isomorphisms among trees, a problem that arises in language pattern matching and parsing applications. A parse tree is often used to describe the structure of a text; two parse trees will be isomorphic if the underlying pair of texts have the same structure.

  Efficient algorithms for tree isomorphism begin with the leaves of both trees and work inward towards the center. Each vertex in one tree is assigned a label representing the set of vertices in the second tree that might possibly be isomorphic to it, based on the constraints of labels and vertex degrees. For example, all the leaves in tree T_1 are initially potentially equivalent to all leaves of T_2. Now, working inward, we can partition the vertices adjacent to leaves in T_1 into classes based on how many leaves and nonleaves they are adjacent to. By carefully keeping track of the labels of the subtrees, we can make sure that we have the same distribution of labeled subtrees for T_1 and T_2. Any mismatch means $T_1 \neq T_2$, while completing the process partitions the vertices into equivalence classes defining all isomorphisms. See the references below for more details.

No polynomial-time algorithm is known for graph isomorphism, but neither is it known to be NP-complete. Along with integer factorization (see Section 8.2.8), it one of the few important algorithmic problems whose rough computational complexity is still not known. The conventional wisdom is that isomorphism is a problem that lies between P and NP-complete if P $\neq$ NP.

Although no worst-case polynomial-time algorithm is known, testing isomorphism in practice is *usually* not very hard. The basic algorithm backtracks through all $n!$ possible relabelings of the vertices of graph h with the names of vertices of graph g, and then tests whether the graphs are identical. Of course, we can prune the search of all permutations with a given prefix as soon as we detect any mismatch between edges both of whose vertices are in the prefix.

However, the real key to efficient isomorphism testing is to preprocess the vertices into "equivalence classes", partitioning them into sets of vertices such that two vertices in different sets cannot possibly be mistaken for each other. All vertices in each equivalence class must share the same value of some invariant that is independent of labeling. Possibilities include:

- *Vertex degree* – This simplest way to partition vertices is based on their degree, the number of edges incident on the vertex. Clearly, two vertices of different degree cannot be identical. This simple partition can often be a big win, but it won't do much for regular graphs, where each vertex has the same degree.
- *Shortest path matrix* – For each vertex v, the all-pairs shortest path matrix (see Section 8.4.4) defines a multiset of $n - 1$ distances (possibly with repeats) representing the distances between v and each of the other vertices. Any two vertices that are identical in isomorphic graphs will define the exact same multiset of distances, so we can partition the vertices into equivalence classes defining identical distance multisets.
- *Counting length-k paths* – Taking the adjacency matrix of G and raising it to the kth power gives a matrix where $G^k[i, j]$ counts the number of (nonsimple) paths from i to j. For each vertex and each k, this matrix defines a multiset of path-counts, which can be used for partitioning as with distances above. You could try all $1 \leq k \leq n$ or beyond, and use any single deviation as an excuse to partition.

Using these invariants, you should be able to partition the vertices of each graph into a large number of small equivalence classes. Finishing the job off with backtracking, using the name of each equivalence class as a label, should usually be quick work. If the sizes of the equivalence classes of both graphs are not identical, then the graphs cannot be isomorphic. It is harder to detect isomorphisms between graphs with high degrees of symmetry than it is for arbitrary graphs, because of the effectiveness of these equivalence-class partitioning heuristics.

Implementations: The world's fastest isomorphism testing program is Nauty, by Brendan D. McKay. Nauty (No AUTomorphisms, Yes?) is a set of very efficient C language procedures for determining the automorphism group of a vertex-colored graph. Nauty is also able to produce a canonically labeled isomorph of the graph, to assist in isomorphism testing. It was the basis of the first program to generate all the 11-vertex graphs without isomorphs, and can test most graphs of fewer than one hundred vertices in well under a second. Nauty has been successfully ported to a variety of operating systems and C compilers. It may be obtained from http://cs.anu.edu.au/people/bdm/. It is free for educational and research applications, but for commercial use contact the author at bdm@cs.anu.edu.au.

Combinatorica [Ski90] provides (slow) Mathematica implementations of graph isomorphism and automorphism testing. See Section 9.1.4 for further information on Combinatorica.

Notes: Graph isomorphism is an important problem in complexity theory. Monographs on isomorphism detection include Hoffmann [Hof82].

Polynomial-time algorithms are known for planar graph isomorphism [HW74] and for graphs where the maximum vertex degree is bounded by a constant [Luk80]. The all-pairs shortest path heuristic is due to [SD76], although there exist nonisomorphic graphs that realize the same set of distances [BH90]. A linear-time tree isomorphism algorithm for both labeled and unlabeled trees is presented in [AHU74].

A problem is said to be *isomorphism-complete* if it is provably as hard as isomorphism. Testing the isomorphism of bipartite graphs is isomorphism-complete, since any graph can be made bipartite by replacing each edge by two edges connected with a new vertex. Clearly, the original graphs are isomorphic if and only if the transformed graphs are.

Related Problems: Shortest path (see page 279), string matching (see page 403).

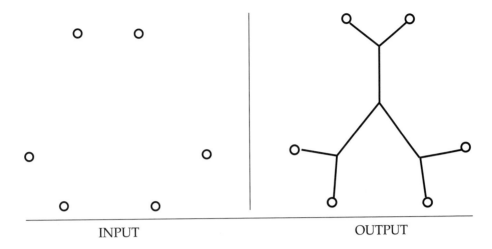

INPUT   OUTPUT

8.5.10 Steiner Tree

Input description: A graph $G = (V, E)$. A subset of vertices $T \in V$.

Problem description: Find the smallest tree connecting all the vertices of T.

Discussion: Steiner tree often arises in network design and wiring layout problems. Suppose we are given a set of sites that must be connected by wires as cheaply as possible. The minimum Steiner tree describes the way to connect them using the smallest amount of wire. Analogous problems arise in designing networks of water pipes or heating ducts in buildings. Similar considerations also arise in VLSI circuit layout, where we seek to connect a set of sites to (say) ground under constraints such as material cost, signal propagation time, or reducing capacitance.

The Steiner tree problem is distinguished from the minimum spanning tree problem (see Section 8.4.3) in that we are permitted to construct or select intermediate connection points to reduce the cost of the tree. Issues in Steiner tree construction include:

- *How many points do you have to connect?* – The Steiner tree of a pair of vertices is simply the shortest path between them (see Section 8.4.4). The Steiner tree of all the vertices, when $S = V$, simply defines the minimum spanning tree of G. Despite these special cases, the general minimum Steiner tree problem is NP-hard and remains so under a broad range of restrictions.

- *Is the input a set of geometric points or a distance graph?* – Geometric versions of Steiner tree take as input a set of points, typically in the plane, and seek the smallest tree connecting the points. A complication is that the set of possible intermediate points is not given as part of the input but must be deduced from the set of points. These possible Steiner points must satisfy several geometric properties, which can be used to reduce the set of candidates down to a finite number. For example, every Steiner point will have degree exactly three in a minimum Steiner tree, and the angles formed between any two of these edges will be exactly 120 degrees.

- *Are there constraints on the edges we can use?* – Many wiring problems correspond to geometric versions of the problem where all edges are restricted to being either horizontal or vertical, which is the so-called *rectilinear Steiner problem*. A different set of angular and degree conditions applies for rectilinear Steiner trees than for Euclidean trees. In particular, all angles must be multiples of 90 degrees, and each vertex is of degree up to four.

- *Do I really need an optimal tree?* – In certain applications, such as minimum cost communications networks, construction costs are high enough to invest large amounts of computation in finding the best possible Steiner tree. This implies an exhaustive search technique such as backtracking or branch-and-bound. There are many opportunities for pruning search based on geometric constraints. For graph instances, network reduction procedures can reduce the problem to a graph typically one-quarter the size of the input graph.

    Still, Steiner tree remains a hard problem. Through exhaustive search methods, instances as large as 32 points for the Euclidean and 30 for the rectilinear problems can be confidently solved to optimality. We recommend experimenting with the implementations described below before attempting your own.

- *How can I reconstruct Steiner vertices I never knew about?* – A very special type of Steiner tree arises in classification and evolution. A *phylogenic tree* illustrates the relative similarity between different objects or species. Each object represents (typically) a terminal vertex of the tree, with intermediate vertices representing branching points between classes of objects. For example, an evolutionary tree of animal species might have leaf nodes including *(human, dog, snake)* and internal nodes corresponding to the taxa *(animal, mammal, reptile)*. A tree rooted at *animal* with *dog* and *human* classified under *mammal* implies that humans are closer to dogs than to snakes.

    Many different phylogenic tree construction algorithms have been developed, which vary in the data they attempt to model and what the desired optimization criterion is. Because they all give different answers, identifying the correct algorithm for a given application is somewhat a matter of faith. A reasonable procedure is to acquire a standard package of implementations, discussed below, and then see what happens to your data under all of them.

Fortunately, there is a good, efficient heuristic for finding Steiner trees that works well on all versions of the problem. Construct a graph modeling your input, with the weight of edge (i, j) equal to the distance from point i to point j. Find a minimum spanning tree of this graph. You are guaranteed a provably good approximation for both Euclidean and rectilinear Steiner trees.

The worst case for a minimum spanning tree approximation of the Euclidean distance problem is three points forming an equilateral triangle. The minimum spanning tree will contain two of the sides (for a length of 2), whereas the minimum Steiner tree will connect the three points using an interior point, for a total length of $\sqrt{3}$. This ratio of $\sqrt{3}/2 \approx 0.866$ is always achieved, and in practice the easily-computed minimum spanning tree is usually within a few percent of the optimal Steiner tree. For rectilinear Steiner trees, the ratio with rectilinear minimum spanning trees is always $\geq 2/3 \approx 0.667$.

Such a minimum spanning tree can be refined by inserting a Steiner point whenever the edges of the minimum spanning tree incident on a vertex form an angle of less than 120 degrees between them. Inserting these points and locally readjusting the tree edges can move the solution a few more percent towards the optimum. Similar optimizations are possible for rectilinear spanning trees.

An alternative heuristic for graphs is based on shortest path. Start with a tree consisting of the shortest path between two terminals. For each remaining terminal t, find the shortest path to a vertex v in the tree and add this path to the tree. The time complexity and quality of this heuristic depend upon the insertion order of the terminals and how the shortest-path computations are performed, but something simple and fairly effective is likely to result.

Implementations: Salowe and Warme [SW95] have developed a program for computing exact rectilinear Steiner minimal trees. It is available by anonymous ftp from ftp.cs.virginia.edu in pub/french/salowe/newsteiner.tar.Z. It should be capable of handling up to 30 points routinely. A heuristic program by Robins and Zhang is available from the algorithm repository http://www.cs.sunysb.edu/~algorith.

PHYLIP is an extensive and widely used package of programs for inferring phylogenic trees. It contains over twenty different algorithms for constructing phylogenic trees from data. Although many of them are designed to work with molecular sequence data, several general methods accept arbitrary distance matrices as input. With versions written in C and Pascal, it is available on the WWW from http://evolution.genetics.washington.edu/phylip.html.

Notes: The most complete reference on the Steiner tree problem is the monograph by Hwang, Richards, and Winter [HRW92]. Surveys on the problem include [Kuh75]. Steiner tree problems arising in VLSI design are discussed in [KPS89, Len90]. Empirical results on Steiner tree heuristics include [SFG82, Vos92].

The Euclidean Steiner problem dates back to Fermat, who asked how to find a point p in the plane minimizing the sum of the distances to three given points. This was solved by Torricelli before 1640. Steiner was apparently one of several mathematicians who worked the general problem for n points, and he was mistakenly credited with the problem. An interesting, more detailed history appears in [HRW92].

Gilbert and Pollak [GP68] first conjectured that the ratio of the length of the minimum Steiner tree over the minimum spanning tree is always $\geq \sqrt{3}/2 \approx 0.866$. After twenty years of active research, the Gilbert-Pollak ratio was finally proven by Du and Hwang [DH92]. The Euclidean minimum spanning tree for n points in the plane can be constructed in $O(n \lg n)$ time [PS85].

Expositions on the proof that the Steiner tree problem for graphs is hard [Kar72] include [Eve79a]. Expositions on exact algorithms for Steiner trees in graphs include [Law76]. The hardness of Steiner tree for Euclidean and rectilinear metrics was established in [GGJ77, GJ77]. Euclidean Steiner tree is not known to be in NP, because of numerical issues in representing distances.

Analogies can be drawn between minimum Steiner trees and minimum energy configurations in certain physical systems. The case that such analog systems, including the behavior of soap films over wire frames, "solve" the Steiner tree problem is discussed in [Mie58].

Related Problems: Minimum spanning tree (see page 275), shortest path (see page 279).

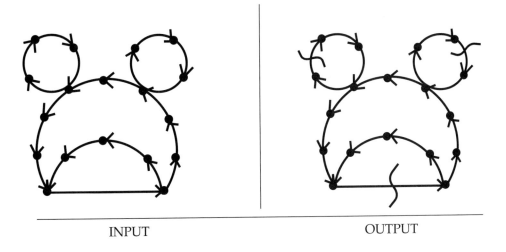

INPUT   OUTPUT

8.5.11   Feedback Edge / Vertex Set

Input description: A (directed) graph $G = (V, E)$.

Problem description: What is the smallest set of edges E' or vertices V' whose deletion leaves an acyclic graph?

Discussion: Feedback set problems arise because many algorithmic problems are much easier or much better defined on directed acyclic graphs than on arbitrary digraphs. Topological sorting (see Section 8.4.2) can be used to test whether a graph is a DAG, and if so, to order the vertices so as to respect the edges as precedence scheduling constraints. But how can you design a schedule if there are cyclic constraints, such as A must be done before B, which must be done before C, which must be done before A?

By identifying a feedback set, we identify the smallest number of constraints that must be dropped so as to permit a valid schedule. In the *feedback edge* (or arc) set problem, we drop precedence constraints (job A must come before job B). In the *feedback vertex set* problem, we drop entire jobs and any associated constraints. It is also referred to in the literature as the *maximum acyclic subgraph problem*.

- *Do any constraints have to be dropped?* – Not if the graph is a DAG, which can be tested via topological sort in linear time, Further, topological sorting also gives a simple way to find a feedback set if we modify the algorithm to delete the edge or vertex whenever a contradiction is found instead of simply printing a warning. The catch is that this feedback set might be much larger than needed, no surprise since both feedback edge set and feedback vertex set are NP-complete on directed graphs.

- *How can I find a good feedback edge set?* – A simple but effective linear-time heuristic constructs a vertex ordering, just as in the topological sort heuristic above, and deletes any arc going from right to left. This heuristic builds up the ordering from the outside in based on the in- and out-degrees of each vertex. Any vertex of in-degree 0 is a source and can be placed first in the ordering. Any vertex of out-degree 0 is a sink and can be placed last in the ordering, again without violating any constraints. If not, we find the vertex with the maximum difference between in- and out-degree, and place it on the side of the permutation that will salvage the greatest number of constraints. Delete any vertex from the DAG after positioning it and repeat until the graph is empty.

- *How can I find a good feedback vertex set?* – The following variant of the above heuristic should be effective. Keep any source or sink we encounter. If none exist, add to the feedback set a vertex v that maximizes $\min(v_{in}, v_{out})$, since this vertex is furthest from becoming either a source or sink. Again, delete any vertex from the DAG after positioning it and repeat until the graph is empty.

- *What if I want to break all cycles in an undirected graph?* – The problem of finding feedback sets from undirected graphs is different for digraphs, and in one case actually easier. An undirected graph without cycles is a tree. It is well known that any tree on n vertices has exactly $n - 1$ edges. Thus the smallest feedback edge set of any undirected graph is $|E| - (n - 1)$, and it can be found by deleting all the edges not in any given spanning tree. Spanning trees can be most efficiently constructed using depth-first search, as discussed in Section 8.4.1. The feedback vertex set problem remains NP-complete for undirected graphs, however.

Finding an optimal feedback set is NP-complete, and for most applications it is unlikely to be worth searching for the smallest set. However, in certain applications it would pay to try to refine the heuristic solutions above via randomization or simulated annealing. To move between states, we modify the vertex permutation by swapping pairs in order or inserting/deleting vertices into the feedback set.

Implementations: The econ_order program of the Stanford GraphBase (see Section 9.1.3) permutes the rows and columns of a matrix so as to minimize the sum of the numbers below the main diagonal. Using an adjacency matrix as the input and deleting all edges below the main diagonal leaves an acyclic graph.

Notes: The feedback set problem first arose in [Sla61]. Heuristics for feedback set problems include [BYGNR94, ELS93, Fuj96]. Expositions of the proofs that feedback minimization is hard [Kar72] include [AHU74, Eve79a]. Both feedback vertex and edge set remain hard even if no vertex has in-degree or out-degree greater than two [GJ79].

An interesting application of feedback arc set to economics is presented in [Knu94]. For each pair A, B of sectors of the economy, we are given how much money flows from A to B. We seek to order the sectors to determine which sectors are primarily producers to other sectors, and which deliver primarily to consumers.

Related Problems: Bandwidth reduction (see page 202), topological sorting (see page 273), scheduling (see page 263).

8.6 Computational Geometry

Computational geometry is the algorithmic study of geometric problems and objects. Compared to the other topics in this book, computational geometry emerged as a field quite recently, with Shamos's Ph.D. thesis [Sha78] typically cited as its founding event. Its emergence coincided with the explosion of computer graphics and windowing systems, which directly or indirectly provide much of the motivation for geometric computing. The past twenty years have seen enormous growth in computational geometry, resulting in a significant body of useful algorithms, software, textbooks, and research results.

Good books on computational geometry include:

- *Preparata and Shamos* [PS85] – Although aging a bit, this book remains the best general introduction to computational geometry, stressing algorithms for convex hulls, Voronoi diagrams, and intersection detection.

- *O'Rourke* [O'R94] – Perhaps the best practical introduction to computational geometry. The emphasis is on careful and correct implementation (in C language) of the fundamental algorithms of computational geometry. These implementations are available from http://grendel.csc.smith.edu/~orourke/.

- *Edelsbrunner* [Ede87] – This is the definitive book on arrangements, a topic that runs through most of computational geometry. Although not appropriate for beginners, it provides an important perspective for advanced geometers.

- *Mulmuley* [Mul94] – An approach to computational geometry through randomized incremental algorithms. Very interesting, but likely too narrow to serve as a general introduction.

- *Nievergelt and Hindrichs* [NH93] – This idiosyncratic algorithms text focuses on problems in graphics and geometry. Good coverage of line drawing, intersection algorithms, and spatial data structures, but with too many topics touched on too lightly to serve as an effective reference.

The leading conference in computational geometry is the *ACM Symposium on Computational Geometry*, held annually in late May or early June. Although the primary results presented at the conference are theoretical, there has been a concerted effort on the part of the research community to increase the presence of applied, experimental work through video reviews and poster sessions. The other major annual conference is the *Canadian Conference on Computational Geometry* (CCCG), typically held in early August. Useful literature surveys include [Yao90].

A unique source of computational geometry information is geom.bib, a community effort to maintain a complete bibliography on computational geometry. It references over eight thousand books, papers, and reports and includes detailed abstracts for many of them. Grep-ing through the geom.bib is an amazingly efficient way to find out about previous work without leaving your office. It is available via anonymous ftp from ftp.cs.usask.ca, in the file pub/geometry/geombib.tar.Z, and included on the enclosed CD-ROM.

There is a growing body of implementations of geometric algorithms. We point out specific implementations where applicable in the catalog, but the reader should be aware of three specific WWW sites:

- The Geometry Center's directory of computational geometry software, maintained by Nina Amenta, is the "official" site for computational geometry software. Check here first to see what is available: http://www.geom.umn.edu/software/cglist/.

- *Graphics Gems* is a series of books dedicated to collecting small codes of interest in computer graphics. Many of these programs are geometric in nature. All associated codes are available from ftp://ftp-graphics.stanford.edu/pub/Graphics/GraphicsGems.

- CGAL (Computational Geometry Algorithms Library) is a joint European project now underway to produce a comprehensive library of geometric algorithms. This will likely become the definitive geometric software project. Check out its progress at http://www.cs.ruu.nl/CGAL/.

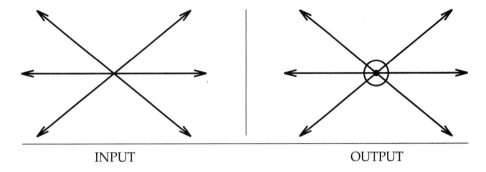

INPUT	OUTPUT

8.6.1 Robust Geometric Primitives

Input description: A point p and a line segment l, or two line segments l_1 and l_2.

Problem description: Does p lie over, under, or on l? Does l_1 intersect l_2?

Discussion: Implementing basic geometric primitives is a task fraught with peril, even for such simple tasks as returning the intersection point of two lines. What should you return if the two lines are parallel, meaning they don't intersect at all? What if the lines are identical, so the intersection is not a point but the entire line? What if one of the lines is horizontal, so that in the course of solving the equations for the intersection point you are likely to divide by zero? What if the two lines are almost parallel, so that the intersection point is so far from the origin as to cause arithmetic overflows? These issues become even more complicated for intersecting line segments, since there are a bunch of other special cases that must be watched for and treated specially.

If you are new to implementing geometric algorithms, I suggest that you study O'Rourke's *Computational Geometry in C* [O'R94] for practical advice and complete implementations of basic geometric algorithms and data structures. You are likely to avoid many headaches by following in his footsteps.

There are two different issues at work here: geometric degeneracy and numerical stability. Degeneracy refers to annoying special cases that must be treated in substantially different ways, such as when two lines intersect in more or less than a single point. There are three primary approaches to dealing with degeneracy:

- *Ignore it* – Make as an operating assumption that your program will work correctly only if no three points are collinear, no three lines meet at a point, no intersections happen at the endpoints of line segments, etc. This is probably the most common approach, and what I would recommend for short-term projects if you can live with frequent crashes. The drawback is that interesting data often comes from points sampled on a grid and is inherently very degenerate.

- *Fake it* – Randomly or symbolically perturb your data so that it seems non-degenerate. By moving each of your points a small amount in a random direction, you can break many of the existing degeneracies in the data, hopefully without creating too many new problems. This probably should be the first thing to try as soon as you decide that your program is crashing too often. One problem with random perturbations is that they do change the shape of your data in subtle ways, which may be intolerable for your application. There also exist techniques to "symbolically" perturb your data to remove degeneracies in a consistent manner, but these require serious study to apply correctly.

- *Deal with it* – Geometric applications can be made more robust by writing special code to handle each of the special cases that arise. This can work well if done with care at the beginning, but not so well if kludges are added whenever the system crashes. Expect to expend a lot of effort if you are determined to do it right.

Geometric computations often involve floating-point arithmetic, which leads to problems with overflows and numerical precision. There are three basic approaches to the issue of numerical stability:

- *Integer arithmetic* – By forcing all points of interest to lie on a fixed-size integer grid, you can perform exact comparisons to test whether any two points are equal or two line segments intersect. The cost is that the intersection point of two lines may not be exactly representable. This is likely to be the simplest and best method, if you can get away with it.

- *Double precision reals* – By using double-precision floating point numbers, you may get lucky and avoid numerical errors. Your best bet might be to keep all the data as single-precision reals, and use double-precision for intermediate computations.

- *Arbitrary precision arithmetic* – This is certain to be correct, but also to be slow. This approach seems to be gaining favor in the research community with the observation that careful analysis can minimize the need for high-precision arithmetic, and thus the performance penalty. Still, you should expect high-precision arithmetic to be several orders of magnitude slower than standard floating-point arithmetic.

The difficulties associated with producing robust geometric software are still under attack by researchers. The best practical technique is to base your applications on a small set of geometric primitives that handle as much of the low-level geometry as possible. These primitives include:

- *Area of a triangle* – Although it is well-known that the area $A(t)$ of a triangle $t = (a, b, c)$ is half the base times the height, computing the length of the base and altitude is messy work with trigonometric functions. It is better to use the determinant formula for *twice* the area:

$$2 \cdot A(t) = \begin{vmatrix} a_x & a_y & 1 \\ b_x & b_y & 1 \\ c_x & c_y & 1 \end{vmatrix} = a_x b_y - a_y b_x + a_y c_x - a_x c_y + b_x c_y - c_x b_y$$

This formula generalizes to compute $d!$ times the volume of a simplex in d dimensions. Thus $3! = 6$ times the volume of a tetrahedron $t = (a, b, c, d)$ in three dimensions is

$$6 \cdot A(t) = \begin{vmatrix} a_x & a_y & a_z & 1 \\ b_x & b_y & b_z & 1 \\ c_x & c_y & c_z & 1 \\ d_x & d_y & d_z & 1 \end{vmatrix}$$

Note that these are signed volumes and can be negative, so take the absolute value first. Section 8.2.4 explains how to compute determinants.

The conceptually simplest way to compute the area of a polygon (or polyhedron) is to triangulate it and then sum up the area of each triangle. An implementation of a slicker algorithm that avoids triangulation is discussed in [O'R94].

- *Above-below-on test* – Does a given point c lie above, below, or on a given line l? A clean way to deal with this is to represent l as a directed line that passes through point a before point b, and ask whether c lies to the left or right of the directed line l. It is up to you to decide whether left means above or below.

  This primitive can be implemented using the sign of the area of a triangle as computed above. If the area of $t(a, b, c) > 0$, then c lies to the left of $\overline{ab}$. If the area of $t(a, b, c) = 0$, then c lies on $\overline{ab}$. Finally, if the area of $t(a, b, c) < 0$, then c lies to the right of $\overline{ab}$. This generalizes naturally to three dimensions, where the sign of the area denotes whether d lies above or below the oriented plane (a, b, c).

- *Line segment intersection* – The above-below primitive can be used to test whether a line intersects a line segment. It does iff one endpoint of the segment is to the left of the line and the other is to the right. Segment intersection is similar but more complicated, and we refer you to implementations described below. The decision whether two segments intersect if they share an endpoint depends upon your application and is representative of the problems of degeneracy.

- *In-circle test* – Does the point d lie inside or outside the circle defined by points a, b, and c in the plane? This primitive occurs in all Delaunay triangulation algorithms and can be used as a robust way to do distance comparisons. Assuming that a, b, c are labeled in counterclockwise order around the circle, compute the determinant:

$$\text{incircle}(a, b, c, d) = \begin{vmatrix} a_x & a_y & a_x^2 + a_y^2 & 1 \\ b_x & b_y & b_x^2 + b_y^2 & 1 \\ c_x & c_y & c_x^2 + c_y^2 & 1 \\ d_x & d_y & d_x^2 + d_y^2 & 1 \end{vmatrix}$$

Incircle will return 0 if all four points are cocircular, a positive value if d is inside the circle, and negative if d is outside.

Check out the implementations described below before you attempt to build your own.

Implementations: LEDA (see Section 9.1.1) provides a very complete set of geometric primitives for planar geometry, written in C++. If you are writing a significant geometric application, you should consider basing it on LEDA. At least check them out before you try to write your own.

O'Rourke [O'R94] provides implementations in C of most of the primitives discussed in this section. See Section 9.1.6.5. These primitives were implemented primarily for exposition rather than production use, but they should be quite reliable and might be more appropriate than LEDA for small applications.

A robust implementation of the basic geometric primitives in C++ using exact arithmetic, by Jonathan Shewchuk, is available at http://www.cs.cmu.edu/~quake/robust.html. Don't expect them to be very fast.

Pascal implementations of basic geometric primitives appear in [MS91]. Sedgewick [Sed92] provides fragments of the basic primitives in C++. See Section 9.1.6.6 for both of them.

An alternative C++ library of geometric algorithms and data structures (although you are almost certainly better off sticking to LEDA) is *Geolab*, written by Pedro J. de Rezende, Welson R. Jacometti, Cesar N. Gon, and Laerte F. Morgado, Universidade Estadual de Campinas, Brazil. Geolab requires the SUN C++ compiler, but a Sparc binary and visualization environment is included along with all source code. Geolab appears to be primarily for the brave, since its robustness is uncertain and it contains little documentation, but it does provide 40 algorithms, including such advanced topics as farthest point Voronoi diagrams, nearest neighbor search, and ray shooting.

Notes: O'Rourke [O'R94] provides an implementation-oriented introduction to computational geometry, which stresses robust geometric primitives and is recommended reading.

Shewchuk [She96] and Fortune and van Wyk [FvW93] present careful studies on the costs of using arbitrary-precision arithmetic for geometric computation. By being careful about when to use it, reasonable efficiency can be maintained while achieving complete robustness. Other approaches to achieving robustness include [DS88, Hof89, Mil89].

Related Problems: Intersection detection (see page 370), maintaining arrangements (see page 392).

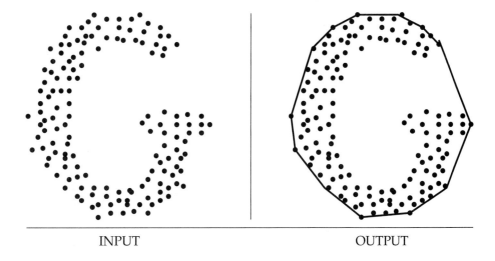

<table>
<tr><td>INPUT</td><td>OUTPUT</td></tr>
</table>

8.6.2 Convex Hull

Input description: A set S of n points in d-dimensional space.

Problem description: Find the smallest convex polygon containing all the points of S.

Discussion: Finding the convex hull of a set of points is *the* most elementary interesting problem in computational geometry, just as minimum spanning tree is the most elementary interesting problem in graph algorithms. It arises because the hull quickly captures a rough idea of the shape or extent of a data set.

Convex hull also serves as a first preprocessing step to many, if not most, geometric algorithms. For example, consider the problem of finding the diameter of a set of points, which is the pair of points a maximum distance apart. The diameter will always be the distance between two points on the convex hull. The $O(n \lg n)$ algorithm for computing diameter proceeds by first constructing the convex hull, then for each hull vertex finding which other hull vertex is farthest away from it. This so-called "rotating-calipers" method can be used to move efficiently from one hull vertex to another.

There are almost as many convex hull algorithms as there are sorting algorithms. Answer the following questions to help choose between them:

- *How many dimensions are you working with?* – Convex hulls in two and even three dimensions are fairly easy to work with. However, as the dimension of a space increases, certain assumptions that were valid in lower dimensions break down. For example, any n-vertex polygon in two dimensions has exactly n edges. However, the relationship between the numbers of faces and vertices is more complicated even in three dimensions. A cube has 8 vertices and 6 faces, while an octahedron has 8 faces and 6 vertices. This has implications for data structures that represent hulls – are you just looking for the hull points or do you need the defining polyhedron? The need to find convex hulls in high-dimensional spaces arises in many applications, so be aware of such complications if your problem takes you there.

Gift-wrapping is the basic algorithm for constructing higher-dimensional convex hulls. Observe that a three-dimensional convex polyhedron is composed of two-dimensional faces, or *facets*, which are connected by one-dimensional lines, or *edges*. Each edge joins exactly two facets together. Gift-wrapping starts by finding an initial facet associated with the lowest vertex and then conducting a breadth-first search from this facet to discover new, additional facets. Each edge e defining the boundary of a facet must be shared with one other facet, so by running through each of the n points we can identify which point defines the next facet with e. Essentially, we "wrap" the points one facet at a time by bending the wrapping paper around an edge until it hits the first point.

The key to efficiency is making sure that each edge is explored only once. Implemented properly in d dimensions, gift-wrapping takes $O(n\phi_{d-1} + \phi_{d-2} \lg \phi_{d-2})$, where ϕ_{d-1} is the number of facets and ϕ_{d-2} is the number of edges in the convex hull. Thus gift-wrapping can be very efficient when there is only a constant number of facets on the hull. However, this can be as bad as $O(n^{\lfloor d/2 \rfloor + 1})$ when the convex hull is very complex.

Better convex hull algorithms are available for the important special case of three dimensions, where $O(n \log n)$ time in fact suffices. For three or higher dimensions, I recommend that you use one of the codes described below rather than roll your own.

- *Is your data given as vertices or half-spaces?* – The problem of finding the intersection of a set of n half-spaces in d dimensions is dual to that of computing convex hulls of n points in d dimensions. Thus the same basic algorithm suffices for both problems. The necessary duality transformation is discussed in Section 8.6.15.

- *How many points are likely to be on the hull?* – If your points are selected "randomly", it is likely that most of them lie within the interior of the hull. Planar convex hull programs can be made more efficient in practice using the observation than the leftmost, rightmost, topmost, and bottommost points must all be on the convex hull. Unless the topmost is leftmost and bottommost is rightmost, this gives a set of either three or four distinct points, defining a triangle or quadrilateral. Any point inside this region *cannot* be on the convex hull and can be discarded in a linear sweep through the points. Ideally, only a few points will then remain to run through the full convex hull algorithm.

  This trick can also be applied beyond two dimensions, although it loses effectiveness as the dimension increases.

- *How do I find the shape of my point set?* – Although convex hulls provide a gross measure of shape, any details associated with concavities are lost. For example, the shape of the 'G' would be indistinguishable from the shape of the 'O'. A more general structure, called *alpha-shapes*, can be parameterized so as to retain arbitrarily large concavities. Implementations and references on alpha-shapes are included below.

The primary convex hull algorithm in the plane is the *Graham scan*. Graham scan starts with one point p known to be on the convex hull (say the point with lowest x-coordinate) and sorts the rest of the points in angular order around p. Starting with a hull consisting of p and the point with the smallest angle, we proceed counterclockwise around v adding points. If the angle formed by the new point and the last hull edge is less than 180 degrees, we add this new point to the hull. If the angle formed by the new point and the last "hull" edge is greater than 180 degrees, then a chain of vertices starting from the last hull edge must be deleted to maintain convexity. The total time is $O(n \lg n)$, since the bottleneck is sorting the points around v.

The basic Graham scan procedure can also be used to construct a nonself-intersecting (or *simple*) polygon passing through all the points. Sort the points around v, but instead of testing angles simply connect the points in angular order. Connecting this to v gives a polygon without self-intersection, although it typically has many skinny protrusions.

The gift-wrapping algorithm becomes especially simple in two dimensions, since each "facet" becomes an edge, each "edge" becomes a vertex of the polygon, and the "breadth-first search" simply walks around the hull in a clockwise or counterclockwise order. The 2D gift-wrapping, or *Jarvis march*, algorithm runs in $O(nh)$ time, where h is the number of vertices on the convex hull. I would recommend sticking with Graham scan unless you *really* know in advance that there cannot be too many vertices on the hull.

Implementations: O'Rourke [O'R94] provides a robust implementation of the Graham scan in two dimensions and an $O(n^2)$ implementation of an incremental algorithm for convex hulls in three dimensions. Both are written in C. The latter has been proven capable of solving 10,000-point problems in a few minutes on a modern workstation. See Section 9.1.6.5.

Qhull [BDH97] appears to be the convex hull code of choice for general dimensions (in particular from 2 to about 8 dimensions). It is written in C and can also construct Delaunay triangulations, Voronoi vertices, furthest-site Voronoi vertices, and half-space intersections. Qhull has been widely used in scientific applications and has a well-maintained home page at http://www.geom.umn.edu/software/qhull/.

An alternative higher-dimensional convex hull code in ANSI C is Ken Clarkson's Hull, available at http://www.cs.att.com/netlib/voronoi/hull.html. It does not appear to be as widely used or actively maintained as Qhull, but it also does alpha-shapes. For an excellent alpha-shapes code, originating from the work of Edelsbrunner and Mucke [EM94], check out http://fiaker.ncsa.uiuc.edu/alpha/.

Fukuda's cdd program is the best choice for nonsimplicial polytopes in about 6D and higher. See ftp://ifor13.ethz.ch/pub/fukuda/cdd/. It may be used for computing convex hulls and half-space intersection.

XTango (see Section 9.1.5) provides animations of the Graham scan and Jarvis march algorithms in the plane.

A Pascal implementation of Graham scan appears in [MS91]. See Section 9.1.6.4. C++ implementations of planar convex hulls includes LEDA (see Section 9.1.1). Algorithm 523 [Edd77] of the *Collected Algorithms of the ACM* is a Fortran code for planar convex hulls. It is available from Netlib (see Section 9.1.2).

Notes: Constructing planar convex hulls plays a similar role in computational geometry as sorting does in algorithm theory. Like sorting, convex hull is a fundamental problem for which a wide variety of different algorithmic approaches lead to interesting or optimal algorithms. Preparata and Shamos [PS85] give a good exposition of several such algorithms, including quickhull and mergehull, both inspired by the sorting algorithms. In fact, a simple construction involving points on a parabola reduces sorting to convex hull, so the information-theoretic lower bound for sorting implies that planar convex hull requires $\Omega(n \lg n)$ time to compute. A stronger lower bound is established in [Yao81].

Good expositions of the Graham scan algorithm [Gra72] and the Jarvis march [Jar73] include [CLR90, PS85]. The gift wrapping algorithm was introduced by Chand and Kapur [CK70]. Noteworthy among planar convex hull algorithms is Seidel and Kirkpatrick [KS86], which takes $O(n \lg h)$ time, where h is the number of hull vertices, which captures the best performance of both Graham scan and gift wrapping and is (theoretically) better in between.

Alpha-hulls, introduced in [EKS83], provide a useful notion of the shape of a point set. A generalization to three dimensions, with an implementation, is presented in [EM94].

Reverse-search algorithms for constructing convex hulls are effective in higher dimensions [AF92], although constructions demonstrating the poor performance of convex hull algorithms for nonsimplicial polytopes are presented in [AB95]. Through a clever lifting-map construction [ES86], the problem of building Voronoi diagrams in d-dimensions can be reduced to constructing convex hulls in $(d + 1)$-dimensions. See Section 8.6.4 for more details.

Dynamic algorithms for convex-hull maintenance are data structures that permit inserting and deleting arbitrary points while always representing the current convex hull. The first such dynamic data structure [OvL81] supported insertions and deletions in $O(\lg^2)$ time. Expositions of this result include [PS85].

Related Problems: Sorting (see page 236), Voronoi diagrams (see page 358).

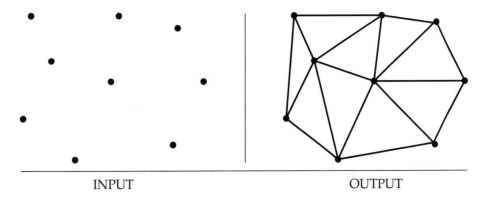

| INPUT | OUTPUT |

8.6.3 Triangulation

Input description: A set of points or a polyhedon.

Problem description: Partition the interior of the point set or polyhedron into triangles.

Discussion: Triangulation is a fundamental problem in computational geometry, because the first step in working with complicated geometric objects is to break them into simple geometric objects. The simplest geometric objects are triangles in two dimensions, and tetrahedra in three. Classical applications of triangulation include finite element analysis and computer graphics.

A particularly interesting application of triangulation is surface or function interpolation. Suppose that we have sampled the height of a mountain at a certain number of points. How can we estimate the height at any point q in the plane? If we project the points on the plane, and then triangulate them, the triangulation completely partitions the plane into regions. We can estimate the height of q by interpolating among the three points of the triangle that contains it. Further, this triangulation and the associated height values define a surface of the mountain suitable for graphics rendering.

In the plane, a triangulation is constructed by adding nonintersecting chords between the vertices until no more such chords can be added. Specific issues arising in triangulation include:

- *Does the shape of the triangles in your triangulation matter?* – There are usually many different ways to partition your input into triangles. Consider a set of n points in convex position in the plane. The simplest way to triangulate them would be to add to the convex hull diagonals from the first point to all of the others. However, this has the tendency to create skinny triangles.

  If the shape of the triangles matters for your application, you will usually want to avoid skinny triangles, or equivalently, small angles in the triangulation. The *Delaunay triangulation* of a point set minimizes the maximum angle over all possible triangulations. This isn't exactly what we are looking for, but it is pretty close, and the Delaunay triangulation has enough other interesting properties (including that it is dual to the Voronoi diagram) to make it the quality triangulation of choice. Further, it can be constructed in $O(n \lg n)$ time, with implementations described below.

- *What dimension are we working in?* – As always, three-dimensional problems are harder than two-dimensional problems. The three-dimensional generalization of triangulation involves partitioning the space into four-vertex tetrahedra by adding nonintersecting faces. One important difficulty is that for certain polyhedra there is no way to tetrahedralize the interior without adding extra vertices. Further, it is NP-complete to decide whether such a tetrahedralization exists, so we should not feel afraid to add extra vertices to simplify our problem.

- *What constraints does the input have?* – When we are triangulating a polygon or polyhedra, we are restricted to adding chords that do not intersect any of the boundary facets. In general, we may have a set of obstacles or constraints that cannot be intersected by inserted chords. The best such triangulation is likely to be the so-called *constrained Delaunay triangulation*. Implementations are described below.

- *Are you allowed to add extra points?* – When the shape of the triangles does matter, it might pay to strategically add a small number of extra "Steiner" points to the data set to facilitate the construction of a triangulation (say) with no small angles. As discussed above, there may be *no* triangulation possible for certain polyhedra without adding Steiner points.

To construct a triangulation of a convex polygon in linear time, just pick an arbitrary starting vertex v and insert chords from v to each other vertex in the polygon. Because the polygon is convex, we can be confident that none of the boundary edges of the polygon will be intersected by these chords and that all of them lie within the polygon. The simplest algorithm for constructing general polygon triangulations tries each of the $O(n^2)$ possible chords and inserts them if they do not intersect a boundary edge or previously inserted chord. There are practical algorithms that run in $O(n \lg n)$ time and theoretically interesting algorithms that run in linear time. See the implementations and notes below for details.

Implementations: Triangle, by Jonathan Shewchuk of Carnegie-Mellon University, is a C language code that generates Delaunay triangulations, constrained Delaunay triangulations (forced to have certain edges), and quality-conforming Delaunay triangulations (which avoid small angles by inserting extra points). It has been widely used for finite element analysis and other applications and is fast and robust. Triangle is the first thing I would try if I needed a two-dimensional triangulation code. Although Triangle is available at http://www.cs.cmu.edu/~quake/triangle.html, it is copyrighted by the author and may not be sold or included in commercial products without a license.

GEOMPACK is a suite of Fortran 77 codes by Barry Joe of the University of Alberta, for 2- and 3-dimensional triangulation and convex decomposition problems. In particular, it does both Delaunay triangulation and convex decompositions of polygonal and polyhedral regions, as well as arbitrary-dimensional Delaunay triangulations. They can be obtained from ftp://ftp.cs.ualberta.ca/pub/geompack.

Steve Fortune is the author of a widely used 2D code for Voronoi diagrams and Delaunay triangulations, written in C. This code is smaller and probably simpler to work with than either of the above, if all you need is the Delaunay triangulation of points in the plane. It is based on Fortune's own sweepline algorithm [For87] for Voronoi diagrams and is available from Netlib (see Section 9.1.2) at http://netlib.bell-labs.com/netlib/voronoi/index.html.

O'Rourke [O'R94] provides asymptotically slow implementations in C of polygon triangulation (in $O(n^3)$) and Delaunay triangulation (in $O(n^4)$). These will be unusable for more than modest numbers of points, but see Section 9.1.6.5 if interested. See Section 9.1.6.5.

Algorithm 624 [Ren84] of the *Collected Algorithms of the ACM* is a Fortran implementation of triangulation for surface interpolation. See Section 9.1.2. A linear-time implementation for triangulating a planar map is included with LEDA (see Section 9.1.1).

Higher-dimensional Delaunay triangulations are a special case of higher-dimensional convex hulls, and Qhull [BDH97] appears to be the convex hull code of choice for general dimensions (i.e. three dimensions and beyond). It is written in C, and it can also construct Voronoi vertices, furthest-site Voronoi vertices, and half-space intersections. Qhull has been widely used in scientific applications and has a well-maintained home page at http://www.geom.umn.edu/software/qhull/.

Notes: After a long search, Chazelle [Cha91] discovered a linear-time algorithm for triangulating a simple polygon. This algorithm is quite hopeless to implement. The first $O(n \lg n)$ algorithm for polygon triangulation was given by [GJPT78]. An $O(n \lg \lg n)$ algorithm by Tarjan and Van Wyk [TW88] followed before Chazelle's result. Expositions on polygon and point set triangulation include [O'R94, PS85].

Linear-time algorithms for triangulating monotone polygons have been long known [GJPT78] and are the basis of algorithms for triangulating simple polygons. A polygon is monotone when there exists a direction d such that any line with slope d intersects the polygon in at most two points.

A heavily studied class of optimal triangulations seeks to minimize the total length of the chords used. The computational complexity of constructing this *minimum weight triangulation* is a long-standing open problem in computational geometry, so the interest has shifted to provably good approximation algorithms. The minimum weight triangulation of a convex polygon can be found in $O(n^3)$ time using dynamic programming, as discused in Section 3.1.5.

Related Problems: Voronoi diagrams (see page 358), polygon partitioning (see page 380).

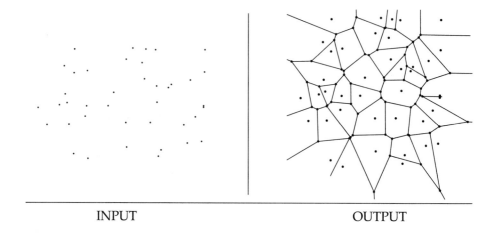

INPUT   OUTPUT

8.6.4   Voronoi Diagrams

Input description: A set S of points $p_1, \ldots, p_n$.

Problem description: Decompose space into regions around each point such that all the points in the region around p_i are closer to p_i than they are to any other point in S.

Discussion: Voronoi diagrams represent the region of influence around each of a given set of sites. If these sites represent the locations of McDonald's restaurants, the Voronoi diagram partitions space into cells around each restaurant. For each person living in a particular cell, the defining McDonald's represents the closest place to get a Big Mac.

Voronoi diagrams have a surprising variety of uses:

- *Nearest neighbor search* – For a query point q, finding its nearest neighbor from a fixed set of points S is simply a matter of determining which cell in the Voronoi diagram of S contains q. See Section 8.6.5 for more details.

- *Facility location* – Suppose McDonald's wanted to open another restaurant. To minimize interference with existing McDonald's, it should be located as far away from the closest restaurant as possible. This location is always at a vertex of the Voronoi diagram, and it can be found in a linear-time search through all the Voronoi vertices.

- *Largest empty circle* – Suppose you needed to obtain a large, contiguous, undeveloped piece of land on which to build a factory. The same condition used for picking McDonald's locations is appropriate for other undesirable facilities, namely that it be as far as possible from any relevant sites of interest. A Voronoi vertex defines the center of the largest empty circle among the points.

- *Path planning* – If the sites of S are the centers of obstacles we seek to avoid, the edges of the Voronoi diagram define the possible channels that maximize the distance to the obstacles. Thus in planning paths among the sites, it will be "safest" to stick to the edges of the Voronoi diagram.

- *Quality triangulations* – In triangulating a set of points, we often desire nice, fat triangles, which avoid small angles and skinny triangles. The *Delaunay triangulation* maximizes the minimum angle over all triangulations and is exactly what we want. Further, it is easily constructed as the dual of the Voronoi diagram. See Section 8.6.3 for details.

Each edge of a Voronoi diagram is part of the perpendicular bisector of two points in S, since this is the line that partitions the plane between the points. The conceptually simplest method to construct Voronoi diagrams is randomized incremental construction. To add another site to the diagram, locate the cell that contains it and add the perpendicular bisectors between the new site and all sites defining impacted regions. When the sites are inserted in random order, only a small number of regions are likely to be impacted.

However, the method of choice is Fortune's sweepline algorithm, especially since robust implementations of it are readily available. Use an existing implementation instead of trying to develop your own. The algorithm works by projecting the set of sites in the plane into a set of cones in three dimensions such that the Voronoi diagram is defined by projecting the cones back onto the plane. The advantages of Fortune's algorithm are (1) it runs in optimal $\Theta(n \log n)$ time, (2) it is reasonable to implement, and (3) we need not store the entire diagram if we can use it as we sweep over it.

There is an interesting relationship between convex hulls in $d + 1$ dimensions and Delaunay triangulations (or equivalently Vornoi diagrams) in d-dimensions, which provides the best way to construct Voronoi diagrams in higher dimensions. By projecting each site in E^d $(x_1, x_2, \ldots, x_d)$ into the point $(x_1, x_2, \ldots, x_d, \sum_{i=1}^d x_i^2)$, taking the convex hull of this $(d + 1)$-dimensional point set and then projecting back into d dimensions we obtain the Delaunay triangulation. Details are given in the references below. Programs that compute higher-dimensional convex hulls are discussed in Section 8.6.2.

Several important variations of standard Voronoi diagrams arise in practice. See the references below for details:

- *Non-Euclidean distance metrics* – Recall that the idea of a Voronoi diagram is to decompose space into regions of influence around each of the given sites. Thus far, we have assumed that Euclidean distance measures influence, but for many applications this is inappropriate. If people drive to McDonald's, the time it takes to get there depends upon where the major roads are. Efficient algorithms are known for constructing Voronoi diagrams under a variety of different metrics, and for curved or constrained objects.

- *Power diagrams* – These structures decompose space into regions of influence around the sites, where the sites are no longer constrained to have all the same power. Imagine a map of the listening range of a set of radio stations operating at a given frequency. The region of influence around a station depends both on the power of its transmitter and the position of neighboring transmitters.

- *kth-order and furthest-site diagrams* – The idea of decomposing space into regions sharing some property can be taken beyond closest-point Voronoi diagrams. Any point within a single cell of the kth-order Voronoi diagram shares the same set of k closest points in S. In furthest-site diagrams, any point within a particular region shares the same furthest point in S. Point location (see Section 8.6.7) on these structures permits fast retrieval of the appropriate points.

Implementations: Steve Fortune is the author of a widely used 2D code for Voronoi diagrams and Delaunay triangulations, written in C. It is based on his own sweepline algorithm [For87] for Voronoi diagrams and is likely to be the right code to try first. It is available from Netlib (see Section 9.1.2) at http://netlib.bell-labs.com/netlib/voronoi/index.html.

LEDA (see Section 9.1.1) provides an implementation of a randomized incremental construction algorithm for planar Voronoi diagrams in C++.

Higher-dimensional and furthest-site Voronoi diagrams can be constructed as a special case of higher-dimensional convex hulls. Qhull [BDH97] appears to be the convex hull code of choice in three dimensions and beyond. It is written in C, and it can also construct Delaunay triangulations and half-space intersections. Qhull has been widely used in scientific applications and has a well-maintained home page at http://www.geom.umn.edu/software/qhull/.

The Stanford GraphBase (see Section 9.1.3) contains an implementation of a randomized incremental algorithm to construct Voronoi diagrams and Delaunay triangulations for use as a generator of planar graph instances.

Algorithm 558 [Che80] of the *Collected Algorithms of the ACM* is a Fortran code for the multifacility location problem. It is based on a network flow approach, instead of using Voronoi diagrams. Interestingly, the network flow code is taken from Nijenhuis and Wilf (see Section 9.1.6.3). See Section 9.1.2.

Notes: Voronoi diagrams were studied by Dirichlet in 1850 and are occasionally referred to as *Dirichlet tessellations*. They are named after G. Voronoi, who discussed them in a 1908 paper. In mathematics, concepts get named after the last person to discover them.

Aurenhammer [Aur91] and Fortune [For92] provide excellent surveys on Voronoi diagrams and associated variants such as power diagrams. The first $O(n \lg n)$ algorithm for constructing Voronoi diagrams was based on divide-and-conquer and is due to Shamos and Hoey [SH75]. Good expositions of Fortune's sweepline algorithm for constructing Voronoi diagrams in $O(n \lg n)$ [For87] include [O'R94]. Good expositions on the relationship between Delaunay triangulations and $(d + 1)$-dimensional convex hulls [ES86] include [O'R94].

In a kth-order Voronoi diagram, we partition the plane such that each point in a region is closest to the same set of k sites. Using the algorithm of [ES86], the complete set of kth-order Voronoi diagrams can be constructed in $O(n^3)$ time. By doing point location on this structure, the k nearest neighbors to a query point can be found in $O(k + \lg n)$. Expositions on kth-order Voronoi diagrams include [O'R94, PS85].

The smallest enclosing circle problem can be solved in $O(n \lg n)$ time using $(n - 1)$st order Voronoi diagrams [PS85]. In fact, there exist linear-time algorithms based on low-dimensional linear programming [Meg83]. A linear algorithm for computing the Voronoi diagram of a convex polygon is given by [AGSS89].

Related Problems: Nearest neighbor search (see page 361), point location (see page 367), triangulation (see page 355).

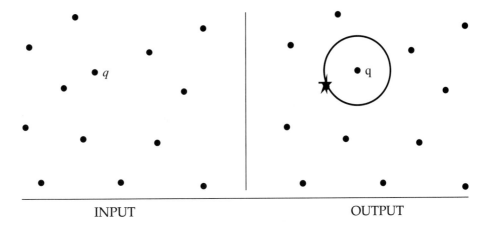

| INPUT | OUTPUT |

8.6.5 Nearest Neighbor Search

Input description: A set S of n points in d dimensions; a query point q.

Problem description: Which point in S is closest to q?

Discussion: The need to quickly find the nearest neighbor to a query point arises in a variety of geometric applications. The classic example in two dimensions is designing a system to dispatch emergency vehicles to the scene of a fire. Once the dispatcher learns the location of the fire, she uses a map to find the firehouse closest to this point so as to minimize transportation delays. This situation occurs in any application mapping customers to service providers.

Nearest-neighbor search is also important in classification. Suppose we are given a collection of data about people (say age, height, weight, years of education, sex, and income level) each of whom has been labeled as Democrat or Republican. We seek a classifier to decide which way a different person is likely to vote. Each of the people in our data set is represented by a party-labeled point in d-dimensional space. A simple classifier can be built by assigning to the new point the party affiliation of its nearest neighbor.

Such nearest-neighbor classifiers are widely used, often in high-dimensional spaces. The vector-quantization method of image compression partitions an image into 8×8 pixel regions. This method uses a predetermined library of several thousand 8×8 pixel tiles and replaces each image region by the most similar library tile. The most similar tile is the point in 64-dimensional space that is closest to the image region in question. Compression is achieved by reporting the identifier of the closest library tile instead of the 64 pixels, at some loss of image fidelity.

Issues arising in nearest-neighbor search include:

- *How many points are you searching?* – When your data set contains only a small number of points (say $n \leq 100$) or if only few queries are ever destined to be performed, the simple approach is best. Compare the query point q against each of the n data points. Only when fast queries are needed for large numbers of points does it pay to consider more sophisticated methods.

- *How many dimensions are you working in?* – Nearest neighbor search gets slowly but progressively harder as the dimensionality increases. The *kd*-tree data structure, presented in Section 8.1.6, does a very good job in moderate-dimensional spaces, even the plane. Still, above 20 or so dimensions, you might as well do a linear search through the data points. Search in high-dimensional spaces becomes hard because a sphere of radius r, representing all the points with distance $\leq r$ from the center, progressively fills up less volume relative to a cube as the dimensionality increases. Thus any data structure based on partitioning points into enclosing volumes will become less and less effective.

  In two dimensions, Voronoi diagrams (see Section 8.6.4) provide an efficient data structure for nearest-neighbor queries. The Voronoi diagram of a point set in the plane decomposes the plane into regions such that the cell containing each data point consists of the part of the plane that is nearer to that point than any other in the set. Finding the nearest neighbor of query point q reduces to finding which cell in the Voronoi diagram contains q and reporting the data point associated with it. Although Voronoi diagrams can be built in higher dimensions, their size rapidly grows to the point of unusability.

- *Is your data set static or dynamic?* – Will there be occasional insertions or deletions of new data points in your application? If these are just rare events, it might pay to build your data structure from scratch each time. If they are frequent, select a version of the kd-tree that supports insertions and deletions.

The nearest neighbor graph on a set S of n points links each vertex to its nearest neighbor. This graph is a subgraph of the Delaunay triangulation and so can be computed in $O(n \log n)$. This is quite a bargain since it takes $\Theta(n \log n)$ time just to discover the closest pair of points in S.

As a lower bound, the closest pair problem in one dimension reduces to sorting. In a sorted set of numbers, the closest pair corresponds to two numbers that lie next to each other in sorted order, so we need only check which is the minimum gap between the $n - 1$ adjacent pairs. The limiting case of this occurs when the closest pair are distance zero apart, meaning that the elements are not unique.

Implementations: *Ranger* is a tool for visualizing and experimenting with nearest-neighbor and orthogonal-range queries in high-dimensional data sets, using multidimensional search trees. Four different search data structures are supported by *Ranger*: naive kd-trees, median kd-trees, nonorthogonal kd-trees, and the vantage point tree.

For each of these, *Ranger* supports queries in up to 25 dimensions under any Minkowski metric. It includes generators for a variety of point distributions in arbitrary dimensions. Finally, *Ranger* provides a number of features to aid in visualizing multidimensional data, best illustrated by the accompanying video

[MS93]. To identify the most appropriate projection at a glance, *Ranger* provides a $d \times d$ matrix of all two-dimensional projections of the data set. *Ranger* is written in C, using Motif. It runs on Silicon Graphics and HP workstations and is available in the algorithm repository http://www.cs.sunysb.edu/~algorith.

See Section 8.6.4 for a complete collection of Voronoi diagram implementations. In particular, LEDA (see Section 9.1.1) provides an implementation of 2D Voronoi diagrams in C++, as well as planar point location to make effective use of them for nearest-neighbor search.

A Pascal implementation of the divide-and-conquer algorithm for finding the closest pair of points in a set of n points appears in [MS91]. See Section 9.1.6.4.

Notes: The best reference on kd-trees and other spatial data structures is two volumes by Samet [Sam90b, Sam90a], where all major variants are developed in substantial detail. Good expositions on finding the closest pair of points in the plane [BS76] include [CLR90, Man89]. These algorithms use a divide-and-conquer approach instead of just selecting from the Delaunay triangulation.

A recent development in higher-dimensional nearest-neighbor search is algorithms that quickly produce a point that, if not the nearest neighbor, lies provably close to the query point. A sparse weighted graph structure is built from the data set, and the nearest neighbor is found by starting at a random point and greedily walking in the graph towards the query point. The closest point found over several random trials is declared the winner. Similar data structures hold promise for other problems in high-dimensional spaces. See [AM93, AMN$^+$94].

Related Problems: Kd-trees (see page 194), Voronoi diagrams (see page 358), range search (see page 364).

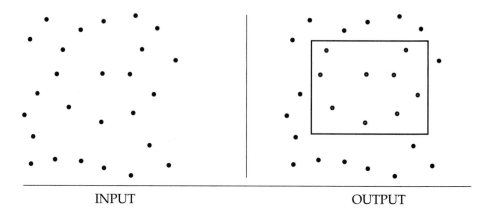

INPUT   OUTPUT

8.6.6 Range Search

Input description: A set S of n points in E^d and a query region Q.

Problem description: Which points from S lie within Q?

Discussion: Range search problems arise in database and geographic information system (GIS) applications. Any data object with d numerical fields, such
as person with height, weight, and income, can be modeled as a point in d-
dimensional space. A *range query* describes a region in space and asks for all
points or the number of points in the region. For example, asking for all people
with income between $0 and $10,000, with height between 6′0″ and 7′0″, and
weight between 50 and 140 lbs. defines a box containing people whose body
and wallet are both thin.

The difficulty of a range search problem depends on several factors:

- *How many range queries are you going to perform?* – The simplest approach to
  range search tests each of the n points one by one against the query polygon
  Q. This works just fine if the number of queries will be small. Algorithms to
  test whether a point is within a given polygon are presented in Section 8.6.7.

- *What shape is your query polygon?* – The easiest type of regions to query against
  are *axis-parallel rectangles*, because the inside/outside test reduces to verifying
  whether each coordinate lies within a prescribed range. The figure above
  illustrates such an *orthogonal range query*.

  If you are querying against a nonconvex polygon, it will pay to partition
  your polygon into convex pieces or (even better) triangles and then query
  the point set against each one of the pieces. This is because testing whether
  a point is inside a convex polygon can be done more quickly and easily
  than for arbitrary polygons. Algorithms for such convex decompositions are
  discussed in Section 8.6.11.

- *How many dimensions?* – The best general-purpose approach to range queries
  builds a kd-tree on the point set, as discussed in Section 8.1.6. To perform a
  query, a depth-first traversal of the kd-tree is performed, with each tree node

expanded only if the associated rectangle intersects the query region. For sufficiently large or misaligned query regions, the entire tree might have to be traversed, but in general; kd-trees lead to an efficient solution. Although algorithms with efficient worst-case performance are known in two dimensions, kd-trees are likely to work even better in the plane. In higher-dimensions, kd-trees provide the only viable solution to the problem.

- *Can I just count the number of points in a region, or do I have to identify them?* – For many applications it suffices to count the number of points in a region instead of returning them. Harkening back to the introductory example, we may want to know whether there are more thin/poor people or rich/fat ones. The need to find the densest or emptiest region in space often naturally arises, and the problem can be solved by counting range queries.

  A data structure for efficiently answering such aggregate range queries can be based on the dominance ordering of the point set. A point x is said to *dominate* point y if y lies both below and to the left of x. Let $DOM(p)$ be a function that counts the number of points in S that are dominated by p. The number of points m in the orthogonal rectangle defined by $x_{min} \leq x \leq x_{max}$ and $y_{min} \leq y \leq y_{max}$ is given by

$$m = DOM(\{x_{max}, y_{max}\}) - DOM(\{x_{max}, y_{min}\})$$
$$- DOM(\{x_{min}, y_{max}\}) + DOM(\{x_{min}, y_{min}\})$$

The second additive term corrects for the points for the lower left-hand corner that have been subtracted away twice.

  To answer arbitrary dominance queries efficiently, partition the space into n^2 rectangles by drawing a horizontal and vertical line through each of the n points. The set of points dominated is identical for each point in each rectangle, so the dominance count of the lower left-hand corner of each rectangle can precomputed, stored, and reported for any query point within it. Queries reduce to binary search and thus take $O(\lg n)$ time. Unfortunately, this data structure takes quadratic space. However, the same idea can be adapted to kd-trees to create a more space-efficient search structure.

Implementations: LEDA (see Section 9.1.1) provides excellent support for maintaining planar subdivisions in C++. In particular, it supports orthogonal range queries in $O(k + \lg^2 n)$ time, where n is the complexity of the subdivision and k is the number of points in the rectangular region.

  Ranger is a tool for visualizing and experimenting with nearest-neighbor and orthogonal-range queries in high-dimensional data sets, using multidimensional search trees. Four different search data structures are supported by *Ranger*: naive kd-trees, median kd-trees, nonorthogonal kd-trees, and the vantage point tree. For each of these, *Ranger* supports queries in up to 25 dimensions under any Minkowski metric. It is available in the algorithm repository.

  A bare bones implementation in C of orthogonal range search using kd-trees appears in [GBY91]. Sedgewick [Sed92] provides code fragments of the grid method for orthogonal range search in C++. See Section 9.1.6.6 for details on both of them.

Notes: Good expositions on data structures with worst-case $O(\lg n + k)$ performance for orthogonal-range searching [Wil85] include [PS85]. An exposition on *kd*-trees for orthogonal range queries in two dimensions appears in [PS85]. Their worst-case performance can be very bad; [LW77] describes an instance in two dimensions requiring $O(\sqrt{n})$ time to report that a rectangle is empty.

The problem becomes considerably more difficult for nonorthogonal range queries, where the query region is not an axis-aligned rectangle. For half-plane intersection queries, $O(\lg n)$ time and linear space suffice [CGL85]; for range searching with simplex query regions (such as a triangle in the plane), lower bounds preclude efficient worst-case data structures. See [Yao90] for a discussion.

Related Problems: Kd-trees (see page 194), point location (see page 367).

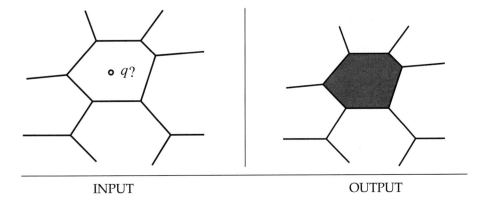

| INPUT | OUTPUT |

8.6.7 Point Location

Input description: A decomposition of the plane into polygonal regions and a query point q.

Problem description: Which region contains the query point q?

Discussion: Point location is a fundamental subproblem in computational geometry, usually needed as an ingredient to solve larger geometric problems. In a dispatch system to assign policemen to the scene of a crime, the city will be partitioned into different precincts or districts. Given a map of regions and a query point (the crime scene), the system must identify which region contains the point. This is exactly the problem of planar point location, variations of which include:

- *Is a given point inside or outside of polygon P?* – The simplest version of point location involves only two regions, inside-P and outside-P, and asks which contains a given query point. For polygons with lots of narrow spirals, this can be surprisingly difficult to tell by inspection. The secret to doing it both by eye or machine is to draw a ray starting from the query point and ending beyond the furthest extent of the polygon. Count the number of times the polygon crosses through an edge. If this number is odd, we must be within the polygon. If it is even, we must be outside. The case of the line passing through a vertex instead of an edge is evident from context, since we are counting the number of times we pass through the boundary of the polygon. Testing each of the n edges for intersection against the query ray takes $O(n)$ time. Faster algorithms for convex polygons are based on binary search and take $O(\lg n)$ time.

- *How many queries will have to be performed?* – When we have a subdivision with multiple regions, it is always possible to repeat the inside-polygon test above on each region in the subdivision. However, this is wasteful if we will be performing many such point location queries on the same subdivision. Instead, we can construct a grid-like or tree-like data structure on top of our subdivision to get us near the correct region quickly. Such search structures are discussed in more detail below.

- *How complicated are the regions of your subdivision?* – More sophisticated inside-outside tests are required when the regions of your subdivision are arbitrary polygons. By triangulating all polygonal regions first, each inside-outside test reduces to testing whether a point is in a triangle. Such a test can be made particularly fast and simple, at the minor cost of recording the full-polygon name for each triangle. An added benefit is that the smaller your regions are, the better grid-like or tree-like superstructures are likely to perform. Some care should be taken when you triangulate to avoid long skinny triangles, as discussed in Section 8.6.3.

- *How regularly sized and spaced are your regions?* – If all resulting triangles are about the same size and shape, the simplest point location method imposes a regularly-spaced $k \times k$ grid of horizontal and vertical lines over the entire subdivision. For each of the k^2 rectangular regions, we maintain a list of all the regions that are at least partially contained within the rectangle. Performing a point location query in such a *grid file* involves a binary search or hash table lookup to identify which rectangle contains query point q and then searching each region in the resulting list to identify the right one.

  Such grid files will perform very well, provided that each triangular region overlaps only a few rectangles (thus minimizing storage space) and each rectangle overlaps only a few triangles (thus minimizing search time). Whether it will perform well is a function of the regularity of your subdivision. Some flexibility can be achieved by spacing the horizontal lines irregularly, as a function of the regions of the subdivision. The *slab method*, discussed below, is a variation on this idea that guarantees worst-case efficient point location at the cost of quadratic space.

- *How many dimensions will you be working in?* – In three or more dimensions, some flavor of kd-tree will almost certainly be the point-location method of choice. They are also likely to be the right answer for planar subdivisions too irregular for grid files.

  Kd-trees, described in Section 8.1.6, decompose the space into a hierarchy of rectangular boxes. At each node in the tree, the current box is split into a small number (typically 2 or 4 or 2^d, where d is the dimension) of smaller boxes. At the leaves of the tree, each box is labeled with the small number of regions that are at least partially contained in the box. The point location search starts at the root of the tree and keeps traversing down the child whose box contains the query point. When the search hits a leaf, we test each of the relevant regions against q to see which one of them contains the point. As with grid files, we hope that each leaf contains a small number of regions and that each region does not cut across too many leaf cells.

The simplest algorithm to guarantee $O(\lg n)$ worst-case access is the *slab* method, which draws horizontal lines through each vertex, thus creating $n + 1$ "slabs" between the lines. Since the slabs are defined by horizontal lines, finding the slab containing a particular query point can be done using a binary search on the y-coordinate of q. Since there can be no vertices within any slab, looking for the region containing a point within a slab can be done by a second binary search on the edges that cross the slab. The catch is that a binary search tree must be maintained for each slab, for a worst-case of $O(n^2)$ space if each region intersects each slab. A more space-efficient approach based on building a hierarchy of triangulations over the regions also achieves $O(\lg n)$ for search and is discussed in the notes below.

Worst-case efficient computational geometry methods either require a lot of storage or are fairly complicated to implement. We identify implementations of worst-case methods below, which are worth at least experimenting with. However, we recommend kd-trees for most general point-location applications.

Implementations: LEDA (see Section 9.1.1) provides excellent support for maintaining planar subdivisions in C++ and, in particular, supports point location in $O(\lg^2 n)$ time.

Arrange is a package for maintaining arrangements of polygons in either the plane or on the sphere. Polygons may be degenerate, and hence represent arrangements of lines. A randomized incremental construction algorithm is used, and efficient point location on the arrangement is supported. Polygons may be inserted but not deleted from the arrangement, and arrangements of several thousand vertices and edges can be constructed in a few seconds. *Arrange* is written in C by Michael Goldwasser and is available from http://theory.stanford.edu/people/wass/wass.html.

A routine in C to test whether a point lies in a simple polygon has been provided by O'Rourke [O'R94], and a Pascal routine for the same problem by [MS91]. For information on both, see Section 9.1.6.4.

Notes: The inside-outside test for convex polygons is described in [PS85], which has a very thorough treatment of deterministic planar point location data structures. Expositions on the inside-outside test for simple polygons include [Man89, PS85].

An experimental study of algorithms for planar point location is described in [EKA84]. The winner was a bucketing technique akin to the grid file.

The elegant triangle refinement method of Kirkpatrick [Kir83] builds a hierarchy of triangulations above the actual planar subdivision such that each triangle on a given level intersects only a constant number of triangles on the following level. Since each triangulation is a fraction of the size of the subsequent one, the total space is obtained by summing up a geometric series and hence is linear. Further, the height of the hierarchy is $O(\lg n)$, ensuring fast query times. An alternative algorithm realizing the same time bounds is [EGS86]. The slab method described above is due to [DL76] and is presented in [PS85].

More recently, there has been interest in dynamic data structures for point location, which support fast incremental updates of the planar subdivision (such as insertions and deletions of edges and vertices) as well as fast point location. Chiang and Tamassia's [CT92] survey is an appropriate place to begin.

Related Problems: Kd-trees (see page 194), Voronoi diagrams (see page 358), nearest neighbor search (see page 361).

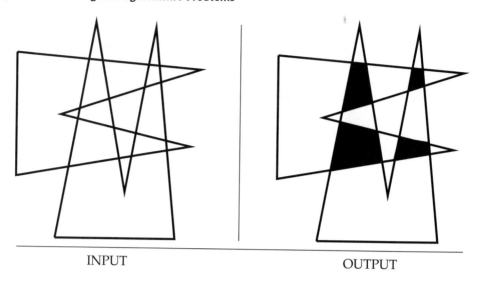

INPUT   OUTPUT

8.6.8   Intersection Detection

Input description: A set S of lines and line segments $l_1, \ldots, l_n$ or a pair of polygons or polyhedra P_1 and P_2.

Problem description: Which pairs of line segments intersect each other? What is the intersection of P_1 and P_2?

Discussion: Intersection detection is a fundamental geometric primitive that arises in many applications. Picture a virtual-reality simulation of an architectural model for a building. The illusion of reality vanishes the instant the virtual person walks through a virtual wall. To enforce such physical constraints, any such intersection between polyhedral models must be immediately detected and the operator notified or constrained.

Another application of intersection detection is design rule checking for VLSI layouts. A minor design mistake resulting in two crossing metal strips can short out the chip, but such errors can be detected before fabrication using programs to find all intersections between line segments.

Issues arising in intersection detection include:

- *Do you want to compute the intersection or just report it?* – We distinguish between intersection detection and computing the actual intersection. The latter problem is obviously harder than the former and is not always necessary. In the virtual-reality application above, it might not be important to know exactly where we hit the wall, just that we hit it.

- *Are you intersecting lines or line segments?* – The big difference here is that any two lines with different slopes must intersect at exactly one point. By comparing each line against every other line, all points of intersections can be found in constant time per point, which is clearly optimal. Constructing the arrangement of the lines provides more information than just intersection points and is discussed in Section 8.6.15.

Finding all the intersections between n line segments is considerably more challenging. Even the basic primitive of testing whether two line segments intersect is not as trivial as it might seem, and this is discussed in Section 8.6.1. To find all intersections, we can explicitly test each line segment against each other line segment and thus find all intersections in $O(n^2)$ time. Each segment can always intersect each other segment, yielding a quadratic number of intersections, so in the worst case, this is optimal. For many applications, however, this worst case is not very interesting.

- *How many intersection points do you expect?* – Sometimes, as in VLSI design rule checking, we expect the set of line segments to have few if any intersections. What we seek is an algorithm whose time is *output sensitive*, taking time proportional to the number of intersection points.

  Such output-sensitive algorithms exist for line-segment intersection, with the fastest algorithm taking $O(n \lg n + k)$ time, where k is the number of intersections. Such algorithms are sufficiently complicated that you should use an existing implementation if you can. These algorithms are based on the sweepline approach, discussed below.

- *Can you see point x from point y?* – Visibility queries ask whether vertex x can see vertex y unobstructed in a room full of obstacles. This can be phrased as the following line-segment intersection problem: does the line segment from x to y intersect any obstacle? Such visibility problems arise in robot motion planning (see Section 8.6.14) and in hidden-surface elimination for computer graphics.

- *Are the intersecting objects convex?* – Better intersection algorithms exist when the line segments form the boundaries of polygons. The critical issue here becomes whether the polygons are convex. Intersecting a convex n-gon with a convex m-gon can be done in $O(n + m)$ time, using a sweepline algorithm as discussed below. This is possible because the intersection of two convex polygons must form another convex polygon using at most $n + m$ vertices.

  However, the intersection of two nonconvex polygons is not so well behaved. Consider the intersection of two "combs" generalizing the Picasso-like frontispiece to this section. As illustrated, the intersection of nonconvex polygons may be disconnected and have quadratic size in the worst case.

  Intersecting two polyhedra is somewhat more complicated than intersecting polygons, because two polyhedra can intersect even when no edges do. Consider the example of a needle piercing the interior of a face. In general, however, the same issues arise for both polygons and polyhedra.

- *Are you searching for intersections repeatedly with the same basic objects?* – In the walk-through application described above, the room and the objects in it don't change between one scene and the next. Only the person moves, and, further, the intersections are rare.

  One common solution is to approximate the objects in the scene by simpler objects that enclose them, such as boxes. Whenever two enclosing boxes intersect, then the underlying objects *might* intersect, and so further work is necessary to decide the issue. However, it is much more efficient to test whether simple boxes intersect than more complicated objects, so we win if collisions are rare. Many variations on this theme are possible, but this idea can lead to large performance improvements for complicated environments.

Planar sweep algorithms can be used to efficiently compute the intersections among a set of line segments, or the intersection/union of two polygons. These algorithms keep track of interesting changes as we sweep a vertical line from left to right over the data. At its leftmost position, the line intersects nothing, but as it moves to the right, it encounters a series of events:

- *Insertion* – the leftmost point of a line segment may be encountered, and it is now available to intersect some other line segment.

- *Deletion* – the rightmost point of a line segment is encountered. This means that we have completely swept over the segment on our journey, and so it can be deleted from further consideration.

- *Intersection* – if we maintain the active line segments that intersect the sweep line as sorted from top to bottom, the next intersection must occur between neighboring line segments. After the intersection, these two line segments swap their relative order.

Keeping track of what is going on requires two data structures. The future is maintained by an *event queue*, a priority queue ordered by the x-coordinate of all possible future events of interest: insertion, deletion, and intersection. See Section 8.1.2 for priority queue implementations. The present is represented by the *horizon*, an ordered list of line segments intersecting the current position of the sweepline. The horizon can be maintained using any dictionary data structure, such as a balanced tree.

To adapt this approach to computing the intersection or union of polygons, we modify the processing of the three event types to keep track of what has occurred to the left of the sweepline. This algorithm can be considerably simplified for pairs of convex polygons, since (1) at most four polygon edges intersect the sweepline, so no horizon data structure is needed and (2) no event-queue sorting is needed, since we can start from the leftmost vertex of each polygon and proceed to the right by following the polygonal ordering.

Implementations: LEDA (see Section 9.1.1) provides a C++ implementation of the Bentley-Ottmann sweepline algorithm [BO79], finding all k intersection points between n line segments in the plane in $O((n + k) \lg n)$ time.

O'Rourke [O'R94] provides a robust program in C to compute the intersection of two convex polygons. See Section 9.1.6.5.

RAPID is a "Rapid and Accurate Polygon Interference Detection library" for large environments composed of polygonal models. It is free for noncommercial use and available from http://www.cs.unc.edu/~geom/OBB/OBBT.html.

Model Pascal subroutines for convexity testing and for finding an intersection in a set of line segments appear in [MS91]. See Section 9.1.6.4 for details. XTango (see Section 9.1.5) includes an animation of polygon clipping against a polygonal window.

Finding the mutual intersection of a collection of half-spaces is a special case of higher-dimensional convex hulls, and Qhull [BDH97] is convex hull code of choice for general dimensions. Qhull has been widely used in scientific applications and has a well-maintained home page at http://www.geom.umn.edu/software/qhull/.

Notes: Good expositions on line segment intersection detection [BO79, SH75] include [CLR90, Man89, NH93, PS85]. Good expositions on polygon and polyhedra intersection [HMMN84, MP78, SH76] include [PS85]. Preparata and Shamos [PS85] provide a good exposition on the special case of finding intersections and unions of axis-oriented rectangles, a problem that arises often in VLSI design.

An optimal $O(n \lg n + k)$ algorithm for computing line segment intersections is due to Chazelle and Edelsbrunner [CE92]. Simpler, randomized algorithms achieving the same time bound are thoroughly presented by Mulmuley [Mul94].

Surveys on hidden-surface removal include [Dor94, SSS74].

Related Problems: Maintaining arrangements (see page 392), motion planning (see page 389).

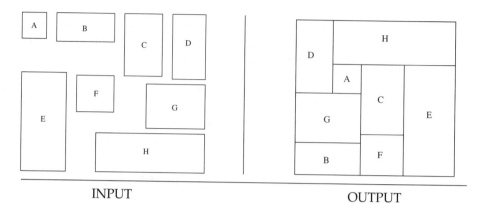

INPUT   OUTPUT

8.6.9 Bin Packing

Input description: A set of n items with sizes $d_1, \ldots, d_n$. A set of m bins with capacity $c_1, \ldots, c_m$.

Problem description: How can you store all the items using the smallest number of bins?

Discussion: Bin packing arises in a variety of packaging and manufacturing problems. Suppose that you are manufacturing widgets with parts cut from sheet metal, or pants with parts cut from cloth. To minimize cost and waste, we seek to lay out the parts so as to use as few fixed-size metal sheets or bolts of cloth as possible. Identifying which part goes on which sheet in which location is a bin-packing variant called the *cutting stock* problem. After our widgets have been successfully manufactured, we will be faced with another bin packing problem, namely how best to fit the boxes into trucks to minimize the number of trucks needed to ship everything.
   Even the most elementary-sounding bin-packing problems are NP-complete (see the discussion of integer partition in Section 8.2.10), so we are doomed to think in terms of heuristics instead of worst-case optimal algorithms. Fortunately, relatively simple heuristics tend to work well on most bin-packing problems. Further, many applications have peculiar, problem-specific constraints that would frustrate highly tuned algorithms for the problem. The following factors will affect the choice of heuristic:

- *What are the shapes and sizes of objects?* – The character of the problem depends greatly on the shapes of the objects to be packed. Packing pieces of a standard jigsaw puzzle is a different problem than packing squares into a rectangular box. In one-dimensional bin packing, each object's size is given simply as an integer, making it a version of the knapsack problem of Section 8.2.10. This is equivalent to packing boxes of equal width into a chimney of that width. If all the boxes are of identical size and shape, the optimal packing is simply a cubic lattice, oriented appropriately with the walls of the bin. It is a waste to use the methods described in this section on such simple problems.

- *Are there constraints on the orientation and placement of objects?* – In loading a truck, many of the boxes will say "this side up." Respecting this constraint restricts our flexibility in packing and will likely lead to an increase in the number of trucks needed to send out the shipment. Similarly, boxes with fragile objects may be labeled "do not stack" and thus are constrained to sit on top of a pile of boxes. Most shippers seem to ignore these constraints. Indeed, your task will be simpler if you don't have to worry about the consequences of them.

- *Is the problem on-line or off-line?* – Do we know the complete set of objects that we will be packing at the beginning of the job (an *off-line* problem)? Or will we get them one at a time and have to deal with them as they arrive (an *on-line* problem)? The difference is important, because we can do a better job packing when we can take a global view and plan ahead. For example, we can arrange the objects in an order that will facilitate efficient packing, perhaps by sorting them from biggest to smallest.

The standard heuristics for bin packing order the objects by size or shape (or in an on-line problem, simply the order they arrive in) and then insert them into bins. Typical insertion rules include (1) select the first or leftmost bin the object fits in, (2) select the bin with the most room, (3) select the bin that provides the tightest fit, or (4) select a random bin.

Analytical and empirical results suggest that the best heuristic is *first-fit, decreasing*. Sort the objects in decreasing order of size, so that the biggest object is first and the smallest last. Insert each object one by one into the first bin that has room for it. If no bin has room for it, we must start another bin. In the case of one-dimensional bin packing, this can never require more than 22% more bins than necessary and usually does much better. First-fit decreasing has an intuitive appeal to it; we pack the bulky objects first and hope that the little objects can be used to fill up the cracks. First-fit decreasing is easily implemented in $O(n \lg n + bn)$ time, where $b \leq \min(n, m)$ is the number of bins actually used, by doing a linear sweep through the bins on each insertion. A faster $O(n \lg n)$ implementation is possible by using a binary tree to keep track of the space remaining in each bin.

We can fiddle with the insertion order in such a scheme to deal with problem-specific constraints. For example, it is reasonable to take "do not stack" boxes last (perhaps after artificially lowering the height of the bins to give some room up top to work with) and to place fixed-orientation boxes at the beginning (so we can use the extra flexibility later to stick boxes into cracks).

Packing boxes is much easier than packing arbitrary geometric shapes, enough so that a reasonable approach for general shapes is to pack each part into its own box and then pack the boxes. Finding an enclosing rectangle for a polygonal part is easy; just find the upper, lower, left, and right tangents in a given orientation. A minimum-area enclosing rectangle can be found by determining the orientation that leads to the smallest box.

In the case of nonconvex parts, considerable useful space can be wasted in the holes created by placing the part in a box. One solution is to find the *maximum empty rectangle* within each boxed part and use this to contain other parts if it is sufficiently large. More advanced solutions are discussed in the references.

Implementations: Codes for the one-dimensional version of bin packing, the so-called knapsack problem, are presented in Section 8.2.10.

XTango (see Section 9.1.5) includes an animation of the first-fit bin packing heuristic. Test data for bin packing is available from http://mscmga.ms.ic.ac.uk/info.html.

Notes: See [CGJ96] for a survey of the extensive literature on approximation algorithms for bin packing. Expositions on heuristics for bin packing include [Baa88]. Experimental results on bin-packing heuristics include [BJLM83].

Sphere packing is an important and well-studied special case of bin packing, with applications to error-correcting codes. Particularly notorious is the "Kepler conjecture," the apparently still-open problem of establishing the densest packing of unit spheres in three dimensions. Conway and Sloane [CS93] is the best reference on sphere packing and related problems. Sloane provides an extensive set of tables of the best known packings, available from ftp:/netlib.bell-labs.com.

Victor Milenkovic and his students have worked extensively on two-dimensional bin-packing problems for the apparel industry, minimizing the amount of material needed to manufacture pants and other clothing. Recent reports of this work include [DM97, Mil97].

Related Problems: Knapsack problem (see page 228), set packing (see page 401).

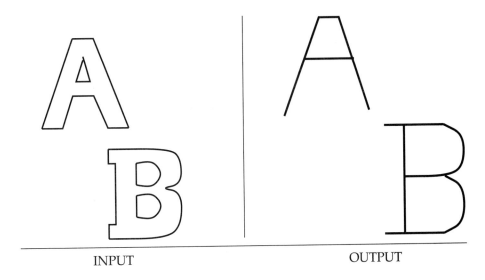

INPUT   OUTPUT

8.6.10 Medial-Axis Transformation

Input description: A polygon or polyhedron P.

Problem description: What is the set of points within P that have more than one closest point on the boundary of P?

Discussion: The medial-axis transformation is useful in *thinning* a polygon, or, as is sometimes said, finding its *skeleton*. The goal is to extract a simple, robust representation of the shape of the polygon. As can be seen from the figures above, the thinned versions of the letters capture the essence of the shape of an 'A' and a 'B', and would be relatively unaffected by changing the thickness of strokes or by adding font-dependent flourishes such as serifs.

The medial-axis transformation of a polygon is always a tree, making it fairly easy to use dynamic programming to measure the "edit distance" between the skeleton of a known model and the skeleton of an unknown object. Whenever the two skeletons are close enough, we can classify the unknown object as an instance of our model. This technique has proven useful in computer vision and in optical character recognition. The skeleton of a polygon with holes (like the 'A' and 'B') is not a tree, but an embedded planar graph, but it remains easy to work with.

There are two distinct approaches to computing medial-axis transforms, depending upon whether your inputs are arbitrary geometric points or pixel images:

- *Geometric data* – Recall that the Voronoi diagram of a point set S decomposes the plane into regions around each point $s_i \in S$ such that each point within the region around s_i is closer to s_i than to any other site in S. Similarly, the Voronoi diagram of a set of line segments L decomposes the plane into regions around each line segment $l_i \in L$ such that each point within the region around l_i is closer to l_i than to any other site in L.

Any polygon is defined by a collection of line segments such that l_i shares a vertex with l_{i+1}. The medial-axis transform of a polygon P is simply the portion of the line-segment Voronoi diagram that lies within P.

Any line-segment Voronoi diagram code thus suffices to do polygon thinning. In the absence of such a code, the most readily implementable thinning algorithm starts at each vertex of the polygon and grows the skeleton inward with an edge bisecting the angle between the two neighboring edges. Eventually, these two edges meet at an internal vertex, a point equally far from three line segments. One of the three is now enclosed within a cell, while a bisector of the two surviving segments grows out from the internal vertex. This process repeats until all edges terminate in vertices.

- *Image data* – Whenever attempting geometric approaches to image processing problems, we must remain aware that images are pixel-based and not continuous. All the pixels sit as lattice points on an integer grid. While we can extract a polygonal description from the boundary and feed it to the geometric algorithms above, the internal vertices of the skeleton will most likely not lie at grid points. This may well make geometric approaches inappropriate for your intended application.

  Algorithms that explicitly manipulate pixels tend to be easy to implement, because they avoid complicated data structures. The basic pixel-based approach for constructing a skeleton directly implements the "brush fire" view of thinning. Imagine a brush fire along all edges of the polygon, burning inward at a constant speed. The skeleton is marked by all points where two or more fires meet. The resulting algorithm traverses all the boundary pixels of the object, deletes all except the extremal pixels, and repeats. The algorithm terminates when all pixels are extreme, leaving an object only 1 or 2 pixels thick. When implemented properly, this takes time linear in the number of pixels in the image.

  The trouble with such pixel-based approaches is that the geometry doesn't work out exactly right. For example, the skeleton of a polygon is no longer always a tree or even necessarily connected, and the points in the skeleton will be close-to-but-not-quite equidistant to two boundary edges. The usual solution is to tweak your implementation until it gives skeletons that look decent on the data that you think is most interesting. Since you are trying to do continuous geometry in a discrete world, there is no way to solve the problem completely. You just have to live with it.

Implementations: MAT [Ogn93] is a medial-axis transform code designed for 2D skeletonization of binary images, written by Robert Ogniewicz. MAT accepts a variety of different input formats, including polygonal representations. This seems to be a solidly built program, and it should be your first stop on seeking a routine for thinning. It available from http://hrl.harvard.edu/people/postdocs/rlo/rlo.dir/rlo-soft.html.

Programs for constructing Voronoi diagrams are discussed in Section 8.6.4.

Notes: For comprehensive surveys of thinning approaches in image processing, see [LLS92, Ogn93]. The medial axis transformation was introduced for shape similarity studies in biology [Blu67]. Applications of the medial-axis transformation in pattern recognition are discussed in [DH73]. Good expositions on the medial-axis transform include [O'R94, Pav82].

The medial-axis of a polygon can be computed in $O(n \lg n)$ time for arbitrary n-gons [Lee82], although linear-time algorithms exist for convex polygons [AGSS89]. An $O(n \lg n)$ algorithm for constructing medial-axis transforms in curved regions was given by Kirkpatrick [Kir79].

Related Problems: Voronoi diagrams (see page 358), Minkowski sum (see page 395).

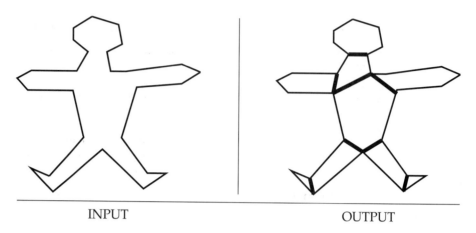

INPUT   OUTPUT

8.6.11   Polygon Partitioning

Input description: A polygon or polyhedron P.

Problem description: How can P be partitioned into a small number of simple (typically convex) pieces?

Discussion: Polygon partitioning is an important preprocessing step for many geometric algorithms, because most geometric problems are simpler and faster on convex objects than on nonconvex ones. We are better off whenever we can partition a nonconvex object into a small number of convex pieces, because it is easier to work with the pieces independently than with the original object.

Several flavors of polygon partitioning arise, depending upon the particular application:

- *Should all the pieces be triangles?* – The mother of all polygon partitioning problems is triangulation, where the interior of the polygon is completely partitioned into triangles. Triangles are always convex and have only three sides, so any geometric operation performed on a triangle is destined to be as simple as it can be.

  Because triangles all have three sides, all triangulations of polygons contain exactly the same number of pieces. Therefore, triangulation cannot be the answer if we seek a small number of convex pieces. The goal of finding "nice" triangulations revolves around the shape of the triangles. See Section 8.6.3 for a thorough discussion of triangulation.

- *Do I want to cover or partition my polygon?* – *Partitioning* a polygon means completely dividing the interior into nonoverlapping pieces. *Covering* a polygon means that our decomposition is permitted to contain mutually overlapping pieces. Both can be useful in different situations. In decomposing a complicated query polygon in preparation for range search (Section 8.6.6), we seek a partitioning, so that each point we locate occurs in exactly one piece. In decomposing a polygon for painting purposes, a covering suffices, since there is no difficulty with filling in a region twice. We will concentrate here on partitioning, since it is simpler to do right, and any application needing a covering will accept a partitioning. The only potential drawback is that partitions can be somewhat larger than coverings.

- *Am I allowed to add extra vertices?* – A final issue associated with polygon decomposition is whether we are allowed to add Steiner vertices (either by splitting edges or adding interior points) or whether we are restricted to adding chords between two existing vertices. The former may result in a smaller number of pieces, at the cost of more complicated algorithms and perhaps messier results.

The Hertel-Mehlhorn heuristic for convex decomposition using diagonals is simple, efficient, and always produces no more than four times the minimum number of convex pieces. It starts with an arbitrary triangulation of the polygon and then deletes a chord that leaves only convex pieces. The decision of whether a chord deletion will create a nonconvex piece can be made locally from the chords and edges surrounding the chord, in constant time. A vertex in a polygon is *reflex* if the internal angle is greater than 180 degrees. We can delete any chord that does not create a reflex vertex.

I recommend using this heuristic unless it is critical for you to absolutely minimize the number of pieces. By experimenting with different triangulations and various deletion orders, you may be able to obtain somewhat better decompositions.

Dynamic programming may be used to minimize the number of diagonals used in the decomposition. The simplest implementation, which maintains the number of pieces for all $O(n^2)$ subpolygons split by a chord, runs in $O(n^4)$. Faster algorithms use fancier data structures, running in $O(r^2 n \lg n)$ time, where r is the number of reflex vertices. An $O(n^3)$ algorithm that further reduces the number of pieces by adding interior vertices is cited below, although it is complex and presumably difficult to implement.

Implementations: Many triangulation codes start by finding a trapezoidal or monotone decomposition of polygons. Further, a triangulation is a simple form of convex decomposition. Check out the codes in Section 8.6.3 as a starting point for most any decomposition problem.

A triangulation code of particular relevance here is GEOMPACK, a suite of Fortran 77 codes by Barry Joe, for 2- and 3-dimensional triangulation and convex decomposition problems. In particular, it does both Delaunay triangulation and convex decompositions of polygonal and polyhedral regions, as well as arbitrary-dimensional Delaunay triangulations.

Notes: Keil and Sack [KS85] given an excellent survey on what is known about partitioning and covering polygons. Expositions on the Hertel-Mehlhorn heuristic [HM83] include [O'R94]. The $O(r^2 n \lg n)$ dynamic programming algorithm for minimum convex decomposition using diagonals is due to Keil [Kei85]. The $O(r^3 + n)$ algorithm minimizing the number of convex pieces with Steiner points appears in [CD85]. Feng and Pavlidis

[FP75b] give a heuristic algorithm for polygon decomposition and apply it to optical character recognition.

Art gallery problems are an interesting topic related to polygon covering, where we seek to position the minimum number of guards in a given polygon such that every point in the interior of the polygon is watched by at least one guard. This corresponds to covering the polygon with a minimum number of star-shaped polygons. O'Rourke [O'R87] is a beautiful book which presents the art gallery problem and its many variations.

Related Problems: Triangulation (see page 355), set cover (see page 398).

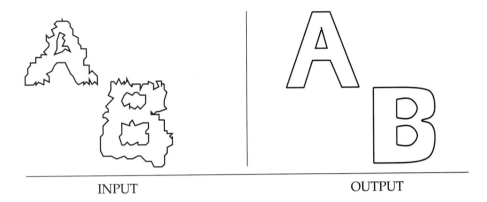

INPUT   OUTPUT

8.6.12   Simplifying Polygons

Input description: A polygon or polyhedron p, with n vertices.

Problem description: Find a polygon or polyhedron p' with n' vertices, where the shape of p' is close to p and $n' < n$.

Discussion: Polygon simplification has two primary applications. The first is in cleaning up a noisy representation of a polygon, perhaps obtained by scanning a picture of an object. By processing it, we hope to remove the noise and reconstruct the original object. The second is in data compression, where given a large and complicated object, we seek to simplify it by reducing detail. Ideally, we obtain a polygon with far fewer vertices that looks essentially the same. This can be a big win in computer graphics, where replacing a large model with a smaller model might have little visual impact but be significantly faster to render.

Several issues arise in shape simplification:

- *Do you want the convex hull?* – The simplest way to simplify a polygon is to take the convex hull of its vertices (see Section 8.6.2). The convex hull removes all internal concavities from the polygon. If you were simplifying a robot model for motion planning, this is a good thing, because you are unlikely to be able to take advantage of the concavities in finding paths. If you were building an OCR system, the convex hull would be disastrous, because the concavities of characters provide most of the interesting features. An 'X' would be identical to an 'I', since both hulls are boxes. Another problem is that if the polygon is already convex, taking the convex hull will do nothing to simplify it further.

- *Am I allowed to insert or just delete points?* – What is typically needed is a way to represent the object as well as possible using only a given number of vertices. The simplest approaches employ local modifications to the boundary of the polygon, in order to reduce the vertex count. For example, if three consecutive vertices form a small-area triangle or define an extremely large angle, the center vertex can be deleted and replaced with an edge without severely distorting the polygon.

Methods that only delete vertices quickly melt the shape into unrecognizability, however. More robust heuristics move vertices around to cover up the gaps that are created by deletions. Such "split-and-merge" heuristics can do a decent job, although nothing is guaranteed. Better results are likely by using the Douglas-Peucker algorithm, described below.

- *Must the resulting polygon be intersection-free?* – A serious drawback of such incremental procedures is that they fail to ensure *simple* polygons, those without self-intersections. Thus the "simplified" polygon may have artifacts that look ugly and that may cause problems for subsequent routines working on the polygon. If simplicity is important, you should test all the line segments of your polygon for any pairwise intersections, as discussed in Section 8.6.8.

  An approach to polygon simplification that guarantees a simple approximation involves computing minimum-link paths. The *link distance* of a path between points s and t is the number of straight segments on the path. An as-the-crow-flies path has a link distance of one. In general, the link distance is one more than the number of turns. The link distance between points s and t in a scene with obstacles is defined by the minimum link distance over all paths from s to t.

  The link distance approach "fattens" the boundary of the polygon by some acceptable error window ϵ (see Section 8.6.16) in order to construct a channel around the polygon and then constructs the minimum-link path through this channel. The minimum-link cycle in this channel represents the simplest polygon that never deviates from the original boundary by more than ϵ. It constructs a globally optimal simplification that will not self-intersect, at the cost of implementation and time complexity.

- *Are you given an image to clean up instead of a polygon to simplify?* – The conventional, nongeometric approach to cleaning up noise from a digital image is to take the Fourier transform of the image, filter out the high-frequency elements, and then take the inverse transform to recreate the image. See Section 8.2.11 for details on the fast Fourier transform.

The Douglas-Plucker algorithm for shape simplification starts with a simple approximation and then refines it, instead of starting with a complicated polygon and trying to simplify it. Start by selecting two vertices v_1 and v_2 of polygon P, and propose the degenerate polygon v_1, v_2, v_1 as a simple approximation P'. Scan through each of the vertices of P, and select the one that is farthest from the corresponding edge of the polygon P'. Inserting this vertex adds the triangle to P' so as to minimize the maximum deviation from P. Points can be inserted until satisfactory results are achieved. This takes $O(kn)$ to insert k points when $|P| = n$.

Simplification becomes considerably more difficult in three dimensions. For example, it is NP-complete to find the minimum-size surface separating two polyhedra. Higher-dimensional analogies of the planar algorithms discussed here can be used to heuristically simplify polyhedra. See the references below.

Implementations: A program for automatically generating level-of-detail hierarchies for polygonal models is available from http://www.cs.unc.edu/~geom/envelope.html and is free for noncommercial use. The user specifies a single error tolerance, and the maximum surface deviation of the simplified model from the original model, and a new, simplified model is generated. This code preserves holes and prevents self-intersection.

Yet another approach to polygonal simplification is based on simplifying and expanding the medial-axis transform of the polygon. The medial-axis transform (see Section 8.6.10) produces a skeleton of the polygon, which can be trimmed before inverting the transform to yield a simpler polygon. MAT [Ogn93] is a medial-axis transform code designed for 2D skeletonization and inversion of binary images, written by Robert Ogniewicz and available from http://hrl.harvard.edu/people/postdocs/rlo/rlo.dir/rlo-soft.html.

Notes: See [HG95] for a thorough survey of algorithms for shape simplification. It is also available from http://www.cs.cmu.edu/afs/cs/user/ph/www/heckbert.html, along with implementations.

The Douglas-Peucker incremental refinement algorithm [DP73] is the basis for most shape simplification schemes, with a faster implementation due to Hershberger and Snoeyink [HS94]. The link distance approach to polygon simplification is presented in [GHMS93]. Shape simplification problems become considerably more complex in three dimensions. Even finding the minimum-vertex convex polyhedron lying between two nested convex polyhedra is NP-complete [DJ92], although approximation algorithms are known [MS95b].

Testing whether a polygon is simple can be performed in linear time, at least in theory, as a consequence of Chazelle's linear-time triangulation algorithm [Cha91].

Related Problems: Fourier transform (see page 232), convex hull (see page 351).

INPUT   OUTPUT

8.6.13   Shape Similarity

Input description: Two polygonal shapes, P_1 and P_2.

Problem description: How similar are P_1 and P_2?

Discussion: Shape similarity is a problem that underlies much of pattern recognition. Consider a system for optical character recognition (OCR). We have a known library of shape models representing letters and the unknown shapes we obtain by scanning a page. We seek to identify an unknown shape by matching it to the most similar shape model.

The problem of shape similarity is inherently ill-defined, because what "similar" means is application dependent. Thus no single algorithmic approach can solve all shape matching problems. Whichever method you select, expect to spend a large chunk of time tweaking it so as to achieve maximum performance. Don't kid yourself – this is a difficult problem.

Among your possible approaches are:

- *Hamming Distance* – Assume that your two polygons have been overlaid one on top of the other. The *Hamming distance* measures the area of symmetric difference between the two polygons, in other words, the area of the regions lying within one of the two polygons but not both of them. If the two polygons are identical, and properly aligned, the Hamming distance will be zero. If the polygons differ only in a little noise at the boundary, then the Hamming distance will be small if the polygons are properly aligned. If the two polygons are completely disjoint, then the Hamming distance is the sum of the areas of the two polygons.

  Computing the area of the symmetric difference reduces to finding the intersection or union of two polygons, as discussed in Section 8.6.8, and then computing areas, as discussed in 8.6.1. The difficult part of computing Hamming distance is finding the right alignment, or overlay, of the two polygons. For certain applications, such as OCR, the overlay problem is simplified because the characters are inherently aligned with the page and thus not free to rotate. Hamming distance is particularly simple and efficient to compute on bit-mapped images, since after alignment all we do is sum the differences of the corresponding pixels.

Although Hamming distance makes sense conceptually and can be simple to implement, it captures only a crude notion of shape and is likely to be ineffective in most applications.

- *Hausdorff distance* – An alternative distance metric is *Hausdorff distance*, which identifies the point on P_1 that is the maximum distance from P_2 and returns this distance. The Hausdorff distance is not symmetrical, for the distance from P_1 to P_2 is not necessarily the distance from P_2 to P_1. Note the difference between Hamming and Hausdorff distance. A large but thin protrusion from one of the models will have a large effect on the Hausdorff distance but little on the Hamming distance. However, a fattening of the entire boundary of one of the models (as is liable to happen with boundary noise) by a small amount will increase the Hamming distance yet have little effect on the Hausdorff distance.

  Which is better, Hamming or Hausdorff? It depends upon your application. As with Hamming distance, computing the right alignment between the polygons can be difficult and time-consuming. Again, Hausdorff distance captures only a crude notion of shape.

- *Comparing Skeletons* – A more powerful approach to shape similarity uses thinning (see Section 8.6.10) to extract a tree-like skeleton for each object. This skeleton captures many aspects of the original shape. The problem now reduces to comparing the shape of two such skeletons, using such features as the topology of the tree and the lengths/slopes of the edges. This comparison can be modeled as a form of subgraph isomorphism (see Section 8.5.9), with edges allowed to match whenever their lengths and slopes are sufficiently similar.

- *Neural Networks* – A final method for shape comparison uses neural networks, which are discussed in Section 5.5.2. Neural nets prove a reasonable approach to recognition problems when you have a lot of data to experiment with and no particular ideas of what to do with it. First, you must identify a set of easily computed features of the shape, such as area, number of sides, and number of holes. Based on these features, a black-box program (the neural network training algorithm) takes your training data and produces a classification function. This classification function accepts as input the values of these features and returns a measure of what the shape is, or how close it is to a particular shape.

  How good are the resulting classifiers? It depends upon the application. Like any ad hoc method, neural networks usually take a fair amount of tweaking and tuning to realize their full potential.

  One caveat. Because your classifier was developed by a black box, you never really know why your classifier is making its decisions, so you can't know when it will fail. An interesting case was a system built for the military to distinguish between images of cars and tanks. It performed very well on test images but disastrously in the field. Eventually, someone realized that the car images had been filmed on a sunnier day than the tanks, and the program was classifying solely on the presence of clouds in the background of the image!

Implementations: The *Stuttgart Neural Network Simulator* supports many types of networks and training algorithms, as well as sophisticated graphical visualization tools under X11. It has been ported to many flavors of UNIX. It is available for ftp from ftp.informatik.uni-stuttgart.de [129.69.211.2] in directory /pub/SNNS as SNNSv4.1.tar.gz (1.4 MB, Source code) and SNNSv4.1.Manual.ps.gz (1 MB, Documentation). It may be best to first have a look at the file SNNSv4.1.Readme. More information can be found in the WWW under http://www.informatik.uni-stuttgart.de/ipvr/bv/projekte/snns/snns.html

An alternate distance metric between polygons can be based on its angle turning function [ACH+91]. An implementation in C of this turning function metric by Eugene K. Ressler is provided on the algorithm repository http://www.cs.sunysb.edu/~algorith.

Notes: General books on pattern classification algorithms include [DH73, JD88]. A wide variety of computational geometry approaches to shape similarity testing have been proposed, including [AMWW88, ACH+91, Ata84, AE83, BM89, OW85].

A linear-time algorithm for computing the Hausdorff distance between two convex polygons is given in [Ata83], with algorithms for the general case reported in [HK90].

Related Problems: Graph isomorphism (see page 335), thinning (see page 377).

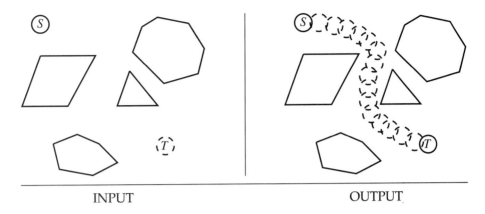

| INPUT | OUTPUT |

8.6.14 Motion Planning

Input description: A polygonal-shaped robot s in a given starting position in a room containing polygonal obstacles, with a desired ending position t.

Problem description: Find the shortest path in the room taking s to t without intersecting any of the obstacles.

Discussion: The difficulty of motion planning will be obvious to anyone who has ever had to move a large piece of furniture into a small apartment. The problem of motion planning also arises in systems for molecular docking. Many drugs are small molecules that act by binding to a given target model. The problem of identifying which binding sites are accessible to a candidate drug is clearly an instance of motion planning. Plotting paths for mobile robots is another canonical motion-planning application.

There is a wide range in the complexity of motion planning problems, with many factors to consider:

Motion planning also provides a tool for computer animation. Given a set of object models that appear in two different scenes s_1 and s_2, a motion planning algorithm can construct a short sequence of intermediate motions to transform s_1 to s_2. These motions can serve to fill in the intermediate scenes between s_1 and s_2, with such scene interpolation greatly reducing the amount of work the animator has to do.

There is a wide range in the complexity of motion planning problems, with many factors to consider:

- *Is your robot a point?* – When the robot is a point, the problem becomes finding the shortest path from s to t around the obstacles, also known as geometric shortest path. The most readily implementable approach constructs the *visibility graph* of the polygonal obstacles, plus the points s and t. This visibility graph has a vertex for each obstacle vertex and an edge between two obstacle vertices if they "see" each other without being blocked by some obstacle edge.

  A brute-force algorithm to construct the visibility graph tests each candidate edge against the $\Theta(n)$ obstacle edges for a total time of $O(n^3)$ time. By weighting each edge of the visibility graph with its length and using Dijkstra's shortest-path algorithm (see Section 8.4.4) on this graph, we can find the shortest path from s to t in time bounded by the time to construct the graph.

- *How is your robot free to move?* – Motion planning becomes considerably more difficult when the robot becomes a polygon instead of a point. Now we must make sure that all of the corridors we use are wide enough to permit the robot to pass through. The complexity depends upon the number of *degrees of freedom* the robot has to move. Is it free to rotate as well as to translate? Does the robot have links that are free to bend or to rotate independently, as in an arm with a hand? Each degree of freedom corresponds to a dimension in the search space. Therefore, the more freedom, the harder it is to compute a path between two locations, although it becomes more likely that such a path exists.

- *Can you simplify the shape of your robot?* – Motion planning algorithms tend to be complex and time-consuming. Anything you can do to simplify your environment would be a win. In particular, consider replacing your robot by an enclosing disk. If there is a path for the disk, there will be a path for whatever is inside of it. Further, since any orientation of a disk is equivalent to any other orientation, rotation provides no help in finding a path Therefore, all movements are limited to translation only.

- *Are motions limited to translation only?* – When rotation is not allowed, the *expanded obstacles* approach can be used to reduce the problem to that of motion planning for a point robot, which is simply the shortest path in a graph. Pick a reference point on the robot, and replace each obstacle by the Minkowski sum of the object and the robot (see Section 8.6.16). This creates a larger obstacle, defined as the robot walks a loop around the obstacle while maintaining contact with it. Finding a path from the initial position of the reference point to the goal point amidst these fattened obstacles defines a legal path for the polygonal robot.

- *Are the obstacles known in advance?* – Thus far we have assumed that the robot starts out with a map of its environment. This is not always true, or even possible, in applications where the obstacles move. There are two approaches to solving motion planning problems without a map. The first approach explores the environment, building a map of what has been seen, and then uses this map to plan a path to the goal. A simpler strategy, which will fail in environments of sufficient complexity, proceeds like a sightless man with a compass. Walk in the direction towards the goal until progress is blocked by an obstacle, and then trace a path along the obstacle until the robot is again free to proceed directly towards the goal.

The most practical approach to motion planning involves randomly sampling the configuration space of the robot. The configuration space defines the set of legal positions for the robot and has one dimension for each degree of freedom. For a planar robot capable of translation and rotation, the degrees of freedom are the x- and y-coordinates of a reference point on the robot and the angle θ relative to this point. Certain points in this space represent legal positions, while others intersect obstacles.

Construct a set of legal configuration-space points by random sampling. For each pair of points p_1 and p_2, decide whether there exists a direct, nonintersecting path between them. Construct a graph with vertices for each legal point and edges for each such traversable pair. The problem of finding a motion between two arbitrary positions reduces to seeing if there is a direct path from the initial/final position to some vertex in the graph, and then solving a shortest-path problem in the graph.

There are lots of ways to enhance this basic technique for specific applications, such as adding additional vertices to regions of particular interest. This is a nice, clean approach for solving problems that would get very messy otherwise.

Implementations: An implementation of collision detection (not really motion planning) is the I_COLLIDE collision detection library. For more information, check out the I_COLLIDE WWW page: http://www.cs.unc.edu/~geom/I_COLLIDE.html.

O'Rourke [O'R94] gives a toy implementation of an algorithm to plot motion for a two-jointed robot arm in the plane. See Section 9.1.6.5.

Notes: Motion planning was originally studied by Schwartz and Sharir as the "piano mover's problem." Their solution constructs the complete free space of robot positions which do not intersect obstacles, and then finds the shortest path within the proper connected component. These free space descriptions are very complicated, involving arrangements of higher-degree algebraic surfaces. The fundamental papers on the piano mover's problem appear in [HSS87], with [SS90] being a survey of current results. The best general result for this free space approach to motion planning is due to Canny [Can87], who showed that any problem with d degrees of freedom can be solved in $O(n^d \lg n)$, although faster algorithms exist for special cases of the general motion planning problem.

Latombe's book [Lat91] describes practical approaches to motion planning, including the random sampling method described above. The expanded obstacle approach to motion planning is due to Lozano-Perez and Wesley [LPW79]. The heuristic, sightless man's approach to motion planning discussed above has been studied by Lumelski [LS87].

The time complexity of algorithms based on the free-space approach to motion planning depends intimately on the combinatorial complexity of the arrangement of surfaces defining the free space. Algorithms for maintaining arrangements are presented in Section 8.6.15. Davenport-Schintzel sequences often arise in the analysis of such arrangements. Sharir and Agarwal [SA95] provide a comprehensive treatment of Davenport-Schintzel sequences and their relevance to motion planning.

Kedem and Sharir [KS90] give an $O(kn \lg(kn))$ time algorithm to find a path (not necessarily shortest) to translate a convex k-gon from s to t. Vegter [Veg90] gives an optimal $\Theta(n^2)$ algorithm for moving a line segment (often called a ladder) in the plane with both translation and rotation.

Related Problems: Shortest path (see page 279), Minkowski sum (see page 395).

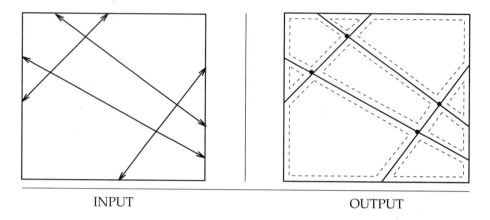

INPUT   OUTPUT

8.6.15   Maintaining Line Arrangements

Input description: A set of lines and line segments $l_1, \ldots, l_n$.

Problem description: What is the decomposition of the plane defined by $l_1, \ldots, l_n$?

Discussion: One of the most fundamental problems in computational geometry is constructing arrangements of lines, that is, explicitly building the regions formed by the intersections of a set of n lines. Algorithms for a surprising number of problems are based on constructing and analyzing the arrangement of a specific set of lines:

- *Degeneracy testing* – Given a set of n lines in the plane, do any three of them pass through the same point? Brute-force testing of all triples takes $O(n^3)$ time. Instead, we can construct the arrangement of the lines and then walk over each vertex and explicitly count its degree, all in quadratic time.

- *Satisfying the maximum number of linear constraints* – Suppose that we are given a set of n linear constraints, each of the form $y \leq a_i x + b_i$. Which point in the plane satisfies the largest number of them? Construct the arrangement of the lines. All points in any region or *cell* of this arrangement satisfy exactly the same set of constraints, so we need to test only one point per cell in order to find the global maximum.

Thinking of geometric problems in terms of the appropriate features in an arrangement can be very useful in formulating algorithms. Unfortunately, it must be admitted that arrangements are not as popular in practice as might be supposed. First, a certain depth of understanding is required to apply them correctly. Second, there have been few available implementations of the fundamental algorithms, a situation that is partially addressed below.

- *What do you want to do with the arrangement?* – Given an arrangement and a query point, we are often interested in identifying which cell of the arrangement contains the point. This is the problem of point location, discussed in

Section 8.6.7. Given an arrangement of lines or line segments, we are often interested in computing all points of intersection of the lines. The problem of intersection detection is discussed in Section 8.6.8.

- *How big will your arrangement be?* – Algorithms for constructing arrangements are incremental. Beginning with an arrangement of one or two lines, subsequent lines are inserted into the arrangement one at a time, building larger and larger arrangements. To insert a new line, we start on the leftmost cell containing the line and walk over the arrangement to the right, moving from cell to neighboring cell and splitting into two pieces those cells that contain the new line.

  A geometric fact called the *zone theorem* implies that the kth line inserted cuts through k cells of the arrangement, and further that $O(k)$ total edges form the boundary of these cells. This means that we can scan through each edge of every cell we encounter on our insertion walk, confident that linear total work will be performed while inserting the line into the arrangement. Therefore, the total time to insert all n lines in constructing the full arrangement is $O(n^2)$.

- *Does your input consist of points instead of lines?* – Although lines and points seem to be different geometric objects, such appearances can be misleading. Through the use of *duality transformations*, we can turn line L into point p and vice versa:

$$L : y = 2ax - b \leftrightarrow p : (a, b)$$

Duality is important because we can now apply line arrangements to point problems, often with surprising results.

For example, suppose we are given a set of n points, and we want to know whether any three of them all lie on the same line. This sounds similar to the degeneracy testing problem discussed above. Not only is it similar, it is *exactly the same*, with only the role of points and lines exchanged. The answer follows from taking our points, dualizing them into lines as above, constructing the arrangement as above, and then searching for a vertex with three lines passing through it. The dual of this vertex gives the line on which the three initial vertices lie.

Once we have constructed an arrangement through incremental methods, it often becomes useful to traverse each face of the arrangement exactly once. Such traversals are called *sweepline algorithms* and are discussed in some detail in Section 8.6.8. The basic procedure is to sort the intersection points by x-coordinate and then walk from left to right while keeping track of all we have seen.

Implementations: *Arrange* is a package written in C by Michael Goldwasser for maintaining arrangements of polygons in either the plane or on the sphere. Polygons may be degenerate and hence represent arrangements of lines. A randomized incremental construction algorithm is used and efficient point location on the arrangement supported. Polygons may be inserted but not deleted from the arrangement, and arrangements of several thousand vertices and edges can be constructed in a few seconds. Arrange is available from http://theory.stanford.edu/people/wass/wass.html.

LEDA (see Section 9.1.1) provides a function that constructs an embedded planar graph from a set of line segments, essentially constructing their arrangement.

Notes: Edelsbrunner [Ede87] provides a comprehensive treatment of the combinatorial theory of arrangements, plus algorithms on arrangements with applications. It is an essential reference for anyone seriously interested in the subject. Good expositions on constructing arrangements include [O'R94].

Arrangements generalize naturally beyond two dimensions. Instead of lines, the space decomposition is defined by planes (or beyond 3-dimensions, *hyperplanes*). In general dimensions, the zone theorem states that any arrangement of n d-dimensional hyperplanes has total complexity $O(n^d)$, and any single hyperplane intersects cells of complexity $O(n^{d-1})$. This provides the justification for the incremental construction algorithm for arrangements. Walking around the boundary of each cell to find the next cell that the hyperplane intersects takes time proportional to the number of cells created by inserting the hyperplane.

The history of the zone theorem has become somewhat muddled, because the original proofs were later found to be in error in higher dimensions. See [ESS93] for a discussion and a correct proof. The theory of Davenport-Schintzel sequences is intimately tied into the study of arrangements. It is presented in [SA95].

The naive algorithm for sweeping an arrangement of lines sorts the n^2 intersection points by x-coordinate and hence requires $O(n^2 \lg n)$ time. The *topological sweep* [EG89, EG91] eliminates the need to sort, and it traverses the arrangement in quadratic time. This algorithm is readily implementable and can be applied to speed up many sweepline algorithms.

Related Problems: Intersection detection (see page 370), point location (see page 367).

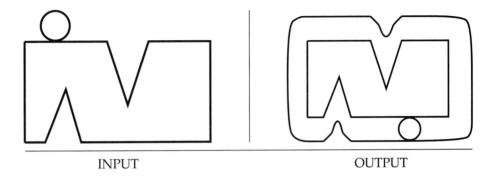

INPUT	OUTPUT

8.6.16 Minkowski Sum

Input description: Point sets or polygons A and B, with n and m vertices, respectively.

Problem description: What is the convolution of A and B, i.e. the Minkowski sum $A + B = \{x + y \mid x \in A, y \in B\}$?

Discussion: Minkowski sums are useful geometric operations that can be used to *fatten* objects in appropriate ways. For example, a popular approach to motion planning for polygonal robots in a room with polygonal obstacles (see Section 8.6.14) fattens each of the obstacles by taking the Minkowski sum of them with the shape of the robot. This reduces the problem to moving a point from the start to the goal using a standard shortest-path algorithm. Another application is in shape simplification (see Section 8.6.12), where we fatten the boundary of an object to create a channel and then define as the shape the minimum link path lying within this channel. Similarly, convolving an irregular object with a small circle will help smooth out the boundaries by eliminating minor nicks and cuts.

The definition of Minkowski sum assumes that the polygons A and B have been positioned on a coordinate system:

$$A + B = \{x + y | x \in A, y \in B\}$$

where $x + y$ is the vector sum of two points. Thinking of this in terms of translation, the Minkowski sum is the union of all translations of A by a point defined within B. Issues arising in computing Minkowski sums include:

- *Are your objects rasterized images or explicit polygons?* – The definition of Minkowski summation suggests a simple algorithm if A and B are rasterized images, and thus contain a number of pixels proportional to their area. Initialize a sufficiently large matrix of pixels by determining the size of the convolution of the bounding boxes of A and B. For each pair of points in A and B, sum up their coordinates and darken the appropriate pixel. These algorithms get somewhat more complicated if an explicit polygonal representation of the Minkowski sum is needed.

- *Are your objects convex or nonconvex?* – The complexity of computing Minkowski sum depends in a serious way on the shape of the polygons. If both A and B are convex, the Minkowski sum can be found in $O(n + m)$ time by tracing the boundary of one polygon with another. If one of them is nonconvex, the *size* of the sum alone can be as large as $\Theta(nm)$. Even worse is when both A and B are nonconvex, in which case the *size* of the sum can be as large as $\Theta(n^2m^2)$. Be aware that the Minkowski sum of nonconvex polygons can have a certain ugliness to it. For example, holes can be either created or destroyed.

Although more efficient algorithms exist, a straightforward approach to computing the Minkowski sum is based on triangulation and union. First, triangulate both polygons, then compute the Minkowski sum of each triangle of A against each triangle of B. The sum of a triangle against another triangle is easy to compute and is a special case of convex polygons, discussed below. The union of these $O(nm)$ convex polygons will be $A + B$. Algorithms for computing the union of polygons are based on plane sweep, as discussed in Section 8.6.8.

Computing the Minkowski sum of two convex polygons is easier than the general case, because the sum will always be convex. For convex polygons, it is easiest to slide A along the boundary of B and compute the sum edge by edge. This is the best approach for triangles against triangles as well.

Implementations: To date, we have not uncovered a suitable Minkowski sum code. When we do, it will be made available on http://www.cs.sunysb.edu/~algorith, the Algorithm Repository site.

Notes: Good expositions on algorithms for Minkowski sums include [O'R94]. The fastest algorithms for various cases of Minkowski sums include [KOS91, Sha87]. Kedem and Sharir [KS90] present an efficient algorithm for translational motion planning for polygonal robots, based on Minkowski sums.

Related Problems: Thinning (see page 377), motion planning (see page 389), simplifying polygons (see page 383).

8.7 Set and String Problems

Sets and strings both represent collections of objects. The primary difference is whether order matters. Sets are collections of symbols whose order is assumed to carry no significance, while the arrangement of symbols is exactly what defines a string.

The assumption of a canonical order makes it possible to solve string problems much more efficiently than set problems, through techniques such as dynamic programming and advanced data structures like suffix trees. The interest in and importance of string processing algorithms have been increasing, largely due to biological and text-processing applications. A product of this interest are three recent books on string algorithms:

- *Crochemore and Rytter* [CR94] – A comprehensive book on advanced string algorithms, but somewhat formal and fairly difficult to follow.
- *Stephen* [Ste94] – A reasonably gentle introduction to basic string algorithmics. Possibly the best available book for the beginner.
- *Gusfield* [Gus97] – This is now the most comprehesive introduction to string algorithms. It contains a thorough discussion on suffix trees, with new, clear formulations of classical exact string-matching algorithms.

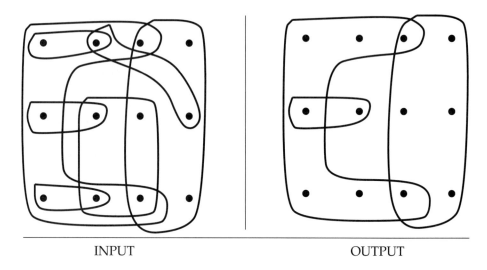

INPUT   OUTPUT

8.7.1 Set Cover

Input description: A set of subsets $S = \{S_1, \ldots, S_m\}$ of the universal set $U = \{1, \ldots, n\}$.

Problem description: What is the smallest subset T of S such that $\cup_{i=1}^{|T|} T_i = U$?

Discussion: Set cover arises when you try to efficiently acquire or represent items that have been packaged in a fixed set of lots. You want to obtain all the items, while buying as few lots as possible. Finding *a* cover is easy, because you can always buy one of each lot. However, by finding a small set cover you can do the same job for less money.

An interesting application of set cover is Boolean logic minimization. We are given a particular Boolean function of k variables, which for each of the 2^k possible input vectors describes whether the desired output is 0 or 1. We seek the simplest circuit that exactly implements this function. One approach is to find a disjunctive normal form (DNF) formula on the variables and their complements, such as $x_1\bar{x}_2 + \bar{x}_1\bar{x}_2$. We could build one "and" term for each input vector and then "or" them all together, but we might save considerably by factoring out common subsets of variables. Given a set of feasible "and" terms, each of which covers a subset of the vectors we need, we seek to "or" together the smallest number of terms that realize the function. This is exactly the set cover problem.

There are several variations of set cover problems to be aware of:

- *Are you allowed to cover any element more than once?* – The distinction here is between set covering and set packing, the latter of which is discussed in Section 8.7.2. If we are allowed to cover elements more than once, as in the logic minimization problem above, we should take advantage of this freedom, as it usually results in a smaller covering.

- *Are your sets derived from the edges or vertices of a graph?* – Set cover is a very general problem, and it includes several useful graph problems as special cases. Suppose instead that you seek the smallest set of edges in a graph that covers each vertex at least once. The solution is to find a maximum matching in the graph (see Section 8.4.6), and then add arbitrary edges to cover the remaining vertices. Suppose you seek the smallest set of vertices in a graph that covers each edge at least once. This is the vertex cover problem, discussed in Section 8.5.3.

    It is instructive to model vertex cover as an instance of set cover. Let the universal set U be the set of edges $\{e_1, \ldots, e_m\}$. Construct n subsets, with S_i consisting of the edges incident on vertex v_i. Although vertex cover is just a set cover problem in disguise, you should take advantage of the fact that better algorithms exist for vertex cover.

- *Do your subsets contain only two elements each?* – When all of your subsets have at most two elements each, you are in luck. This is about the only special case that you can solve efficiently to optimality, by using the matching technique described above. Unfortunately, as soon as your subsets get to have three elements each, the problem becomes NP-complete.

- *Do you want to cover elements with sets, or sets with elements?* – In the *hitting set* problem, we seek a small number of items that together represent an entire population. Formally, we are given a set of subsets $S = \{S_1, \ldots, S_m\}$ of the universal set $U = \{1, \ldots, n\}$, and we seek the smallest subset $T \subset U$ such that each subset S_i contains at least one element of T. Thus $S_i \cap T \neq \emptyset$ for all $1 \leq i \leq m$. Suppose we desire a small Congress with at least one representative of each ethnic group. If each ethnic group is represented as a subset of people, the minimum hitting set is the smallest possible politically correct Congress. Hitting set is illustrated in Figure 8-6.

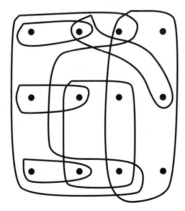 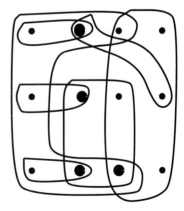

Figure 8-6. Hitting set is dual to set cover

Hitting set is, in fact, *dual* to set cover, meaning it is exactly the same problem in disguise. Replace each element of U by a set of the names of the subsets that contain it. Now S and U have exchanged roles, for we seek a set of subsets from U to cover all the elements of S. This is is exactly the set cover problem. Thus we can use any of the set cover codes below to solve hitting set after performing this simple translation.

Since the vertex cover problem is NP-complete, the set cover problem must be at least as hard. In fact, it is somewhat harder. Approximation algorithms do no worse than twice optimal for vertex cover, but only a $\Theta(\lg n)$ times optimal approximation algorithm exists for set cover.

The greedy heuristic is the right approach for set cover. Begin by placing the largest subset in the set cover, and then mark all its elements as covered. We will repeatedly add the subset containing the largest number of uncovered elements, until all elements are covered. This heuristic always gives a set cover using at most $\ln n$ times as many sets as optimal, and in practice it usually does a lot better.

The simplest implementation of the greedy heuristic sweeps through the entire input instance of m subsets for each greedy step. However, by using such data structures as linked lists and a bounded-height priority queue (see Section 8.1.2), the greedy heuristic can be implemented in $O(S)$ time, where $S = \bigcup_{i=1}^{m} |S_i|$ is the size of the input representation.

It pays to check whether there are certain elements that exist in only a few subsets, ideally only one. If so, we should select the biggest subsets containing these elements at the very beginning. We will have to take them eventually, and they carry with them extra elements that we might have to pay to cover by waiting until later.

Simulated annealing is likely to produce somewhat better set covers than these simple heuristics, if that is important for your application. Backtracking can be used to guarantee you an optimal solution, but typically it is not worth the computational expense.

Implementations: Pascal implementations of an exhaustive search algorithm for set packing, as well as heuristics for set cover, appear in [SDK83]. See Section 9.1.6.1.

Notes: Survey articles on set cover include [BP76]. See [CK75] for a computational study of set cover algorithms. An excellent exposition on algorithms and reduction rules for set cover is presented in [SDK83].

Good expositions of the greedy heuristic for set cover include [CLR90]. An example demonstrating that the greedy heuristic for set cover can be as bad as $\lg n$ is presented in [Joh74, PS82]. This is not a defect of the heuristic. Indeed, it is provably hard to approximate set cover to within an approximation factor better than $(1/4) \lg n$ [LY93].

Related Problems: Matching (see page 287), vertex cover (see page 317), set packing (see page 401).

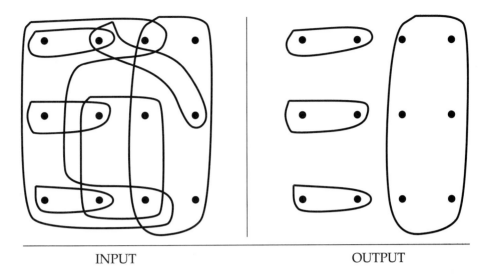

INPUT   OUTPUT

8.7.2 Set Packing

Input description: A set of subsets $S = \{S_1, \ldots, S_m\}$ of the universal set $U = \{1, \ldots, n\}$.

Problem description: What is the largest number of mutually disjoint subsets from S?

Discussion: Set packing problems arise in partitioning applications, where we need to partition elements under strong constraints on what is an allowable partition. The key feature of packing problems is that no elements are permitted to be covered by more than one set. Consider the problem of finding the maximum independent set in a graph G, discussed in Section 8.5.2. We seek a large subset of vertices such that each edge is adjacent to at most one of the selected vertices. To model this as set packing, let the universal set consist of all edges of G, and subset S_i consist of all edges incident on vertex v_i. Any set packing corresponds to a set of vertices with no edge in common, in other words, an independent set.

Scheduling airline flight crews to airplanes is another application of set packing. Each airplane in the fleet needs to have a crew assigned to it, consisting of a pilot, copilot, and navigator. There are constraints on the composition of possible crews, based on their training to fly different types of aircraft, as well as any personality conflicts. Given all possible crew and plane combinations, each represented by a subset of items, we need an assignment such that each plane and each person is in exactly one chosen combination. After all, the same person cannot be on two different planes, and every plane needs a crew. We need a perfect packing given the subset constraints.

Set packing is used here to represent a bunch of problems on sets, all of which are NP-complete and all of which are quite similar:

- *Must every element from the universal set appear in one selected subset?* – In the *exact cover* problem, we seek some collection of subsets such that each element is covered exactly once. The airplane scheduling problem above has the flavor of exact covering, since every plane and crew has to be employed.

    Unfortunately, exact cover is similar to that of Hamiltonian cycle in graphs. If we really must cover all the elements exactly once, and this existential problem is NP-complete, then all we can do is exponential search. This will be prohibitive unless there are so many solutions that we will stumble upon one quickly.

    Things will be far better if we can be content with a partial solution, say by adding each element of U as a singleton subset of S. Thus we can expand any set packing into an exact cover by mopping up the unpacked elements of U with singleton sets. Now our problem is reduced to finding a minimum-cardinality set packing, which can be attacked via heuristics, as discussed below.

- *What is the penalty for covering elements twice?* – In set cover (see Section 8.7.1), there is no penalty for elements existing in many selected subsets. In pure set packing, any such violation is forbidden. For many such problems, the truth lies somewhere in between. Such problems can be approached by charging the greedy heuristic more to select a subset that contains previously covered elements than one that does not.

The right heuristics for set packing are greedy, and similar to those of set cover (see Section 8.7.1). If we seek a packing with many sets, then we repeatedly select the smallest subset, delete all subsets from S that clash with it, and repeat. If we seek a packing with few subsets, we do the same but always pick the largest possible subset. As usual, augmenting this approach with some exhaustive search or randomization (in the form of simulated annealing) is likely to yield better packings at the cost of additional computation.

Implementations: Since set cover is a more popular and more tractable problem than set packing, it might be easier to find an appropriate implementation to solve the cover problem. Many such implementations should be readily modifiable to support certain packing constraints.

Pascal implementations of an exhaustive search algorithm for set packing, as well as heuristics for set cover, appear in [SDK83]. See Section 9.1.6.1 for details on ftp-ing these codes.

Notes: An excellent exposition on algorithms and reduction rules for set packing is presented in [SDK83], including the airplane scheduling application discussed above. Survey articles on set packing include [BP76].

Related Problems: Independent set (see page 315), set cover (see page 398).

" You will always have my love,
my love,   for the love I love is
lovely as love itself."

$\boxed{\text{love}}$?

" You will always have my $\boxed{\text{love}}$,
my $\boxed{\text{love}}$, for the $\boxed{\text{love}}$ I $\boxed{\text{love}}$
is $\boxed{\text{love}}$ ly as $\boxed{\text{love}}$ itself."

INPUT	OUTPUT

8.7.3 String Matching

Input description: A text string t of length n. A pattern string p of length m.

Problem description: Find the first (or all) instances of the pattern in the text.

Discussion: String matching is fundamental to database and text processing applications. Every text editor must contain a mechanism to search the current document for arbitrary strings. Pattern matching programming languages such as Perl and Awk derive much of their power from their built-in string matching primitives, making it easy to fashion programs that filter and modify text. Spelling checkers scan an input text for words in the dictionary and reject any strings that do not match.

Several issues arise in identifying the right string matching algorithm for the job:

- *Are your search patterns and/or texts short?* – If your strings are sufficiently short and your queries sufficiently infrequent, the brute-force $O(mn)$-time search algorithm will likely suffice. For each possible starting position $1 \leq i \leq n - m + 1$, it tests whether the m characters starting from the ith position of the text are identical to the pattern.

  For very short patterns, say $m \leq 4$, you can't hope to beat brute force by much, if at all, and you shouldn't try. Further, since we can reject the possibility of a match at a given starting position the instant we observe a text/pattern mismatch, we expect much better than $O(mn)$ behavior for typical strings. Indeed, the trivial algorithm *usually* runs in linear time. But the worst case certainly can occur, as with pattern $p = a^m$ and text $t = (a^{m-1}b)^{n/m}$.

- *What about longer texts and patterns?* – By being more clever, string matching can be performed in worst-case linear time. Observe that we need not begin the search from scratch on finding a character mismatch between the pattern and text. We know something about the subsequent characters of the string as a result of the partial match from the previous position. Given a long partial match from position i, we can jump ahead to the first character position in the pattern/text that will provide new information about the text starting from position $i + 1$. The Knuth-Moris-Pratt algorithm preprocesses the search pattern to construct such a jump table efficiently. The details are tricky to get correct, but the resulting implementations yield short, simple programs.

Even better in practice is the Boyer-Moore algorithm, although it offers similar worst-case performance. Boyer-Moore matches the pattern against the text from right to left, in order to avoid looking at large chunks of text on a mismatch. Suppose the pattern is *abracadabra*, and the eleventh character of the text is x. This pattern cannot match in any of the first eleven starting positions of the text, and so the next necessary position to test is the 22nd character. If we get very lucky, only n/m characters need ever be tested. The Boyer-Moore algorithm involves two sets of jump tables in the case of a mismatch: one based on pattern matched so far, the other on the text character seen in the mismatch. Although somewhat more complicated than Knuth-Morris-Pratt, it is worth it in practice for patterns of length $m > 5$.

- *Will you perform multiple queries on the same text?* – Suppose you were building a program to repeatedly search a particular text database, such as the Bible. Since the text remains fixed, it pays build a data structure to speed up search queries. The suffix tree and suffix array data structures, discussed in Section 8.1.3, are the right tools for the job.

- *Will you search many texts using the same patterns?* – Suppose you were building a program to screen out dirty words from a text stream. Here, the set of patterns remains stable, while the search texts are free to change. In such applications, we may need to find all occurrences of each of k different patterns, where k can be quite large.

  Performing a linear-time search for each of these patterns yields an $O(k(m + n))$ algorithm. If k is large, a better solution builds a single finite automaton that recognizes each of these patterns and returns to the appropriate start state on any character mismatch. The Aho-Corasick algorithm builds such an automaton in linear time. Space savings can be achieved by optimizing the pattern recognition automaton, as discussed in Section 8.7.7. This algorithm was used in the original version of *fgrep*.

  Sometimes multiple patterns are specified not as a list of strings, but concisely as a regular expression. For example, the regular expression $a + a(a + b + c)^*a$ matches any string on (a, b, c) that begins and ends with an a, including a itself. The best way to test whether input strings are described by a given regular expression R is to construct the finite automaton equivalent to R and then simulate the machine on the string. Again, see Section 8.7.7 for details on constructing automata from regular expressions.

  When the patterns are specified by context-free grammars instead of regular expressions, the problem becomes one of parsing, discussed in books on compilers and programming languages.

- *What if our text or pattern contains a spelling error?* – Finally, observe that the algorithms discussed here work only for exact string matching. If you must allow some tolerance for spelling errors, your problem is approximate string matching, which is thoroughly discussed in Section 8.7.4.

Implementations: SPARE Parts is a string pattern recognition toolkit, written in C++ by Bruce Watson. It provides production-quality implementations of all major variants of the classical string matching algorithms for single patterns (both Knuth-Morris-Pratt and Boyer-Moore) and multiple patterns (both Aho-Corasick and Commentz-Walter). SPARE Parts is available by anonymous ftp from ftp.win.tue.nl in /pub/techreports/pi/watson.phd/. A greatly improved commercial version is available from www.RibbitSoft.com.

XTango (see Section 9.1.5) provides animations for both the Boyer-Moore and Knuth-Morris-Pratt algorithms. The C source code for each animation is included.

Implementations in C and Pascal of several algorithms for exact and approximate string matching appear in [GBY91]. Sedgewick provides similar implementations of Knuth-Morris-Pratt, Rabin-Karp, and Boyer-Moore in C++. See Section 9.1.6.6 for details on both codes.

Implementations in C of the Boyer-Moore, Aho-Corasick, and regular expression matching algorithms appear in [BR95]. The code for these algorithms is printed in the text and available on disk for a modest fee.

Notes: All books on string algorithms contain thorough discussions of exact string matching, including [CR94, Ste94, Gus97]. Good expositions on the Boyer-Moore [BM77] and Knuth-Morris-Pratt algorithms [KMP77] include [Baa88, CLR90, Man89].

Aho [Aho90] provides a good survey on algorithms for pattern matching in strings, particularly where the patterns are regular expressions instead of strings, and for the Aho-Corasick algorithm for multiple patterns [AC75]. An algorithm merging Aho-Corasick and Boyer-Moore can be faster for small numbers of patterns [CW79], but the window where it wins is apparently fairly narrow.

Empirical comparisons of string matching algorithms include [DB86, Hor80, dVS82]. Which algorithm performs best depends upon the properties of the strings and the size of the alphabet. For long patterns and texts, I recommend that you use the best implementation of Boyer-Moore that you can find.

An interesting classical problem is determining the minimum number of comparisons needed to perform exact string matching. One version of Boyer-Moore never makes more than $2n$ comparisons independent of the number of occurrences of the pattern in the text [AG86]. More recent results are very technical, depending upon the details of the model and the alphabet size. There is a lower bound of $n - m + 1$ text characters that any algorithm must be examine in the worst case [Riv77]. The history of string matching algorithms is somewhat checkered because several published proofs were incorrect or incomplete [Gus97].

The Karp-Rabin algorithm [KR87] uses a hash function to perform string matching in linear expected time. Its worst-case time remains quadratic, and its performance in practice appears somewhat worse than the character comparison methods described above.

Related Problems: Suffix trees (see page 183), approximate string matching (see page 406).

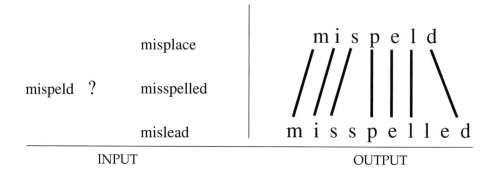

misplace	
mispeld ? misspelled	
mislead	
INPUT	OUTPUT

8.7.4 Approximate String Matching

Input description: A text string t and a pattern string p. An edit cost bound k.

Problem description: Can we transform t to p using at most k insertions, deletions, and substitutions?

Discussion: Approximate string matching is fundamental to text processing, because we live in an error-prone world. Any spelling correction program must be able to identify the closest match for any text string not found in a dictionary. Genbank has become a fundamental tool for molecular biology by supporting homology (similarity) searches on DNA sequences. Suppose you were to sequence a new gene in man, and you discovered that it is similar to the hemoglobin gene in rats. It is likely that this new gene also produces hemoglobin, and any differences are the result of genetic mutations during evolution.

I once encountered approximate string matching in evaluating the performance of an optical character recognition system that we built. After scanning and recognizing a test document, we needed to compare the correct answers with those produced by our system. To improve our system, it was important to count how often each pair of letters were getting confused and to identify gibberish when the program was trying to make out letters where none existed. The solution was to do an alignment between the two texts. Insertions and deletions corresponded to gibberish, while substitutions signaled errors in our recognizers. This same principle is used in file difference programs, which identify the lines that have changed between two versions of a file.

When no errors are permitted, the problem becomes that of exact string matching, which is presented in Section 8.7.3. Here, we restrict our discussion to dealing with errors.

Dynamic programming provides the basic approach to approximate string matching, as discussed in Section 3.1.3. Let $D[i, j]$ denote the cost of editing the first i characters of the text string t into the first j characters of the pattern string p. The recurrence follows because we must have done *something* with t_i and p_j. The only options are matching them, substituting one for the other, deleting t_i, or inserting a match for p_j. Thus $D[i, j]$ is the minimum of the costs of these possibilities:

1. If $t_i = p_j$ then $D[i-1, j-1]$ else $D[i-1, j-1]$ + substitution cost.
2. $D[i-1, j]$ + deletion cost of p_j.
3. $D[i, j-1]$ + deletion cost of t_i.

Several issues remain before we can make full use of this recurrence:

- *Do I want to match the pattern against the full text, or against a substring? –* Appropriately initializing the boundary conditions of the recurrence distinguishes between the algorithms for string matching and substring matching. Suppose we want to align the full text against the full pattern. Then the cost of $D[i, 0]$ must be that of deleting the first i characters of the text, so $D[i, 0] = i$. Similarly, $D[0, j] = j$.

  Now suppose that the pattern may occur anywhere within the text. The proper cost of $D[i, 0]$ is now 0, since there should be no penalty for starting the alignment in the ith position. The cost of $D[0, j]$ remains j, however, since the only way to match the first j pattern characters with nothing is to delete all of them. The cost of the best pattern match against the text will be given by $\min_{k=1}^{n} D[k, m]$.

- *How should I select the substitution and insertion/deletion costs? –* The basic algorithm can be easily modified to use different costs for insertion, deletion, and the substitutions of specific pairs of characters. Which costs you use depends upon what you are planning to do with the alignment.

  The most common cost assignment charges the same for insertion, deletion, or substitution. Charging a substitution cost of more than insertion + deletion is a good way to ensure that substitutions never get performed, since it will be cheaper to edit both characters out of the string. With just insertion and deletion to work with, the problem reduces to *longest common subsequence*, discussed in Section 8.7.8. Often, it pays to tweak the edit distance costs and study the resulting alignments until you find a set that does the job.

- *How do I get the actual alignment of the strings, instead of just the cost? –* To obtain the transcript of the editing operations performed in achieving the minimum cost alignment, we can work backwards from the complete cost matrix D. To get to cell $D[n, m]$, we had to come from one of $D[n-1, m]$ (insertion), $D[n, m-1]$ (deletion), or $D[n-1, m-1]$ (substitution/match). Which of the three options was chosen can be reconstructed given these costs and the characters t_n and p_m. By continuing to work backwards from the previous cell, we can trace the entire path and thus reconstruct the alignment.

- *What if the two strings are very similar to each other? –* The dynamic programming algorithm can be thought of as finding a shortest path across an $n \times n$ grid, where the cost of each edge depends upon which operation it represents. If we seek an alignment involving a combination of at most d insertions, deletions, and substitutions, we need only traverse the band of $O(dn)$ cells within a distance d of the central diagonal. If no such low-cost alignment exists within this band, then no low-cost alignment can exist in the full cost matrix.

- *How can I minimize the required storage? –* Dynamic programming takes both quadratic time and space. For many applications, the space required to store the dynamic programming table is a much more serious problem. Observe that only $O(\min(m, n))$ space is needed to compute $D[m, n]$, since we need only maintain two active rows (or columns) of the matrix in order to compute the final value. The entire matrix is required only if we need to reconstruct the actual sequence alignment.

To save space, we can use Hirschberg's clever recursive algorithm. During one pass of the linear-space algorithm above to compute $D[m, n]$, we maintain all the values for the $(m/2)$nd column and identify which middle-element cell $D[m/2, x]$ was used to optimize $D[m, n]$. This reduces our problem to finding the best paths from $D[1, 1]$ to $D[m/2, x]$ and from $D[m/2, x]$ to $D[m/2, n]$, both of which can be solved recursively. Each time we throw away half of the matrix elements from consideration, and so the total time remains $O(mn)$. This linear-space algorithm proves to be a big win in practice on long strings, although it is somewhat more difficult to program.

- *Does string similarity mean that the strings sound alike?* – Other models of approximate pattern matching become more appropriate than edit distance in certain applications. Particularly interesting is *Soundex*, a hashing scheme that attempts to pair up English words that sound alike. This can be useful in testing whether two names that have been spelled differently are likely to be the same. For example, my last name is often spelled "Skina", "Skinnia", "Schiena", and occasionally "Skiena". All of these hash to the same Soundex code, *S25*.

  The algorithm works by dropping vowels and silent letters, removing doubled letters, and then assigning the remaining letters numbers from the following classes: *BFPV* gets a 1, *CGJKQSXZ* gets a 2, *DT* gets a 3, *L* gets a 4, *MN* gets a 5, and *R* gets a 6. The code starts with the first letter and contains at most three digits. This all sounds very hokey, but experience shows that it works reasonably well. Experience indeed: Soundex has been used since the 1920's.

Implementations: The best available software tools for approximate pattern matching are *glimpse* and *agrep* [WM92a, WM92b], developed by Manber and Wu at the University of Arizona and available from http://glimpse.cs.arizona.edu/. Glimpse is a tool for building and using an index to search through file systems, while agrep (approximate general regular expression pattern matcher) is a tool supporting text search with spelling errors. Both programs are widely used and respected.

ht://Dig is an alternative WWW text search engine from Andrew Scherpbier, which contains implementations of Soundex and Metaphone. It is available from http://htdig.sdsu.edu/ and is released under the GNU general public license.

Implementations in C of the Soundex and dynamic programming edit-distance algorithms appear in [BR95]. The code for these algorithms is printed in the text and is available on disk for a modest fee.

Bare bones implementations in C and Pascal of several algorithms for exact and approximate string matching appear in [GBY91]. See Section 9.1.6.2 for further details.

Notes: The wide range of applications for approximate string matching was made apparent in Sankoff and Kruskal's book [SK83], which remains a useful historical reference for the problem. Surveys on approximate pattern matching include [HD80]. The basic dynamic programming alignment algorithm is attributed to [WF74], although it is apparently folklore. The edit distance between two strings is sometimes referred to as the *Levenshtein distance*. Expositions of dynamic programming to compute Levenshtein distance include [Baa88, CLR90, Man89]. Expositions of Hirschberg's linear-space algorithm [Hir75] include [CR94, Gus97].

Masek and Paterson [MP80] compute the edit distance between m- and n-length strings in time $O(mn/\log(\min\{m, n\}))$ for constant-sized alphabets, using ideas from the four Russians algorithm for Boolean matrix multiplication [ADKF70]. The shortest path formulation leads to a variety of algorithms that are good when the edit distance is small, including an $O(n \lg n + d^2)$ algorithm due to Myers [Mye86]. Longest increasing subsequence can be done in $O(n \lg n)$ time [HS77], as presented in [Man89].

Soundex was invented and patented by M. K. Odell and R. C. Russell. Expositions on Soundex include [BR95, Knu73b]. Metaphone is a recent attempt to improve on Soundex [BR95, Par90].

Related Problems: String matching (see page 403), longest common substring (see page 422).

Fourscore and seven years ago our father brought forth on
this continent a new nation conceived in Liberty and dedicated
to the proposition that all men are created equal. Now we are
engaged in a great civil war testing whether that nation or any
nation so conceived and so dedicated can long endure. We are
met on a great battlefield of that war. We have come to
dedicate a portion of that field as a final resting place for those
who here gave their lives that the nation might live. It is altogether
fitting and we can not consecrate we can not hallow this groud.
The brave men living and dead who struggled here have consecrated
it for above our poor power to add or detract. The world will little
note nor long remember what we say here but it can never forget
what they did here. It is for us the living here have thus far so nobly
advanced. It is rather for us to be here dedicated to the great task
remaining before us that from these honored dead we take increased
devotion to that cause for which they here gave the last full measure
of devotion that we here highly resolve that these dead shall not have
died in vain that this nation under God shall have a new birth of
freedom and that government of the people by the people for the
people shall not perish from the earth.

Fourscore and seven years ago our father brought forth on
this continent a new nation conceived in Liberty and dedicated
to the proposition that all men are created equal. Now we are
engaged in a great civil war testing whether that nation or any
nation so conceived and so dedicated can long endure. We are
met on a great battlefield of that war. We have come to
dedicate a portion of that field as a final resting place for those
who here gave their lives that the nation might live. It is altogether
fitting and we can not consecrate we can not hallow this groud.
The brave men living and dead who struggled here have consecrated
it for above our poor power to add or detract. The world will little
note nor long remember what we say here but it can never forget
what they did here. It is for us the living here have thus far so nobly
advanced. It is rather for us to be here dedicated to the great task
remaining before us that from these honored dead we take increased
devotion to that cause for which they here gave the last full measure
of devotion that we here highly resolve that these dead shall not have
died in vain that this nation under God shall have a new birth of
freedom and that government of the people by the people for the
people shall not perish from the earth.

| INPUT | OUTPUT |

8.7.5 Text Compression

Input description: A text string S.

Problem description: A shorter text string S' such that S can be correctly reconstructed from S'.

Discussion: Secondary storage devices fill up quickly on every computer system, even though their capacity doubles each year. Decreasing storage prices have only increased interest in data compression, since there is now more data to compress than ever before. *Data compression* is the algorithmic problem of finding alternative, space-efficient encodings for a given data file. With the rise of computer networks, a new mission for data compression has arisen, that of increasing the effective bandwidth of networks by reducing the number of bits before transmission.

Data compression is a problem for which practical people like to invent ad hoc methods, designed for their particular applications. Sometimes these outperform general methods, but often they do not. The following issues arise in selecting the right data compression algorithm:

- *Must we exactly reconstruct the input text after compression?* – A primary issue in data compression algorithms is the question of lossy versus lossless encodings. Text applications typically demand *lossless* encodings, because users become disturbed whenever their data files get corrupted. However, fidelity is not such an issue in image or video compression, where the presence of small artifacts will be imperceptible to the viewer. Significantly greater compression ratios can be obtained using lossy compression, which is why most image/video/audio compression algorithms take advantage of this freedom.

- *Can I simplify my data before I compress it?* – The most effective way to free up space on a disk is to delete files you don't need. Likewise, any preprocessing you can do to a file to reduce its information content before compression will pay off later in better performance. For example, is it possible to eliminate extra blank spaces or lines from the file? Can the document be converted entirely to uppercase characters or have formatting information removed?

- *Does it matter whether the algorithm is patented?* – One concern is that many data compression algorithms are patented, in particular the LZW variation of the Lempel-Ziv algorithm discussed below. Further, Unisys, the owner of the patent, makes periodic attempts to collect. My personal (although not legal) recommendation is to ignore them, unless you are in the business of selling text compression software. If this makes you uncomfortable, note that there are other variations on the Lempel-Ziv algorithm that are not under patent protection and perform about as well. See the notes and implementations below.

- *How do I compress image data* – Run-length coding is the simplest lossless compression algorithm for image data, where we replace runs of identical pixel values with one instance of the pixel and an integer giving the length of the run. This works well on binary images with large regions of similar pixels (like scanned text) and terribly on images with many quantization levels and a little noise. It can also be applied to text with many fields that have been padded by blanks. Issues like how many bits to allocate to the count field and the traversal order converting the two-dimensional image to a stream of pixels can have a surprisingly large impact on the compression ratio.

  For serious image and video compression applications, I recommend that you use a lossy coding method and not fool around with implementing it yourself. JPEG is the standard high-performance image compression method, while MPEG is designed to exploit the frame-to-frame coherence of video. Encoders and decoders for both are provided in the implementation section.

- *Must compression and decompression both run in real time?* – For many applications, fast decompression is more important than fast compression, and algorithms such as JPEG exist to take advantage of this. While compressing video for a CD-ROM, the compression will be done only once, while decompression will be necessary anytime anyone plays it. In contrast, operating systems that increase the effective capacity of disks by automatically compressing each file will need a symmetric algorithm with fast compression times as well.

Although there are literally dozens of text compression algorithms available, they are characterized by two basic approaches. In *static algorithms*, such as Huffman codes, a single coding table is built by analyzing the entire document. In *adaptive algorithms*, such as Lempel-Ziv, a coding table is built on the fly and adapts to the local character distribution of the document. An adaptive algorithm will likely prove to be the correct answer:

- *Huffman codes* – Huffman codes work by replacing each alphabet symbol by a variable-length code string. ASCII uses eight bits per symbol in English text, which is wasteful, since certain characters (such as 'e') occur far more often than others (such as 'q'). Huffman codes compress text by assigning 'e' a short code word and 'q' a longer one.

Optimal Huffman codes can be constructed using an efficient greedy algorithm. Sort the symbols in increasing order by frequency. We will merge the two least frequently used symbols x and y into a new symbol m, whose frequency is the sum of the frequencies of its two child symbols. By replacing x and y by m, we now have a smaller set of symbols, and we can repeat this operation $n - 1$ times until all symbols have been merged. Each merging operation defines a node in a binary tree, and the left or right choices on the path from root-to-leaf define the bit of the binary code word for each symbol. Maintaining the list of symbols sorted by frequency can be done using priority queues, which yields an $O(n \lg n)$-time Huffman code construction algorithm.

Although they are widely used, Huffman codes have three primary disadvantages. First, you must make two passes over the document on encoding, the first to gather statistics and build the coding table and the second to actually encode the document. Second, you must explicitly store the coding table with the document in order to reconstruct it, which eats into your space savings on short documents. Finally, Huffman codes exploit only nonuniformity in symbol distribution, while adaptive algorithms can recognize the higher-order redundancy in strings such as $0101010101\ldots$.

- *Lempel-Ziv algorithms* – Lempel-Ziv algorithms, including the popular LZW variant, compress text by building the coding table on the fly as we read the document. The coding table available for compression changes at each position in the text. A clever protocol between the encoding program and the decoding program ensures that both sides of the channel are always working with the exact same code table, so no information can be lost.

  Lempel-Ziv algorithms build coding tables of recently-used text strings, which can get arbitrarily long. Thus it can exploit frequently-used syllables, words, and even phrases to build better encodings. Further, since the coding table alters with position, it adapts to local changes in the text distribution, which is important because most documents exhibit significant locality of reference.

  The truly amazing thing about the Lempel-Ziv algorithm is how robust it is on different types of files. Even when you know that the text you are compressing comes from a special restricted vocabulary or is all lowercase, it is very difficult to beat Lempel-Ziv by using an application-specific algorithm. My recommendation is not to try. If there are obvious application-specific redundancies that can safely be eliminated with a simple preprocessing step, go ahead and do it. But don't waste much time fooling around. No matter how hard you work, you are unlikely to get significantly better text compression than with *gzip* or *compress*, and you might well do worse.

Implementations: A complete list of available compression programs is provided in the comp.compression FAQ (frequently asked questions) file, discussed below. This FAQ will likely point you to what you are looking for, if you don't find it in this section.

The best general-purpose program for text compression is *gzip*, which implements a public domain variation of the Lempel-Ziv algorithm. It is distributed under the GNU software licence and can by obtained from ftp://prep.ai.mit.edu/pub/gnu/gzip-1.2.4.tar. Unix *compress* is another popular compression program based on the patented LZW algorithm. It is available from ftp://wuarchive.wustl.edu/packages/compression/compress-4.1.tar.

A JPEG implementation is available from ftp://ftp.uu.net/graphics/jpeg/jpegsrc.v6a.tar.gz. MPEG can be found at ftp://havefun.stanford.edu/pub/mpeg/MPEGv1.2.2.tar.Z.

Algorithm 673 [Vit89] of the *Collected Algorithms of the ACM* is a Pascal implementation of dynamic Huffman codes, which is a one-pass, adaptive text compression algorithm. See Section 9.1.2 for details on fetching this program.

Notes: Many books on data compression are available, but we highly recommend Bell, Cleary, and Witten [BCW90] and Storer [Sto88]. Another good source of information is the USENET newsgroup comp.compression. Check out its particularly comprehensive FAQ (frequently asked questions) compendium at location ftp://rtfm.mit.edu/pub/usenet/news.answers/compression-faq.

Good expositions on Huffman codes [Huf52] include [AHU83, BR95, CLR90, Eve79a, Man89]. Expositions on the LZW [Wel84, ZL78] algorithm include [BR95].

There is an annual IEEE Data Compression Conference, the proceedings of which should be studied seriously before attempting to develop a new data compression algorithm. On reading the proceedings, it will become apparent that this is a mature technical area, where much of the current work (especially for text compression) is shooting for fairly marginal improvements on special applications. On a more encouraging note, we remark that this conference is held annually at a world-class ski resort in Utah.

Related Problems: Shortest common superstring (see page 425), cryptography (see page 414).

The magic words are Squeamish Ossifrage.

I5&AE<&UA9VEC'=0

<F1s"F%R92!3<75E96UI<V

V@*3W-S:69R86=E+@K_

| INPUT | OUTPUT |

8.7.6 Cryptography

Input description: A plaintext message T or encrypted text E, and a key k.

Problem description: Encode T using k giving E, or decode E using k back to T.

Discussion: Cryptography has grown substantially in importance in recent years, as computer networks have made confidential documents more vulnerable to prying eyes. Cryptography is a way to increase security by making messages difficult to read if they fall into the wrong hands. Although the discipline of cryptography is at least two thousand years old, its algorithmic and mathematical foundations have recently solidified to the point where there can now be talk of provably secure cryptosystems.

There are three classes of cryptosystems everyone should be aware of:

- *Caesar shifts* – The oldest ciphers involve mapping each character of the alphabet to a different letter. The weakest such ciphers rotate the alphabet by some fixed number of characters (often 13), and thus have only 26 possible keys. Better is to use an arbitrary permutation of the letters, so there are 26! possible keys. Even so, such systems can be easily attacked by counting the frequency of each symbol and exploiting the fact that 'e' occurs more often than 'z'. While there are variants that will make this more difficult to break, none will be as secure as DES or RSA.

- *Data Encryption Standard (DES)* – This algorithm is based on repeatedly shuffling the bits of your text as governed by the key. The standard key length for DES (56 bits) is now considered too short for applications requiring the highest level of security. However, a simple variant called *triple DES* permits an effective key length of 112 bits by using three rounds of DES with two 56-bit keys. In particular, first encrypt with key1, then *decrypt* with key2, before finally encrypting with key1. There is a mathematical reason for using three rounds instead of two, and the encrypt-decrypt-encrypt pattern is used so that the scheme is equivalent to single DES when $key1 = key2$.

- *Rivest-Shamir-Adelman (RSA)* – RSA is a *public key* cryptosystem, meaning that different keys are used to encode and decode messages. Since the encoding key is of no help in decoding, it can be made public at no risk to security. The security of RSA is based on the difference in the computational complexity of factoring and primality testing (see Section 8.2.8). Encoding is (relatively) fast because it relies on primality testing to construct the key, while the hardness of decryption follows from that of factoring. Still, RSA is slow relative to other cryptosystems, roughly 100 to 1,000 times slower than DES.

The key issue in selecting a cryptosystem is identifying your paranoia level, i.e. deciding how much security you need. Who are you trying to stop from reading your stuff: your grandmother, local thieves, the Mafia, or the NSA? If you can use an accepted implementation of RSA, such as PGP discussed below, you should feel safe against just about anybody.

If there is an implementation of DES on your machine, that will likely be good enough for most applications. For example, I use DES to encrypt my final exam each semester, and it proved more than sufficient the time an ambitious student broke into my office looking for it. If the NSA had been breaking in, the story might have been different, although it is important to understand that *the most serious security holes are human, not algorithmic*. Making sure your password is long enough, hard to guess, and not written down is far more important than obsessing about the encryption algorithm.

Simple ciphers like the Caesar shift are fun and easy to program. For this reason, it is perhaps healthy to use them for applications needing only a casual level of security (such as hiding the punchlines of jokes). Since they are easy to break, they should never be used for serious security applications.

One thing that you should *never* do is mess around with developing your own novel cryptosystem. The security of DES and RSA is accepted largely because these systems have both survived over twenty years of public scrutiny. Over this period, many other cryptosystems have been proposed, proven vulnerable to attack, and then abandoned. This is not a field for amateurs. If you are charged with implementing a cryptosystem, carefully study a respected program such as PGP (discussed below) to see how its author, Philip Zimmermann, skillfully handled such issues as key selection and key distribution. Any cryptosystem is as strong as its weakest link.

There are several problems related to cryptography that arise often in practice:

- *How can I validate the integrity of data against corruption?* – In any communications application, there is need to validate that the transmitted data is identical to that which has been received. One solution is for the receiver to transmit the data back to the source and have the original sender confirm that the two texts are identical. This fails when the exact inverse of an error is made in the retransmission, but a more serious problem is that your available bandwidth is cut in half with such a scheme.

  A more efficient if less reliable method is to use a *checksum*, a simple mathematical function that hashes a long text down to a simple number or digit, and then transmit the checksum along with the text. The checksum can be recomputed on the receiving end and bells set off if the computed checksum is not identical to what was received. Perhaps the simplest checksum scheme just adds up the byte or character values and takes the sum modulo some constant, say $2^8 = 256$. Unfortunately, an error transposing two or more characters would go undetected under such a scheme, since addition is commutative.

Cyclic-redundancy check (CRC) provides a more powerful method for computing checksums and is used in most communications systems and internally in computers to validate disk drive transfers. These codes compute the remainder in the ratio of two polynomials, the numerator of which is a function of the input text. The design of these polynomials involves considerable mathematical sophistication and ensures that all reasonable errors are detected. The details of efficient computation are complicated enough that we recommend that you start from an existing implementation, described below.

- *How can I prove that a file has not been changed?* – If I send you a contract in electronic form, what is to stop you from editing the file and then claiming that your version was what we had really agreed to? I need a way to prove that any modification to a document is fraudulent. *Digital signatures* are a cryptographic way for me to stamp my document as genuine.

  Given a file, I can compute a checksum for it, and then encrypt this checksum using my own private key. I send you the file and the encrypted checksum. You can now edit the file, but to fool the judge you must also edit the encrypted checksum such that it can be decrypted to the correct checksum. With a suitably good checksum function, designing a file that yields the same checksum becomes an insurmountable problem.

- *How can I prove that I am me?* – *Authentication* is the process of one party convincing another that they are who they say they are. The historical solutions have involved passwords or keys, so I prove that I am who I say I am because I know my credit card number or carry an ID card. The problem with such schemes is that anyone who eavesdrops on this conversation or who steals my physical key can now successfully impersonate me.

  What we need is some way for me to convince you that I know my key, without actually telling you the key. One such method to do so is for you to send me a random number or text, and I use my key to encrypt your text and send it back to you. If you then decrypt this text and compare it to the message you sent me, you gain confidence that I am who I say I am. We can repeat this exercise on several random texts until sufficient authentication has been agreed upon. If we use a secure enough cryptosystem, we can be confident that an eavesdropper will not be able to deduce my key even given several plain and encrypted texts.

  Such *authentication protocols* of back-and-forth messages often involve the use of randomness to frustrate eavesdroppers. Different protocols satisfy particular needs and constraints about who has to know what. It is important to do some reading before attempting to design your own protocols. References are provided below.

Implementations: The USENET FAQ (frequently asked questions) file on cryptography provides a wealth of information, including pointers to implementations. Check it out at ftp://rtfm.mit.edu/pub/usenet/news.answers/cryptography-faq/.

Distributing cryptographic software is complicated by United States export restrictions, which make it illegal to export encryption software. PGP (Pretty Good Privacy) is such a good implementation of RSA that its author Philip Zimmerman was charged with export violations by federal authorities. PGP may be obtained from the Electronic Frontier Foundation (EFF) at http://www.eff.org/pub/Net_info/Tools/Crypto/PGP/.

A good discussion on checksums and cyclic-redundancy codes, with implementations in C, appear in [BR95]. The code for these algorithms is printed in the text and is available on disk for a modest fee.

The Stanford Graphbase (see Section 9.1.3) uses checksums to ensure that data files remain unmodified from the original distribution. Algorithm 536 [Kno79] of the *Collected Algorithms of the ACM* is an encryption function for passwords, written in Fortran. See Section 9.1.2 for further information.

Notes: Kahn [Kah67] presents the fascinating history of cryptography from ancient times to 1967 and is particularly noteworthy in light of the secretive nature of the subject. More recent and more technical works on cryptography include Denning [Den82] and Schneier [Sch94], the latter of which provides a through overview of different cryptographic algorithms, including implementations for sale. Rawlins [Raw92] provides a good introduction to cryptographic algorithms, from Caesar shift to public key to zero-knowledge proofs. An algorithm for breaking simple substitution ciphers appears in [PR79].

Expositions on the RSA algorithm [RSA78] include [CLR90]. The RSA Laboratories home page http://www.rsa.com/rsalabs/ is very informative. See [Sta95] for an excellent guide to PGP and its underlying algorithms.

The history of DES is well presented in [Sch94]. Particularly controversial was the decision by the NSA to limit key length to 56 bits, presumably short enough to be cracked by special-purpose computers costing on the order of several million dollars. Despite some theoretical progress in breaking DES analytically [BS93], the most significant threat remains special-purpose hardware.

MD5 [Riv92] is the secure hashing function used by PGP to compute digital signatures. Expositions include [Sch94, Sta95].

Related Problems: Factoring and primality testing (see page 221), text compression (see page 410)).

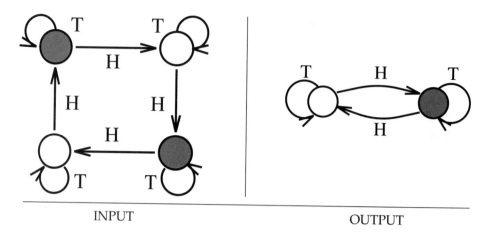

INPUT   OUTPUT

8.7.7   Finite State Machine Minimization

Input description: A deterministic finite automaton M.

Problem description: The smallest deterministic finite automaton M' such that M' behaves identically to M.

Discussion: Problems associated with constructing and minimizing finite state machines arise repeatedly in software and hardware design applications. Finite state machines are best thought of as pattern recognizers, and minimum-size machines correspond to recognizers that require less time and space. Complicated control systems and compilers are often built using finite state machines to encode the current state and associated actions, as well as the set of possible transitions to other states. Minimizing the size of this machine minimizes its cost.

Finite state machines are best thought of as edge-labeled directed graphs, where each vertex represents one of n states and each edge a transition from one state to the other on receipt of the alphabet symbol that labels the edge. The automaton above analyzes a given sequence of coin tosses, with the dark states signifying that an even number of heads have been observed. Automata can be represented using any graph data structure (see Section 8.1.4), or by using an $n \times |\Sigma|$ *transition matrix*, where $|\Sigma|$ is the size of the alphabet.

Finite state machines are often used to specify search patterns in the guise of *regular expressions*, which are patterns formed by and-ing, or-ing, and looping over smaller regular expressions. For example, the regular expression $a + a(a + b + c)^*a$ matches any string on (a, b, c) that begins and ends with an a (including a itself). The best way to test whether a string is described by a given regular expression (especially if many strings will be tested) is to construct the equivalent finite automaton and then simulate the machine on the string. See Section 8.7.3 for alternative approaches to string matching.

We consider three different problems on finite automata:

- *Minimizing deterministic finite state machines* – Transition matrices for finite automata quickly become prohibitively large for sophisticated machines, thus fueling the need for tighter encodings. The most direct approach is to eliminate redundant states in the automaton. As the example above illustrates, automata of widely varying sizes can compute the same function.

    Algorithms for minimizing the number of states in a deterministic finite automaton (DFA) appear in any book on automata theory. The basic approach works by partitioning the states into gross equivalence classes and then refining the partition. Initially, the states are partitioned into accepting, rejecting, and other classes. The transitions from each node branch to a given class on a given symbol. Whenever two states s, t in the same class C branch to elements of different classes, the class C must be partitioned into two subclasses, one containing s, the other containing t.

    This algorithm makes a sweep though all the classes looking for a new partition, and repeating the process from scratch if it finds one. This yields an $O(n^2)$ algorithm, since at most $n - 1$ sweeps need ever be performed. The final equivalence classes correspond to the states in the minimum automaton. In fact, a more efficient, $O(n \log n)$ algorithm is known with available implementations discussed below.

- *Constructing deterministic machines from nondeterministic machines* – DFAs are simple to understand and work with, because the machine is always in exactly one state at any given time. *Nondeterministic automata* (NFAs) can be in more than one state at a time, so its current "state" represents a subset of all possible machine states.

    In fact, any NFA can be mechanically converted to an equivalent DFA, which can then be minimized as above. However, converting an NFA to a DFA can cause an exponential blowup in the number of states, which perversely might later be eliminated in the minimization. This exponential blowup makes automaton minimization problems NP-hard whenever you do not start with a DFA.

    The proofs of equivalence between NFAs, DFAs, and regular expressions are elementary enough to be covered in undergraduate automata theory classes. However, they are surprisingly nasty to implement, for reasons including but not limited to the exponential blowup of states. Implementations are discussed below.

- *Constructing machines from regular expressions* – There are two approaches to converting a regular expression to an equivalent finite automaton, the difference being whether the output automaton is to be a nondeterministic or deterministic machine. The former is easier to construct but less efficient to simulate.

    The nondeterministic construction uses ϵ-moves, which are optional transitions that require no input to fire. On reaching a state with an ϵ-move, we must assume that the machine can be in either state. Using such ϵ-moves, it is straightforward to construct an automaton from a depth-first traversal of the parse tree of the regular expression. This machine will have $O(m)$ states, if m

is the length of the regular expression. Further, simulating this machine on a string of length n takes $O(mn)$ time, since we need consider each state/prefix pair only once.

The deterministic construction starts with the parse tree for the regular expression, observing that each leaf represents one of the alphabet symbols in the pattern. After recognizing a prefix of the text, we can be left in some subset of these possible positions, which would correspond to a state in the finite automaton. The *derivatives* method builds up this automaton state by state as it is needed. Even so, some regular expressions of length m require $O(2^m)$ states in any DFA implementing them, such as $(a+b)^*a(a+b)(a+b)\ldots(a+b)$. There is no way to avoid this exponential blowup in the space required. Note, however, that it takes linear time to simulate an input string on any automaton, regardless of the size of the automaton.

Implementations: FIRE Engine is a finite automaton toolkit, written in C++ by Bruce Watson. It provides production-quality implementations of finite automata and regular expression algorithms. Several finite automaton minimization algorithms have been implemented, including Hopcroft's $O(n \lg n)$ algorithm. Both deterministic and nondeterministic automata are supported. FIRE Engine has been used for compiler construction, hardware modeling, and computational biology applications. It is strictly a computing engine and does not provide a graphical user interface. FIRE Engine is available by anonymous ftp from ftp.win.tue.nl in the directory /pub/techreports/pi/watson.phd/. A greatly improved commercial version is available from www.RibbitSoft.com.

Grail is a C++ package for symbolic computation with finite automata and regular expressions, from Darrell Raymond and Derrick Wood. Grail enables one to convert between different machine representations and to minimize automata. It can handle machines with 100,000 states and dictionaries of 20,000 words. All code and documentation are accessible from the WWW site http://www.csd.uwo.ca/research/grail, as well as pointers to a variety of other automaton packages. Commercial use of Grail is not allowed without approval, although it is freely available to students and educators.

An implementation in C of a regular-expression matching algorithm appears in [BR95]. The source code for this program is printed in the text and is available on disk for a modest fee. A bare bones implementation in C of a regular-expression pattern matching algorithm appears in [GBY91]. See Section 9.1.6.2.

XTango (see Section 9.1.5) includes a simulation of a DFA. Many of the other animations (but not this one) are interesting and quite informative to watch.

FLAP (Formal Languages and Automata Package) is a tool by Susan Rodger for drawing and simulating finite automata, pushdown automata, and Turing machines. Using FLAP, one can draw a graphical representation (transition diagram) of an automaton, edit it, and simulate the automaton on some input. FLAP was developed in C++ for X-Windows. See http://www.cs.duke.edu:80/~rodger/tools/tools.html.

Notes: Aho [Aho90] provides a good survey on algorithms for pattern matching, and a particularly clear exposition for the case where the patterns are regular expressions. The technique for regular expression pattern matching with ϵ-moves is due to Thompson [Tho68]. Other expositions on finite automaton pattern matching include [AHU74].

Hopcroft [Hop71] gave an optimal $O(n \lg n)$ algorithm for minimizing the number of states in DFAs. The derivatives method of constructing a finite state machine from a regular expression is due to Brzozowski [Brz64] and has been expanded upon in [BS86]. Expositions on the derivatives method includes Conway [Con71]. Testing the equivalence of two nondeterministic finite state machines is PSPACE-complete [SM73].

Related Problems: Satisfiability (see page 266). string matching (see page 403).

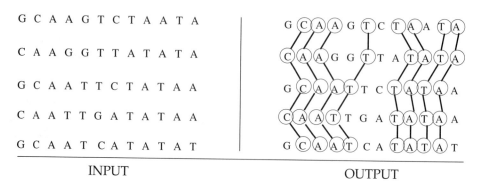

INPUT	OUTPUT

8.7.8 Longest Common Substring

Input description: A set S of strings $S_1, \ldots, S_n$.

Problem description: What is the longest string S' such that for each S_i, $1 \leq i \leq n$, the characters of S appear as a subsequence of S_i?

Discussion: The problem of longest common subsequence arises whenever we search for similarities across multiple texts. A particularly important application is in finding a consensus among DNA sequences. The genes for building particular proteins evolve with time, but the functional regions must remain consistent in order to work correctly. By finding the longest common subsequence of the same gene in different species, we learn what has been conserved over time.

The longest common substring problem is a special case of edit distance (see Section 8.7.4), when substitutions are forbidden and only exact character match, insert, and delete are allowable edit operations. Under these conditions, the edit distance between p and t is $n + m - 2|lcs(p, t)|$, since we can delete the missing characters from p to the $lcs(p, t)$ and insert the missing characters from t to transform p to t. This is particularly interesting because the longest common subsequence can be faster to compute than edit distance.

Issues arising include:

- *Are you looking for a common substring or scattered subsequence?* – In detecting plagiarism or attempting to identify the authors of anonymous works, we might need to find the longest phrases shared between several documents. Since phrases are strings of consecutive characters, we need the longest common *substring* between the texts.

    The longest common substring of a set of strings can be identified in linear time using suffix trees, discussed in Section 8.1.3. The trick is to build a suffix tree containing all the strings, label each leaf with the set of strings that contain it, and then do a depth-first traversal to identify the deepest node that has descendents from each input string.

    For the rest of this discussion, we will restrict attention to finding common scattered subsequences. Dynamic programming can be used to find the longest common subsequence of two strings, S and T, of n and m char-

acters each. This algorithm is a special case of the edit-distance computation of Section 8.7.4. Let $M[i, j]$ denote the number of characters in the
longest common substring of $S[1], \ldots, S[i]$ and $T[1], \ldots, T[j]$. In general, if
$S[i] \neq T[j]$, there is no way the last pair of characters could match, so $M[i, j] =
\max(M[i, j-1], M[i-1, j])$. If $S[i] = T[j]$, we have the option to select this char-
acter for our substring, so $M[i, j] = \max(M[i-1, j-1]+1, M[i-1, j], M[i, j-1])$.
This gives a recurrence that computes M, and thus finds the length of the
longest common subsequence in $O(nm)$ time. We can reconstruct the actual common substring by walking backward from $M[n, m]$ and establishing
which characters were actually matched along the way.

- *What if there are relatively few sets of matching characters?* – For strings that do
  not contain too many copies of the same character, there is a faster algorithm.
  Let r be the number of pairs of positions (i, j) such that $S_i = T_j$. Thus r can be
  as large as mn if both strings consist entirely of the same character, but $r = n$
  if the two strings are permutations of $\{1, \ldots, n\}$. This technique treats the pairs
  of r as defining points in the plane.

  The complete set of r such points can be found in $O(n + m + r)$ time by
  bucketing techniques. For each string, we create a bucket for each letter of the
  alphabet and then partition all of its characters into the appropriate buckets.
  For each letter c of the alphabet, create a point (s, t) from every pair $s \in S_c$
  and $t \in T_c$, where S_c and T_c are buckets for c.

  A common substring represents a path through these points that moves
  only up and to the right, never down or to the left. Given these points, the
  longest such path can be found in $O((n + r) \lg n)$ time. We will sort the points
  in order of increasing x-coordinate (breaking ties in favor of increasing y-
  coordinate). We will insert these points one by one in this order, and for each
  $k, 1 \le k \le n$, maintain the minimum y-coordinate of any path going through
  exactly k points. Inserting a new point will change exactly one of these paths
  by reducing the y-coordinate of the path whose last point is barely greater
  than the new point.

- *What if the strings are permutations?* – If the strings are permutations, then
  there are exactly n pairs of matching characters, and the above algorithm
  runs in $O(n \lg n)$ time. A particularly important case of this occurs in finding
  the longest *increasing* subsequence of a sequence of numbers. Sorting this
  sequence and then replacing each number by its rank in the total order gives
  us a permutation p. The longest common subsequence of p and $\{1, 2, 3, \ldots, n\}$
  gives the longest increasing subsequence.

- *What if we have more than two strings to align?* – The basic dynamic programming algorithm can be generalized to k strings, taking $O(kn^k)$ time, where n is
  the length of the longest string. This algorithm is exponential in the number
  of strings k, and so it will likely be too expensive for more than 3 to 4 strings.
  Further, the problem is NP-complete, so no better exact algorithm is destined
  to come along.

  This problem of multiple sequence alignment has received considerable
  attention, and numerous heuristics have been proposed. Many heuristics

begin by computing the pairwise alignment between each of the $\binom{k}{2}$ pairs of strings, and then work to merge these alignments. One approach is to build a graph with a vertex for each character of each string. There will be an edge between S_i and T_j if the corresponding characters are matched in the alignment between S and T. Any k-clique (see Section 8.5.1) in this graph describes a commonly aligned character, and all such cliques can be found efficiently because of the sparse structure of this graph.

Although these cliques will define a common subsequence, there is no reason to believe that it will be the longest such substring. Appropriately weakening the clique requirement provides a way to increase it, but still there can be no promises.

Implementations: MAP (Multiple Alignment Program) [Hua94] by Xiaoqiu Huang is a C language program that computes a global multiple alignment of sequences using an iterative pairwise method. Certain parameters will need to be tweaked to make it accommodate non-DNA data. It is available by anonymous ftp from cs.mtu.edu in the pub/huang directory.

Combinatorica [Ski90] provides a Mathematica implementation of an algorithm to construct the longest increasing subsequence of a permutation, which is a special case of longest common subsequence. This algorithm is based on Young tableaux rather than dynamic programming. See Section 9.1.4.

Notes: Good expositions on longest common subsequence include [AHU83, CLR90]. A survey of algorithmic results appears in [GBY91]. The algorithm for the case where all the characters in each sequence are distinct or infrequent is due to Hunt and Szymanski [HS77]. Expositions of this algorithm include [Aho90, Man89]. Multiple sequence alignment for computational biology is treated in [Wat95].

Certain problems on strings become easier when we assume a constant-sized alphabet. Masek and Paterson [MP80] solve longest common subsequence in $O(mn/\log(\min\{m,n\}))$ for constant-sized alphabets, using the four Russians technique.

Related Problems: Approximate string matching (see page 406), shortest common superstring (see page 425).

```
            A B R A C A D A B R A
A B R A C   A B R A C
A C A D A       R A C A D
A D A B R           A C A D A
D A B R A               A D A B R
R A C A D                   D A B R A
```

INPUT	OUTPUT

8.7.9 Shortest Common Superstring

Input description: A set of strings $S = \{S_1, \ldots, S_m\}$.

Problem description: Find the shortest string S' that contains each element of S as a substring.

Discussion: Shortest common superstring arises in a variety of applications, including sparse matrix compression. Suppose we have an $n \times m$ matrix with most of the elements being zero. We can partition each row into m/k runs of k elements each and construct the shortest common superstring S' of these runs. We now have reduced the problem to storing the superstring, plus an $n \times m/k$ array of pointers into the superstring denoting where each of the runs starts. Accessing a particular element $M[i, j]$ still takes constant time, but there is a space savings when $|S| << mn$.

Another application arises in DNA sequencing. It happens to be easy to sequence small fragments of DNA, say up to about 500 base pairs or characters. However, the real interest is in sequencing large molecules. The standard approach to large-scale, "shotgun" sequencing clones many copies of the target molecule, breaks them randomly into small fragments, sequences the fragments, and then proposes the shortest superstring of the fragments as the correct sequence. While it is an article of faith that the shortest superstring will be the most likely sequence, this seems to work reasonably well in practice.

Finding a superstring of all the substrings is not difficult, as we can simply concatenate them all together. It is finding the shortest such string that is problematic. Indeed, shortest common superstring remains NP-complete under all reasonable classes of strings.

The problem of finding the shortest common superstring can easily be reduced to that of the traveling salesman problem (see Section 8.5.4). Create an overlap graph G where vertex v_i represents string S_i. Edge (v_i, v_j) will have weight equal to the length of S_i minus the overlap of S_j with S_i. The path visiting all the vertices of minimum total weight defines the shortest common superstring. The edge weights of this graph are not symmetric, after all, the overlap of $S_i = abc$ and $S_j = bcd$ is not the same as the overlap of S_j and S_i. Thus only programs capable of solving asymmetric TSPs can be applied to this problem.

The greedy heuristic is the standard approach to approximating the shortest common superstring. Find the pair of strings with the maximum number of characters of overlap. Replace them by the merged string, and repeat until only one string remains. Given the overlap graph above, this heuristic can be efficiently implemented by inserting all of the edge weights into a heap (see Section 8.1.2) and then merging if the appropriate ends of the two strings have not yet been used, which can be maintained with an array of Boolean flags.

The potentially time-consuming part of this heuristic is in building the overlap graph. The brute-force approach to finding the maximum overlap of two length-l strings takes $O(l^2)$, which must be repeated $\binom{n}{2}$ times. Faster times are possible by appropriately using suffix trees (see Section 8.1.3). Build a tree containing all suffixes of all reversed strings of S. String S_i overlaps with S_j if a suffix of S_i matches a suffix of the reverse of S_j. The longest overlap for each fragment can be found in time linear in its length.

How well does the greedy heuristic perform? If we are unlucky with the input, the greedy heuristic can be fooled into creating a superstring that is at least twice as long as optimal. Usually, it will be a lot better in practice. It is known that the resulting superstring can never be more than 2.75 times optimal.

Building superstrings becomes more difficult with positive and negative substrings, where negative substrings *cannot* be substrings of the superstring. The problem of deciding whether *any* such consistent substring exists is NP-complete, unless you are allowed to add an extra character to the alphabet to use as a spacer.

Implementations: CAP (Contig Assembly Program) [Hua92] by Xiaoqiu Huang is a C language program supporting DNA shotgun sequencing by finding the shortest common superstring of a set of fragments. As to performance, CAP took 4 hours to assemble 1,015 fragments of a total of 252,000 characters on a Sun SPARCstation SLC. Certain parameters will need to be tweaked to make it accommodate non-DNA data. It is available by anonymous ftp from cs.mtu.edu in the pub/huang directory.

Notes: The shortest common superstring problem and its application to DNA shotgun assembly is ably surveyed in [Wat95]. Kececioglu and Myers [KM95] report on an algorithm for the more general version of shortest common superstring, where the strings may have errors. Their paper is recommended reading to anyone interested in fragment assembly.

Blum et al. [BJL+91] gave the first constant-factor approximation algorithms for shortest common superstring, with a variation of the greedy heuristic. More recent research has beaten the constant down to 2.75, progress towards the expected factor-two result.

Related Problems: Suffix trees (see page 183), text compression (see page 410).

CHAPTER 9

Algorithmic Resources

This chapter describes resources that the practical algorithm designer should be familiar with. Although some of this information has appeared at various points in the catalog, the most important pointers have been collected here for general reference. These resources take the form of software systems and libraries, books, and other bibliographic sources. Many of the most interesting resources are available on-line and mirrored on the enclosed CD-ROM.

9.1 Software Systems

In this section, we describe several particularly comprehensive implementations of combinatorial algorithms, all of which are available over the Internet. Although these codes are discussed in the relevant sections of the catalog, they are substantial enough to warrant further attention.

A good algorithm designer does not reinvent the wheel, and a good programmer does not rewrite code that other people have written. Picasso put it best: "Good artists borrow. Great artists steal."

However, a word of caution about stealing. Many of the codes described below (and throughout this book) have been made available for research or educational use, although commercial use requires a licensing arrangement with the author. I urge you to respect this. Licensing terms from academic institutions are usually surprisingly modest. The recognition that industry is using a particular code is important to the authors, often more important than the money involved. This can lead to enhanced support or future releases of the software. Do the right thing and get a license. Information about terms or whom to contact is usually available embedded within the documentation, or available at the source ftp/WWW site.

Although the bulk of the systems we describe here are available by accessing our algorithm repository, http://www.cs.sunysb.edu/~algorith (as well as on the enclosed CD-ROM), we encourage you to get them from the original sites instead of Stony Brook. There are three reasons. First, the version on the original site is much more likely to be maintained. Second, there are often supporting files and documentation that we for whatever reason did not download, and which may be of interest to you. Finally, by ftp-ing from the original sites, you will keep traffic down at the algorithm repository site, which will minimize the complaints from our system staff that I anticipate if this service becomes very popular.

9.1.1   LEDA

LEDA, for Library of Efficient Data types and Algorithms, is perhaps the best single resource available to support combinatorial computing. It has been under development since 1988 by a group at Max-Planck-Instutut in Saarbrücken, Germany, including Kurt Mehlhorn, Stefan Näher, Stefan Schirra, Christian Uhrig, and Christoph Burnikel. LEDA is unique because of (1) the algorithmic sophistication of its developers and (2) the level of continuity and resources invested in the project.

LEDA is implemented in C++ using templates, and it should compile on most new compilers, but not some old ones. LEDA is available by anonymous ftp from ftp.mpi-sb.mpg.de in directory /pub/LEDA, or at http://www.mpi-sb.mpg.de/LEDA/leda.html. The distribution contains all sources, installation instructions, and a substantial users manual [NU95]. An active Usenet newsgroup comp.lang.c++.leda is inhabited by users of LEDA. A good article on LEDA is available [MN95], and a book is promised soon. LEDA is not in the public domain, but it can be used freely for research and teaching. Commerical licenses are also available.

What LEDA offers is a complete collection of well-implemented data structures and types. Particularly useful is the graph type, which supports all the basic operations one needs in an intelligent way, although this generality comes at some cost in size and speed over handcrafted implementations.

9.1.2   Netlib

Netlib is an on-line repository of mathematical software that contains a large number of interesting codes, tables, and papers. Netlib is a compilation of resources from a variety of places, with fairly detailed indices and search mechanisms to help you find what is there. Netlib is important because of its breadth and ease of access. Whenever you need a specialized piece of mathematical software, you should look here first.

There are three ways to access netlib: by e-mail, ftp, or WWW:

- *E-mail* — Netlib provides an email server to send indices and sources on demand. To get an index, send e-mail to netlib@netlib.org with the words *send index* on its own line in the message. The index will provide a list of other files you can send for. The e-mail server and netlib in general are discussed in [DG87].

- *FTP* — Connect by ftp to ftp.netlib.org. Log in as anonymous and use your e-mail address as password. Use "ls" to see the contents of a directory, "cd" to move to a different directory, and "get" to fetch the desired file. Type "binary" before "get" in order to ensure uncorrupted transmission, and "quit" to quit. Obtaining an index first can make it easier to move around.

- *WWW* — With your favorite browser, open the URL address http:// www.netlib.org/ and prowl around to your heart's content. There is a forms index that permits searching based on keywords.

GAMS, the Guide to Available Mathematical Software, is an indexing service for Netlib and other related software repositories that can help you find what you want. Check it out at http://gams.nist.gov. GAMS is a service of the National Institute of Standards and Technology (NIST).

9.1.2.1   Collected Algorithms of the ACM

An early mechanism for the distribution of useful algorithm implementations was *CALGO*, the *Collected Algorithms of the ACM*. It first appeared in *Communications of the ACM* in 1960, covering such famous algorithms as Floyd's linear-time build heap algorithm. More recently, it has been the province of the *ACM Transactions on Mathematical Software*. Each algorithm/implementation is described in a brief journal article, with the implementation validated and collected. These implementations are maintained at http://www.acm.org/calgo/ and at Netlib.

Over 750 algorithms have appeared to date. Most of the codes are in Fortran and are relevant to numerical computing, although several interesting combinatorial algorithms have slithered their way into CALGO. Since the implementations have been refereed, they are presumably more reliable than most readily available software.

9.1.3   The Stanford GraphBase

The Stanford GraphBase is an interesting program for several reasons. First, it was composed as a "literate program," meaning that it was written to be read. If anybody's programs deserve to be read, it is Knuth's, and [Knu94] contains the full source code of the system. The programming language/environment is CWEB, which permits the mixing of text and code in particularly expressive ways.

The GraphBase contains implementations of several important combinatorial algorithms, including matching, minimum spanning trees, and Voronoi diagrams, as well as specialized topics like constructing expander graphs and generating combinatorial objects. Finally, it contains programs for several recreational problems, including constructing word ladders (flour-floor-flood-blood-brood-broad-bread) and establishing dominance relations among football teams.

Although the GraphBase is more fun to play with than LEDA, it is not really suited for building general applications on top of. The GraphBase is perhaps most useful as an instance generator for constructing a wide variety of graphs to serve as test data. It incorporates graphs derived from interactions of characters in famous novels, Roget's thesaurus, the Mona Lisa, and the economy of the United States. Further, because of its machine-independent random number generators, the GraphBase provides a way to construct random graphs that can be reconstructed elsewhere, making them perfect for experimental comparisons of algorithms.

The Stanford GraphBase can be obtained by anonymous ftp from labrea.stanford.edu in the directory pub/sgb. It may be used freely, but the files may not be modified. Installing the GraphBase requires CWEB, which can be obtained by anonymous ftp from labrea.stanford.edu in the directory pub/cweb.

9.1.4   Combinatorica

Combinatorica [Ski90] is a collection of over 230 algorithms for combinatorics and graph theory written in Mathematica. These routines have been designed to work together, enabling one to experiment with discrete structures and build prototype applications. Combinatorica has been widely used for both research and education.

Although (in my totally unbiased opinion) Combinatorica is more comprehensive and better integrated than other libraries of combinatorial algorithms, it is also the slowest such system available. Credit for all of these properties is largely due to Mathematica, which provides a very high-level, functional, interpreted, and thus inefficient programming language. Combinatorica is best for finding quick solutions to small problems, and (if you can read Mathematica code) as a terse exposition of algorithms for translation into other languages.

Combinatorica is included with the standard Mathematica distribution in the directory Packages/DiscreteMath/Combinatorica.m. It can also be obtained by anonymous ftp from ftp.cs.sunysb.edu in the pub/Combinatorica directory. Included on this site are certain extensions to Combinatorica and data sources such as the graphs of Graffiti.

9.1.5   Algorithm Animations with XTango

XTango [Sta90] is a general-purpose algorithm animation system, developed by John Stasko of Georgia Tech, that helps programmers develop color, real-time animations of their own algorithms and programs. Creating animations of your own implementations is useful both for pedagogical and debugging purposes.

Included with the XTango distribution is a large collection of animations, several of which are quite interesting and enlightening to watch. The C language source for each animation is included. My favorites include animations of:

- *Data structures* — including AVL/red-black trees, and binary/Fibonacci heaps.
- *Sorting algorithms* — including bubblesort, radixsort, quicksort, and shellsort.

- *Backtracking* — including graph coloring, and both the eight-queens and knight's tour problems.

- *Geometric algorithms* — including bin packing heuristics and the Graham scan/Jarvis march convex hull algorithms.

- *Graph algorithms* — including minimum spanning trees and shortest paths.

- *String algorithms* — including the Knuth-Morris-Pratt and Boyer-Moore algorithms.

Anybody studying these algorithms might well profit from playing with the animations. The source code for the animations contains C language implementations of these algorithms, although of uncertain quality.

XTango is implemented on top of UNIX and X-windows. It is available via anonymous ftp from ftp.cc.gatech.edu in directory pub/people/stasko. Also available there is Polka, a C++ algorithm animation system that is particularly suited to animating parallel computations. POLKA provides its own high-level abstractions to make the creation of animations easier and faster than with many other systems. POLKA also includes an interactive front end called SAMBA that can be used to generate animations from any type of program that can generate ASCII.

9.1.6   Programs from Books

Several books on algorithms include working implementations of the algorithms in a real programming language. Although these implementations are intended primarily for exposition, they can also be useful for computation. Since they are typically small and clean, they can prove the right foundation for simple applications.

The most useful codes of this genre are described below. Most are available from the algorithm repository, http://www.cs.sunysb.edu/~algorith.

9.1.6.1   Discrete Optimization Algorithms in Pascal

This is a collection of 28 programs for solving discrete optimization problems, appearing in the book by Syslo, Deo, and Kowalik [SDK83]. The package includes programs for integer and linear programming, the knapsack and set cover problems, traveling salesman, vertex coloring, and scheduling as well as standard network optimization problems. They have been made available from the algorithm repository WWW site, http://www.cs.sunysb.edu/~algorith.

This package is noteworthy for the operations-research flavor of the problems and algorithms selected. The algorithms have been selected to solve problems, as opposed to for purely expository purposes. In [SDK83], a description of each algorithm and extensive set of references for each problem is provided, as well as execution times for each program on several instances on an early 1980s mainframe, an Amdahl 470 V/6.

9.1.6.2   Handbook of Data Structures and Algorithms

The *Handbook of Data Structures and Algorithms*, by Gonnet and Baeza-Yates [GBY91], provides a comprehensive reference on fundamental data structures for searching and priority queues, and algorithms for sorting and text searching. The book covers this relatively small number of topics comprehensively, presenting most of the major and minor variations that appear in the literature. Perusing the book makes one aware of the tremendous range of data structures that have been developed and the intense level of analysis many of them have been subjected to.

For each data structure or algorithm, a brief description is provided along with its asymptotic complexity and an extensive set of references. More distinctively, an implementation in C and/or Pascal is usually provided, along with experimental results comparing the performance of these implementations. The programs in [GBY91] are available at a very slick WWW site, http://www.dcc.uchile.cl/rbaeza/handbook/hbook.html.

Since many of the elementary data structures and sorting algorithms can be implemented concisely, most of the programs are very short. They are perhaps most useful as models, or as part of an experimental study to compare the performance on different data structures for a particular application.

9.1.6.3   Combinatorial Algorithms for Computers and Calculators

Nijenhuis and Wilf [NW78] specializes in algorithms for constructing basic combinatorial objects such as permutations, subsets, and partitions. Such algorithms are often very short, but they are hard to locate and usually surprisingly subtle. Fortran programs for all of the algorithms are provided, as well as a discussion of the theory behind each of them. The programs are usually short enough that it is reasonable to translate them directly into a more modern programming language, as I did in writing *Combinatorica* (see Section 9.1.4). Both random and sequential generation algorithms are provided. Descriptions of more recent algorithms for several problems, without code, are provided in [Wil89].

These programs are now available from our algorithm repository WWW site. We tracked them down from Neil Sloane, who had them on a magnetic tape, while the original authors did not! In [NW78], Nijenhuis and Wilf set the proper standard of statistically testing the output distribution of each of the random generators to establish that they really appear uniform. We encourage you to do the same before using these programs, to verify that nothing has been lost in transit.

9.1.6.4   Algorithms from P to NP

This algorithms text [MS91] distinguishes itself by including Pascal implementations of many algorithms, with careful experimental comparisons of different algorithms for such problems as sorting and minimum spanning tree, and heuristics for the traveling salesman problem. It provides a useful model for how to properly do empirical algorithm analysis.

The programs themselves are probably best used as models. Interesting implementations include the eight-queens problem, plus fundamental graph and geometric algorithms. The programs in [MS91] have been made available by anonymous ftp from cs.unm.edu in directory /pub/moret_shapiro.

9.1.6.5   Computational Geometry in C

O'Rourke [O'R94] is perhaps the best practical introduction to computational geometry available, because of its careful and correct C language implementations of the fundamental algorithms of computational geometry. Fundamental geometric primitives, convex hulls, triangulations, Voronoi diagrams, and motion planning are all included. Although they were implemented primarily for exposition rather than production use, they should be quite reliable. The codes are available from http://grendel.csc.smith.edu/~orourke/.

9.1.6.6   Algorithms in C++

Sedgewick's popular algorithms text [Sed92] comes in several different language editions, including C, C++, and Modula-3. This book distinguishes itself through its use of algorithm animation and in its broad topic coverage, including numerical, string, and geometric algorithms.

The language-specific parts of the text consist of many small code fragments, instead of full programs or subroutines. Thus they are best thought of as models, instead of working implementations. Still, the program fragments from the C++ edition have been made available from http://heg-school.aw.com/cseng/authors/sedgewick/algo-in-c++/algo-in-c++.html

9.2 Data Sources

It is often important to have interesting data to feed your algorithms, to serve as test data to ensure correctness or to compare different algorithms for raw speed.

Finding good test data can be surprisingly difficult. Here are some pointers:

- *Combinatorica graphs* — A modest collection of graphs is available from the Combinatorica ftp site. Particularly interesting are the 190 graphs of Graffiti [Faj87], a program that formulated conjectures in graph theory by studying the properties of these graphs. See Section 9.1.4 for more information on Combinatorica.

- *TSPLIB* — This well-respected library of test instances for the traveling salesman problem is available from Netlib (see Section 9.1.2) and by anonymous ftp from softlib.cs.rice.edu. TSPLIB instances are large, real-world graphs, derived from applications such as circuit boards and networks.

- *Stanford GraphBase* — Discussed in Section 9.1.3, this suite of programs by Knuth provides portable generators for a wide variety of graphs. These include graphs arising from distance matrices, arts, and literature, as well as graphs of more theoretical interest.

- *DIMACS Challenge data* — A series of DIMACS Challenge workshops have focused on evaluating algorithm implementations of graph, logic, and data structure problems. Instance generators for each problem have been developed, with the focus on constructing difficult or representative test data. The products of the DIMACS Challenges are available from http://dimacs.rutgers.edu/.

- *Algorithm Repository* — Located at http://www.cs.sunysb.edu/~algorith, and mirrored on the enclosed CD-ROM, the Algorithm Repository contains data sources for a few of the implementation challenge exercises. In particular, we provide an airline routes data set, and a collection of names labeled by ethnicity.

9.3 Textbooks

There have emerged a number of excellent textbooks in the design and analysis of combinatorial algorithms. Below we point out several of our favorites. In this book, we have shied away from giving a detailed exposition or analysis of many algorithms, for our primary mission is to help the reader identify their problem and point them to solutions. The reader is encouraged to turn to these sources for more detail once they have found the name of what they are looking for.

Only general algorithm books are discussed here. Books on specific subareas of algorithms are reviewed at the head of the relevant catalog chapter.

- *Cormen, Leiserson, and Rivest* [CLR90] — This is the one (other) book on algorithms you must own, with its comprehensive treatment of most of the problems we discuss here, including data structures, graph algorithms, and seminumerical algorithms.

- *Baase* [Baa88] — This book is more accessible than [CLR90] for those without a strong mathematical background. It covers standard sorting, string, and graph algorithms, NP-completeness, and more exotically, an introduction to parallel algorithms.

- *Manber* [Man89] — Built around the unconventional notion that induction is the fundamental paradigm of algorithm design, this book is especially good at teaching techniques for designing algorithms and has an outstanding collection of problems. Highly recommended.

- *van Leeuwen* [vL90b] — Not a textbook, but a collection of in-depth surveys on the state of the art in algorithms and computational complexity. Although the emphasis is on theoretical results, this book is perhaps the best single reference to point you to what is known about any given problem.

- *Syslo, Deo, and Kowalik* [SDK83] — This book includes printed Pascal implementations of 28 algorithms for discrete optimization problems, including mathematical programming, network optimization, and traditional operations research problems such as knapsack and TSP. Each algorithm is described in the book, and experimental timings (on a 1980s vintage machine) are provided. These codes are now available by ftp, as discussed in Section 9.1.6.1. Despite its age, this remains a useful reference, particularly with the programs now available on-line.

- *Moret and Shapiro* [MS91] — This algorithms text distinguishes itself by including Pascal implementations of all algorithms and by its careful experimental comparisons of different algorithms for such problems as sorting and minimum spanning tree. It provides a useful model for how to properly do empirical algorithm analysis. These programs are available by ftp — see Section 9.1.6.4 for details.

- *Knuth* [Knu94] — This book presents the implementation of the Stanford GraphBase, a collection of programs for constructing different graphs and working with them. See Section 9.1.3 for details. It is very intriguing to browse.

- *Aho, Hopcroft, and Ullman* [AHU74] — This was the first modern algorithms book, and it has had an enormous influence on how algorithms should be taught. Although it is now dated, it remains a useful guide to certain topics in vogue in the early 1970s, such as matrix multiplication, the fast Fourier transform, and arithmetic algorithms. A more elementary edition, focusing on data structures, is [AHU83].

- *Rawlins* [Raw92] — This may well be the best self-study book available on algorithms. It is fun and inspiring, with built-in pauses so the reader can make sure they understand what is going on. The only drawback is a somewhat idiosyncratic set of topics, so you will miss certain important topics. But you can get that from here. Rawlins's book can teach you the proper mindset to think about algorithms.

- *Papadimitriou and Steiglitz* [PS82] — This book has more of an operations research emphasis than most algorithms texts, with a good coverage of mathematical programming, network flow, and combinatorial search.

- *Lawler* [Law76] — Particularly useful for its coverage of matroid theory, this book also provides a thorough treatment of the network flow, matching, and shortest path algorithms known by the mid-1970s.

- *Binstock and Rex* [BR95] — Although not a textbook, it includes C language implementations of an idiosyncratic variety of algorithms for programmers. Disks containing the code are available for a modest fee. The most interesting implementations are string and pattern matching algorithms, time and date routines, an arbitrary-precision calculator, and a nice section on checksums and cyclic-redundancy checks.

9.4 On-Line Resources

The Internet has proven to be a fantastic resource for people interested in algorithms, as it has for many other subjects. What follows is a highly selective list of the resources that I use most often, partitioned into references to literature, people, and software. All of these should be in the tool chest of every algorist.

9.4.1   Literature

There are many bibliographic sources available on the WWW, but the following I find indispensable.

- *Computer Science Bibliographies* — This is a collection of over 700,000 references to papers and technical reports in Computer Science, beneath a sophisticated search engine. While there is much duplication, this is my first stop whenever I need to look something up. The primary site is http://liinwww.ira.uka.de/bibliography/index.html, although several mirror sites are in operation around the world. All references are provided in bibtex format.

- *Joel Seiferas' paper.lst* — References to over 55,000 papers and technical reports (at last count), mostly on algorithms and related areas of theoretical computer science. Each paper is reduced to a one-line format, which I find easy to grep through. It is available by anonymous ftp from ftp.cs.rochester.edu in /pub/u/joel, and a copy is included on the enclosed CD-ROM. Strongly recommended.

- *Geom.bib* — The *complete* bibliography on anything related to computational geometry, it references over 8,000 books, papers, and reports and includes detailed abstracts for many of them. Grep-ing through geom.bib is an amazingly efficient way to find out about previous work without leaving your office. It is available via anonymous ftp from ftp.cs.usask.ca, in file pub/geometry/geombib.tar.Z, and a copy is included on the enclosed CD-ROM.

- *Usenet FAQ files* — The straightest, most accurate dope about any topic is likely to be had in a USENET frequently asked questions (FAQ) file. Dedicated volunteers maintain FAQ files for many USENET newsgroups, many of which are quite detailed, and which always emphasize other on-line resources. Excellent FAQ files on algorithm-related topics such as cryptography, linear programming, and data compression are currently available. A collection of current FAQ files is maintained at ftp://rtfm.mit.edu/pub/usenet/news.answers.

9.4.2   People

The easiest way to find the answer to any question is to ask someone who knows. Below, I describe some useful resources of contact information for experts in algorithms. E-mail and the WWW make it very easy to track down experts. In fact, maybe too easy. Please use this information responsibly, for many of them are my friends! Limit your queries to short, specific questions; and contact them only after you have checked the standard textbooks and references described in this section.

- *comp.theory* — The USENET newsgroup comp.theory is the proper forum for on-line discussions on algorithms and related topics. Other appropriate newsgroups include comp.graphics.algorithms and sci.math.

- *TCS Virtual Address Book* — This WWW-page address book is particularly useful to track down the whereabouts of researchers in the design and analysis of algorithms. It is maintained by Ian Parbery and is available from the ACM SIGACT WWW page http://sigact.acm.org/sigact/. An excellent way to learn of recent progress on a particular problem is to ask a researcher who has previously published on the problem, or at least check out their WWW page.

- *Algorithm Courses* — A comprehensive page of pointers to algorithm courses on the WWW is maintained at at http://www.cs.pitt.edu/~kirk/ algorithm-courses/index.html. Lecture notes, problems, and solutions are typically available, with some of the material being amazingly thorough. Software associated with course projects is available from some of the sites.

9.4.3   Software

There are many high-quality implementations of algorithms available on the WWW, if you know where to look. We have placed as many of them as we could on the CD-ROM enclosed with this book. My favorite on-line resources are:

- *The Stony Brook Algorithm Repository* — This is the source of the software on the enclosed CD-ROM, organized according to the problem structure of this book. Local copies of most implementations are maintained at http://www.cs.sunysb.edu/~algorith, along with pointers to the latest version available from the original distribution site.

- *Netlib* — This is the standard resource on mathematical software, primarily numerical computation. An enormous amount of material is available. See Section 9.1.2 for details.

- *Directory of Computational Geometry Software* — Maintained by Nina Amenta, this is the source for anything geometric. See http://www.geom.umn.edu/ software/cglist.

- *Combinatorial Optimization codes* — A very nice list of implementations of combinatorial algorithms is maintained by Jiefeng Xu at http://ucsu.colorado. edu/~xu/software.html. With an emphasis on operations research, there is a lot of useful stuff here.

9.5   Professional Consulting Services

Algorist Technologies is a consulting firm that provides its clients with short-term, expert help in algorithm design and implementation. Typically, an Algorist consultant is called in for 1 to 3 days worth of intensive, on-site discussion and analysis with the client's own development staff. Algorist has built an impressive record of performance improvements with several companies and applications. They provide longer-term consulting and contracting services as well.

Call 212-580-9009 or email info@algorist.com for more information on services provided by Algorist Technologies.

Algorist Technologies
312 West 92nd St. Suite 1A
New York, NY 10025
http://www.algorist.com

Bibliography

[AB95]   D. Avis and D. Bremner. How good are convex hull algorithms? In *Proc. 11th Annual ACM Symposium on Computational Geometry*, pages 20–28, 1995.

[ABCC95]   D. Applegate, R. Bixby, V. Chvatal, and W. Cook. Finding cuts in the TSP (a preliminary report). Technical Report 95-05, DIMACS, Rutgers University, Piscataway NJ, 1995.

[Abd80]   N. N. Abdelmalek. A Fortran subroutine for the L_1 solution of overdetermined systems of linear equations. *ACM Trans. Math. Softw.*, 6(2):228–230, June 1980.

[AC75]   A. Aho and M. Corasick. Efficient string matching: an aid to bibliographic search. *Communications of the ACM*, 18:333–340, 1975.

[AC91]   D. Applegate and W. Cook. A computational study of the job-shop scheduling problem. *ORSA Journal on Computing*, 3:149–156, 1991.

[ACH⁺91]   E. M. Arkin, L. P. Chew, D. P. Huttenlocher, K. Kedem, and J. S. B. Mitchell. An efficiently computable metric for comparing polygonal shapes. *IEEE Trans. PAMI*, 13(3):209–216, 1991.

[ACI92]   D. Alberts, G. Cattaneo, and G. Italiano. An empirical study of dynamic graph algorithms. In *Proc. Seventh ACM-SIAM Symp. Discrete Algorithms (SODA)*, pages 192–201, 1992.

[ADKF70]   V. Arlazarov, E. Dinic, M. Kronrod, and I. Faradzev. On economical construction of the transitive closure of a directed graph. *Soviet Mathematics, Doklady*, 11:1209–1210, 1970.

[Adl94a]   L. Adleman. Algorithmic number theory – the complexity contribution. In *Proc. 35th IEEE Symp. Foundations of Computer Science (FOCS)*, pages 88–113, 1994.

[Adl94b]   L. M. Adleman. Molecular computations of solutions to combinatorial problems. *Science*, 266:1021–1024, November 11, 1994.

[AE83]   D. Avis and H. ElGindy. A combinatorial approach to polygon similarity. *IEEE Trans. Inform. Theory*, IT-2:148–150, 1983.

[AF92]   D. Avis and K. Fukuda. A pivoting algorithm for convex hulls and vertex enumeration of arrangements and polyhedra. *Discrete Comput. Geom.*, 8:295–313, 1992.

[AG86]   A. Apostolico and R. Giancarlo. The Boyer-Moore-Galil string searching strategies revisited. *SIAM J. Computing*, 15:98–105, 1986.

[AGSS89]   A. Aggarwal, L. Guibas, J. Saxe, and P. Shor. A linear-time algorithm for computing the Voronoi diagram of a convex polygon. *Discrete and Computational Geometry*, 4:591–604, 1989.

[AGU72]   A. Aho, M. Garey, and J. Ullman. The transitive reduction of a directed graph. *SIAM J. Computing*, 1:131–137, 1972.

[Aho90]   A. Aho. Algorithms for finding patterns in strings. In J. van Leeuwen, editor, *Handbook of Theoretical Computer Science: Algorithms and Complexity*, volume A, pages 255–300. MIT Press, 1990.

[AHU74]   A. Aho, J. Hopcroft, and J. Ullman. *The Design and Analysis of Computer Algorithms*. Addison-Wesley, Reading MA, 1974.

[AHU83]   A. Aho, J. Hopcroft, and J. Ullman. *Data Structures and Algorithms*. Addison-Wesley, Reading MA, 1983.

[Aig88]   M. Aigner. *Combinatorial Search*. Wiley-Teubner, 1988.

[AK89]   E. Aarts and J. Korst. *Simulated annealing and Boltzman machines: A stochastic approach to combinatorial optimization and neural computing*. John Wiley and Sons, 1989.

[AKD83]   J. H. Ahrens, K. D. Kohrt, and U. Dieter. Sampling from gamma and Poisson distributions. *ACM Trans. Math. Softw.*, 9(2):255–257, June 1983.

[AM93]   S. Arya and D. Mount. Approximate nearest neighbor queries in fixed dimensions. In *Proc. Fourth ACM-SIAM Symp. Discrete Algorithms (SODA)*, pages 271–280, 1993.

[AMN⁺94]   S. Arya, D. Mount, N. Netanyahu, R. Silverman, and A. Wu. Approximate nearest neighbor queries in fixed dimensions. In *Proc. Fifth ACM-SIAM Symp. Discrete Algorithms (SODA)*, pages 573–582, 1994.

[AMO93]   R. Ahuja, T. Magnanti, and J. Orlin. *Network Flows*. Prentice Hall, Englewood Cliffs NJ, 1993.

[AMOT88]   R. Ahuja, K. Mehlhorn, J. Orlin, and R. Tarjan. Faster algorithms for the shortest path problem. Technical Report 193, MIT Operations Research Center, 1988.

[AMWW88]   H. Alt, K. Mehlhorn, H. Wagener, and E. Welzl. Congruence, similarity and symmetries of geometric objects. *Discrete Comput. Geom.*, 3:237–256, 1988.

[And76]   G. Andrews. *The Theory of Partitions*. Addison-Wesley, Reading, Mass., 1976.

[Apo85]   A. Apostolico. The myriad virtues of subword trees. In A. Apostolico and Z. Galil, editors, *Combinatorial algorithms on words*. Springer-Verlag, 1985.

[APT79]   B. Aspvall, M. Plass, and R. Tarjan. A linear-time algorithm for testing the truth of certain quantified boolean formulas. *Info. Proc. Letters*, 8:121–123, 1979.

[Aro96]   S. Arora. Polynomial time approximations schemes for Euclidean TSP and other geometric problems. In *Proc. 37th IEEE Foundations of Computer Science (FOCS '96)*, pages 1–10, 1996.

[AS89]   C. Aragon and R. Seidel. Randomized search trees. In *Proc. 30th IEEE Symp. Foundations of Computer Science*, pages 540–545, 1989.

[Ata83]   M. Atallah. A linear time algorithm for the Hausdorff distance between convex polygons. *Info. Proc. Letters*, 8:207–209, 1983.

[Ata84]   M. Atallah. Checking similarity of planar figures. *Internat. J. Comput. Inform. Sci.*, 13:279–290, 1984.

[Aur91]   F. Aurenhammer. Voronoi diagrams: a survey of a fundamental data structure. *ACM Computing Surveys*, 23:345–405, 1991.

[Baa88]   S. Baase. *Computer Algorithms*. Addison-Wesley, Reading MA, second edition, 1988.

[BCGR92]   D. Berque, R. Cecchini, M. Goldberg, and R. Rivenburgh. The SetPlayer system for symbolic computation on power sets. *J. Symbolic Computation*, 14:645–662, 1992.

[BCW90]   T. Bell, J. Cleary, and I. Witten. *Text Compression*. Prentice Hall, Englewood Cliffs NJ, 1990.

[BDH97]   C. Barber, D. Dobkin, and H. Huhdanpaa. The Quickhull algorithm for convex hulls. *ACM Trans. on Mathematical Software*, 22:469–483, 1997.

[Bel58]   R. Bellman. On a routing problem. *Quarterly of Applied Mathematics*, 16:87–90, 1958.

[Ben75]   J. Bentley. Multidimensional binary search trees used for associative searching. *Communications of the ACM*, 18:509–517, 1975.

[Ben86]   J. Bentley. *Programming Pearls*. Addison-Wesley, Reading MA, 1986.

[Ben90]   J. Bentley. *More Programming Pearls*. Addison-Wesley, Reading MA, 1990.

[Ben92a]   J. Bentley. Fast algorithms for geometric traveling salesman problems. *ORSA J. Computing*, 4:387–411, 1992.

[Ben92b]   J. Bentley. Software exploratorium: The trouble with qsort. *UNIX Review*, 10(2):85–93, February 1992.

[Ber89]   C. Berge. *Hypergraphs*. North-Holland, Amsterdam, 1989.

[BETT94]   G. Di Battista, P. Eades, R. Tamassia, and I. Tollis. Algorithms for drawing graphs: An annotated bibliography. *Computational Geometry: Theory and Applications*, 4, 1994.

[BFP+72]   M. Blum, R. Floyd, V. Pratt, R. Rivest, and R. Tarjan. Time bounds for selection. *J. Computer and System Sciences*, 7:448–461, 1972.

[BG95]   J. Berry and M. Goldberg. Path optimization and near-greedy analysis for graph partitioning: An empirical study. In *Proc. 6th ACM-SIAM Symposium on Discrete Algorithms*, pages 223–232, 1995.

[BGL+95]   C. Di Battista, A. Garg, G. Liotta, R. Tamassia, E. Tassinari, and F. Vargiu. An experimental comparison of three graph drawing algorithms. In *Proc. 11th ACM Symposium on Computational Geometry*, pages 306–315, 1995.

[BGS95]   M Bellare, O. Goldreich, and M. Sudan. Free bits, PCPs, and non-approximability – towards tight results. In *Proc. IEEE 36th Symp. Foundations of Computer Science*, pages 422–431, 1995.

[BH90]   F. Buckley and F. Harary. *Distances in Graphs*. Addison-Wesley, Redwood City, Calif., 1990.

[BJ85]   S. Bent and J. John. Finding the median requires $2n$ comparisons. In *Proc. 17th ACM Symp. on Theory of Computing*, pages 213–216, 1985.

[BJL+91]   A. Blum, T. Jiang, M. Li, J. Tromp, and M. Yanakakis. Linear approximation of shortest superstrings. In *Proc. 23rd ACM Symp. on Theory of Computing*, pages 328–336, 1991.

[BJLM83]   J. Bentley, D. Johnson, F. Leighton, and C. McGeoch. An experimental study of bin packing. In *Proc. 21st Allerton Conf. on Communication, Control, and Computing*, pages 51–60, 1983.

[BL77]   B. P. Buckles and M. Lybanon. Generation of a vector from the lexicographical index. *ACM Trans. Math. Softw.*, 3(2):180–182, June 1977.

[BLS91]   D. Bailey, K. Lee, and H. Simon. Using Strassen's algorithm to accelerate the solution of linear systems. *J. Supercomputing*, 4:357–371, 1991.

[Blu67]   H. Blum. A transformation for extracting new descriptions of shape. In W. Wathen-Dunn, editor, *Models for the Perception of speech and Visual Form*, pages 362–380. MIT Press, 1967.

[BLW76]   N. L. Biggs, E. K. Lloyd, and R. J. Wilson. *Graph Theory 1736-1936*. Clarendon Press, Oxford, 1976.

[BM53]   G. Birkhoff and S. MacLane. *A survey of modern algebra*. Macmillian, New York, 1953.

[BM72]   R. Bayer and E. McCreight. Organization and maintenance of large ordered indexes. *Acta Informatica*, 1:173–189, 1972.

[BM77]   R. Boyer and J. Moore. A fast string-searching algorithm. *Communications of the ACM*, 20:762–772, 1977.

[BM89]   J. Boreddy and R. N. Mukherjee. An algorithm to find polygon similarity. *Inform. Process. Lett.*, 33(4):205–206, 1989.

[BO79]   J. Bentley and T. Ottmann. Algorithms for reporting and counting geometric intersections. *IEEE Transactions on Computers*, C-28:643–647, 1979.

[BO83]   M. Ben-Or. Lower bounds for algebraic computation trees. In *Proc. Fifteenth ACM Symp. on Theory of Computing*, pages 80–86, 1983.

[BP76]   E. Balas and M. Padberg. Set partitioning – a survey. *SIAM Review*, 18:710–760, 1976.

[BR80]   I. Barrodale and F. D. K. Roberts. Solution of the constrained L_1 linear approximation problem. *ACM Trans. Math. Softw.*, 6(2):231–235, June 1980.

[BR95]   A. Binstock and J. Rex. *Practical Algorithms for Programmers*. Addison-Wesley, Reading MA, 1995.

[Bre73]   R. Brent. *Algorithms for minimization without derivatives*. Prentice-Hall, Englewood Cliffs NJ, 1973.

[Bre74]   R. P. Brent. A Gaussian pseudo-random number generator. *Commun. ACM*, 17(12):704–706, December 1974.

[Brè79]   D. Brèlaz. New methods to color the vertices of a graph. *Communications of the ACM*, 22:251–256, 1979.

[Bri74]   E. Brigham. *The Fast Fourier Transform*. Prentice Hall, Englewood Cliffs NJ, 1974.

[Bro74]   F. Brooks. *The Mythical Man-Month*. Addison-Wesley, Reading MA, 1974.

[Bro88]   R. Brown. Calendar queuing method for future event list manipulation. *Comm. ACM*, 30, 1988.

[Brz64]   J. Brzozowski. Derivatives of regular expressions. *J. ACM*, 11:481–494, 1964.

[BS76]   J. Bentley and M. Shamos. Divide-and-conquer in higher-dimensional space. In *Proc. Eighth ACM Symp. Theory of Computing*, pages 220–230, 1976.

[BS81]   I. Barrodale and G. F. Stuart. A Fortran program for solving $\mathbf{a}x = b$. *ACM Trans. Math. Softw.*, 7(3):391–397, September 1981.

[BS86]   G. Berry and R. Sethi. From regular expressions to deterministic automata. *Theoretical Computer Science*, 48:117–126, 1986.

[BS93]   E. Biham and A. Shamir. *Differential Cryptanalysis of the Data Encryption Standard*. Springer-Verlag, Berlin, 1993.

[BS96]   E. Bach and J. Shallit. *Algorithmic Number Theory: Efficient Algorithms*, volume 1. MIT Press, Cambridge MA, 1996.

[BS97]   R. Bradley and S. Skiena. Fabricating arrays of strings. In *Proc. First Int. Conf. Computational Molecular Biology (RECOMB '97)*, pages 57–66, 1997.

[BT92]   J. Buchanan and P. Turner. *Numerical methods and analysis*. McGraw-Hill, New York, 1992.

[Buc94]   A. G. Buckley. A Fortran 90 code for unconstrained nonlinear minimization. *ACM Trans. Math. Softw.*, 20(3):354–372, September 1994.

[BW91]   G. Brightwell and P. Winkler. Counting linear extensions is #P-complete. In *Proc. 23rd ACM Symp. Theory Computing (STOC)*, pages 175–181, 1991.

[BYGNR94] R. Bar-Yehuda, D. Geiger, J. Naor, and R. Roth. Approximation algorithms for the vertex feedback set problem with applications to constraint satisfaction and Bayesian inference. In *Proc. Fifth ACM-SIAM Symp. Discrete Algorithms*, pages 344–354, 1994.

[Can87]   J. Canny. *The complexity of robot motion planning*. MIT Press, Cambridge MA, 1987.

[CC92]   S. Carlsson and J. Chen. The complexity of heaps. In *Proc. Third ACM-SIAM Symp. on Discrete Algorithms*, pages 393–402, 1992.

[CCDG82]   P. Chinn, J. Chvátolvá, A. K. Dewdney, and N. E. Gibbs. The bandwidth problem for graphs and matrices – a survey. *J. Graph Theory*, 6:223–254, 1982.

[CD85]   B. Chazelle and D. Dobkin. Optimal convex decompositions. In G. Toussaint, editor, *Computational Geometry*, pages 63–133. North-Holland, Amsterdam, 1985.

[CDT95]   G. Carpento, M. Dell'Amico, and P. Toth. CDT: A subroutine for the exact solution of large-scale, asymmetric traveling salesman problems. *ACM Trans. Math. Softw.*, 21(4):410–415, December 1995.

[CE92]   B. Chazelle and H. Edelsbrunner. An optimal algorithm for intersecting line segments. *J. ACM*, 39:1–54, 1992.

[CG94]   B. Cherkassky and A. Goldberg. On implementing push-relabel method for the maximum flow problem. Technical Report 94-1523, Department of Computer Science, Stanford University, 1994.

[CGJ96]   E. G. Coffman, M. R. Garey, and D. S. Johnson. Approximation algorithms for bin packing: a survey. In D. Hochbaum, editor, *Approximation algorithms*. PWS Publishing, 1996.

[CGL85]   B. Chazelle, L. Guibas, and D. T. Lee. The power of geometric duality. *BIT*, 25:76–90, 1985.

[CGPS76]   H. L. Crane Jr., N. F. Gibbs, W. G. Poole Jr., and P. K. Stockmeyer. Matrix bandwidth and profile reduction. *ACM Trans. Math. Softw.*, 2(4):375–377, December 1976.

[CGR93]   B. Cherkassky, A. Goldberg, and T. Radzik. Shortest paths algorithms: theory and experimental evaluation. Technical Report 93-1480, Department of Computer Science, Stanford University, 1993.

[Cha71]   J. M. Chambers. Partial sorting. *Commun. ACM*, 14(5):357–358, May 1971.

[Cha91]   B. Chazelle. Triangulating a simple polygon in linear time. *Discrete and Computational Geometry*, 6:485–524, 1991.

[Che80]   T-Y. Cheung. A program for the multifacility location problem with rectilinear distance by the minimum-cut approach. *ACM Trans. Math. Softw.*, 6(3):430–431, September 1980.

[Chr76]   N. Christofides. Worst-case analysis of a new heuristic for the traveling salesman problem. Technical report, Graduate School of Industrial Administration, Carnegie-Mellon University, Pittsburgh PA, 1976.

[CHT90]   J. Cai, X. Han, and R. Tarjan. New solutions to four planar graph problems. Technical report, New York University, 1990.

[Chv83]   V. Chvatal. *Linear Programming*. Freeman, San Francisco, 1983.

[CK70]   D. Chand and S. Kapur. An algorithm for convex polytopes. *J. ACM*, 17:78–86, 1970.

[CK75]   N. Christofides and S. Korman. A computational survey of methods for the set covering problem. *Management Science*, 21:591–599, 1975.

[CK80]   W. Cheney and D. Kincaid. *Numerical Mathematics and Computing*. Brooks/Cole, Monterey CA, 1980.

[CK94]   A. Chetverin and F. Kramer. Oligonucleotide arrays: New concepts and possibilities. *Bio/Technology*, 12:1093–1099, 1994.

[Cla92]   K. L. Clarkson. Safe and effective determinant evaluation. In *Proc. 31st IEEE Symposium on Foundations of Computer Science*, pages 387–395, Pittsburgh, PA, 1992.

[CLR90] T. Cormen, C. Leiserson, and R. Rivest. *Introduction to Algorithms*. MIT Press, Cambridge MA, 1990.

[CM69]   E. Cuthill and J. McKee. Reducing the bandwidth of sparse symmetric matrices. In *Proc. 24th Nat. Conf. ACM*, pages 157–172, 1969.

[Cof76] E. Coffman. *Computer and Job Shop Scheduling*. Wiley, New York, 1976.

[Coh94] E. Cohen. Estimating the size of the transitive closure in linear time. In *35th Annual Symposium on Foundations of Computer Science*, pages 190–200. IEEE, 1994.

[Con71] J. H. Conway. *Regular Algebra and Finite Machines*. Chapman and Hall, London, 1971.

[Coo71] S. Cook. The complexity of theorem proving procedures. In *Proc. Third ACM Symp. Theory of Computing*, pages 151–158, 1971.

[CP90]   R. Carraghan and P. Paradalos. An exact algorithm for the maximum clique problem. In *Operations Research Letters*, volume 9, pages 375–382, 1990.

[CR76]   J. Cohen and M. Roth. On the implementation of Strassen's fast multiplication algorithm. *Acta Informatica*, 6:341–355, 1976.

[CR94]   M. Crochemore and W. Rytter. *Text Algorithms*. Oxford University Press, New York, 1994.

[Cra94] R. Crandall. *Topics in Scientific Computation*. Telos/Springer-Verlag, New York, 1994.

[CS93]   J. Conway and N. Sloane. *Sphere packings, lattices, and groups*. Springer-Verlag, New York, 1993.

[CT65]   J. Cooley and J. Tukey. An algorithm for the machine calculation of complex Fourier series. *Mathematics of Computation*, 19:297–301, 1965.

[CT71]   M. W. Coleman and M. S. Taylor. Circular integer partitioning. *Commun. ACM*, 14(1):48, January 1971.

[CT80]   G. Carpaneto and P. Toth. Solution of the assignment problem. *ACM Trans. Math. Softw.*, 6(1):104–111, March 1980.

[CT92]   Y. Chiang and R. Tamassia. Dynamic algorithms in computational geometry. *Proc. IEEE*, 80:1412–1434, 1992.

[CW79]   B. Commentz-Walter. A string matching algorithm fast on the average. In *Proc. Sixth Int. Coll. on Automata, Languages, and Programming (ICALP)*, pages 118–132. Springer Verlag, Lecture Notes in Computer Science, 1979.

[CW87]   D. Coppersmith and S. Winograd. Matrix multiplication via arithmetic progressions. In *Proc. Nineteenth ACM Symp. Theory of Computing*, pages 1–6, 1987.

[Dan63] G. Dantzig. *Linear programming and extensions*. Princeton University Press, Princeton NJ, 1963.

[Dau92] I. Daubechies. *Ten Lectures on Wavelets*. SIAM, Philadelphia, 1992.

[DB74]   G. Dahlquist and A. Bjorck. *Numerical Methods*. Prentice-Hall, Englewood Cliffs NJ, 1974.

[DB86]   G. Davies and S. Bowsher. Algorithms for pattern matching. *Software – Practice and Experience*, 16:575–601, 1986.

[Den82] D. Denning. *Cryptography and Data Security*. Addison-Wesley, Reading MA, 1982.

[DF79]   E. Denardo and B. Fox. Shortest-route methods: 1. reaching, pruning, and buckets. *Operations Research*, 27:161–186, 1979.

[DFJ54] G. Dantzig, D. Fulkerson, and S. Johnson. Solution of a large-scale traveling-salesman problem. *Operations Research*, 2:393–410, 1954.

[dFPP88]   H. de Fraysseix, J. Pach, and R. Pollack. Small sets supporting Fary embeddings of planar graphs. In *Proc. of the 20th Symposium on the Theory of Computing*, pages 426–433. ACM, 1988.

[DG87]   J. Dongarra and E. Grosse. Distribution of mathematical software via electronic mail. *Communications of the ACM*, 30:403–407, 1987.

[DGKK79]   R. Dial, F. Glover, D. Karney, and D. Klingman. A computational analysis of alternative algorithms and labeling techniques for finding shortest path trees. *Networks*, 9:215–248, 1979.

[DH73]   R. Duda and P. Hart. *Pattern Classification and Scene Analysis*. Wiley-Interscience, New York, 1973.

[DH92]   D. Du and F. Hwang. A proof of Gilbert and Pollak's conjecture on the Steiner ratio. *Algorithmica*, 7:121–135, 1992.

[Dij59]   E. W. Dijkstra. A note on two problems in connection with graphs. *Numerische Mathematik*, 1:269–271, 1959.

[DJ92]   G. Das and D. Joseph. Minimum vertex hulls for polyhedral domains. *Theoret. Comput. Sci.*, 103:107–135, 1992.

[DL76]   D. Dobkin and R. Lipton. Multidimensional searching problems. *SIAM J. Computing*, 5:181–186, 1976.

[DLR79]   D. Dobkin, R. Lipton, and S. Reiss. Linear programming is log-space hard for P. *Info. Processing Letters*, 8:96–97, 1979.

[DM80]   D. Dobkin and J. I. Munro. Determining the mode. *Theoretical Computer Science*, 12:255–263, 1980.

[DM97]   K. Daniels and V. Milenkovic. Multiple translational containment. part I: an approximation algorithm. *Algorithmica*, 1997.

[DMBS79]   J. Dongarra, C. Moler, J. Bunch, and G. Stewart. *LINPACK User's Guide*. SIAM Publications, Philadelphia, 1979.

[DNRT81]   J. J. Ducroz, S. M. Nugent, J. K. Reid, and D. B. Taylor. Solution of real linear equations in a paged virtual store. *ACM Trans. Math. Softw.*, 7(4):537–541, December 1981.

[Dor94]   S. Dorward. A survey of object-space hidden surface removal. *Int. J. Computational Geometry Theory and Applications*, 4:325–362, 1994.

[DP73]   D. H. Douglas and T. K. Peucker. Algorithms for the reduction of the number of points required to represent a digitized line or its caricature. *Canadian Cartographer*, 10(2):112–122, December 1973.

[DP84]   N. Deo and C. Pang. Shortest path algorithms: Taxonomy and annotation. *Networks*, 14:275–323, 1984.

[DR90]   N. Dershowitz and E. Reingold. Calendrical calculations. *Software – Practice and Experience*, 20:899–928, 1990.

[DR97]   N. Dershowitz and E. Reingold. *Calendrical Calculations*. Cambridge University Press, New York, 1997.

[DRR+95]   S. Dawson, C. R. Ramakrishnan, I. V. Ramakrishnan, K. Sagonas, S. Skiena, T. Swift, and D. S. Warren. Unification factoring for efficient execution of logic programs. In *22nd ACM Symposium on Principles of Programming Languages (POPL '95)*, pages 247–258, 1995.

[DS88]   D. Dobkin and D. Silver. Recipes for geometry and numerical analysis. In *Proc. 4th ACM Symp. Computational Geometry*, pages 93–105, 1988.

[Duf81]   S. Duff. Permutations for a zero-free diagonal. *ACM Trans. Math. Softw.*, 7(3):387–390, September 1981.

[dVS82]   G. de V. Smit. A comparison of three string matching algorithms. *Software – Practice and Experience*, 12:57–66, 1982.

[DZ95]   D. Dor and U. Zwick. Selecting the median. In *Proc. Sixth ACM-SIAM Symp. Discrete Algorithms (SODA)*, pages 28–37, 1995.

[Ebe88]   J. Ebert. Computing Eulerian trails. *Info. Proc. Letters*, 28:93–97, 1988.

[Edd77]   W. F. Eddy. CONVEX: a new convex hull algorithm for planar sets. *ACM Trans. Math. Softw.*, 3(4):411–412, December 1977.

[Ede87]   H. Edelsbrunner. *Algorithms for Combinatorial Geometry*. Springer-Verlag, Berlin, 1987.

[Edm65]   J. Edmonds. Paths, trees, and flowers. *Canadian J. Math.*, 17:449–467, 1965.

[Edm71]   J. Edmonds. Matroids and the greedy algorithm. *Mathematical Programming*, 1:126–136, 1971.

[EG60]   P. Erdős and T. Gallai. Graphs with prescribed degrees of vertices. *Mat. Lapok (Hungarian)*, 11:264–274, 1960.

[EG89]   H. Edelsbrunner and L. Guibas. Topologically sweeping an arrangement. *J. Computer and System Sciences*, 38:165–194, 1989.

[EG91]   H. Edelsbrunner and L. Guibas. Corrigendum: Topologically sweeping an arrangement. *J. Computer and System Sciences*, 42:249–251, 1991.

[EGIN92]   D. Eppstein, Z. Galil, G. F. Italiano, and A. Nissenzweig. Sparsification: A technique for speeding up dynamic graph algorithms. In *Proc. 33rd IEEE Symp. on Foundations of Computer Science (FOCS)*, pages 60–69, 1992.

[EGS86]   H. Edelsbrunner, L. Guibas, and J. Stolfi. Optimal point location in a monotone subdivision. *SIAM J. Computing*, 15:317–340, 1986.

[EJ73]   J. Edmonds and E. Johnson. Matching, Euler tours, and the Chinese postman. *Math. Programming*, 5:88–124, 1973.

[EKA84]   M. I. Edahiro, I. Kokubo, and T. Asano. A new point location algorithm and its practical efficiency – comparison with existing algorithms. *ACM Trans. Graphics*, 3:86–109, 1984.

[EKS83]   H. Edelsbrunner, D. Kirkpatrick, and R. Seidel. On the shape of a set of points in the plane. *IEEE Trans. on Information Theory*, IT-29:551–559, 1983.

[ELS93]   P. Eades, X. Lin, and W. F. Smyth. A fast and effective heuristic for the feedback arc set problem. *Info. Proc. Letters*, 47:319–323, 1993.

[EM94]   H. Edelsbrunner and Ernst P. Mücke. Three-dimensional alpha shapes. *ACM Transactions on Graphics*, 13:43–72, 1994.

[ES74]   P. Erdős and J. Spencer. *Probabilistic Methods in Combinatorics*. Academic Press, New York, 1974.

[ES86]   H. Edelsbrunner and R. Seidel. Voronoi diagrams and arrangements. *Discrete and Computational Geometry*, 1:25–44, 1986.

[ESS93]   H. Edelsbrunner, R. Seidel, and M. Sharir. On the zone theorem for hyperplane arrangements. *SIAM J. Computing*, 22:418–429, 1993.

[ESV96]   F. Evans, S. Skiena, and A. Varshney. Optimizing triangle strips for fast rendering. In *Proc. IEEE Visualization '96*, pages 319–326, 1996.

[Eul36]   L. Euler. Solutio problematis ad geometriam situs pertinentis. *Commentarii Academiae Scientiarum Petropolitanae*, 8:128–140, 1736.

[Eve79a]   S. Even. *Graph Algorithms*. Computer Science Press, Rockville MD, 1979.

[Eve79b]   G. Everstine. A comparison of three resequencing algorithms for the reduction of matrix profile and wave-front. *Int. J. Numerical Methods in Engr.*, 14:837–863, 1979.

[F48]   I. Fáry. On straight line representation of planar graphs. *Acta. Sci. Math. Szeged*, 11:229–233, 1948.

[Faj87]   S. Fajtlowicz. On conjectures of Graffiti. *Discrete Mathematics*, 72:113–118, 1987.

[FF62]   L. Ford and D. R. Fulkerson. *Flows in Networks*. Princeton University Press, Princeton NJ, 1962.

[Fis95]   P. Fishwick. *Simulation Model Design and Execution: Building Digital Worlds*. Prentice Hall, Englewood Cliffs, NJ, 1995.

[FJ95]   A. Frieze and M. Jerrum. An analysis of a Monte Carlo algorithm for estimating the permanent. *Combinatorica*, 15, 1995.

[Fle74]   H. Fleischner. The square of every two-connected graph is Hamiltonian. *J. Combinatorial Theory, B*, 16:29–34, 1974.

[Fle80]   R. Fletcher. *Practical Methods of Optimization: Unconstrained Optimization*, volume 1. John Wiley, Chichester, 1980.

[Flo62]   R. Floyd. Algorithm 97 (shortest path). *Communications of the ACM*, 7:345, 1962.

[Flo64]   R. Floyd. Algorithm 245 (treesort). *Communications of the ACM*, 18:701, 1964.

[FM71]   M. Fischer and A. Meyer. Boolean matrix multiplication and transitive closure. In *IEEE 12th Symp. on Switching and Automata Theory*, pages 129–131, 1971.

[For87]   S. Fortune. A sweepline algorithm for Voronoi diagrams. *Algorithmica*, 2:153–174, 1987.

[For92]   S. Fortune. Voronoi diagrams and Delaunay triangulations. In D.-Z. Du and F. Hwang, editors, *Computing in Euclidean Geometry*, volume 1, pages 193–234. World Scientific, 1992.

[FP75a]   D. Fayard and G. Plateau. Resolution of the 0-1 knapsack problem: Comparison of methods. *Math. Programming*, 8:272–307, 1975.

[FP75b]   H. Feng and T. Pavlidis. Decomposition of polygons into simpler components: feature generation for syntactic pattern recognition. *IEEE Transactions on Computers*, C-24:636–650, 1975.

[FR75]   R. Floyd and R. Rivest. Expected time bounds for selection. *Communications of the ACM*, 18:165–172, 1975.

[FR94]   M. Fürer and B. Raghavachari. Approximating the minimum-degree Steiner tree to within one of optimal. *J. Algorithms*, 17:409–423, 1994.

[Fra79]   D. Fraser. An optimized mass storage FFT. *ACM Trans. Math. Softw.*, 5(4):500–517, December 1979.

[Fre62]   E. Fredkin. Trie memory. *Communications of the ACM*, 3:490–499, 1962.

[FT87]   M. Fredman and R. Tarjan. Fibonacci heaps and their uses in improved network optimization algorithms. *J. ACM*, 34:596–615, 1987.

[Fuj96]   T. Fujito. A note on approximation of the vertex cover and feedback vertex set problems. *Info. Proc. Letters*, 59:59–63, 1996.

[FvW93]   S. Fortune and C. van Wyk. Efficient exact arithmetic for computational geometry. In *Proc. 9th ACM Symp. Computational Geometry*, pages 163–172, 1993.

[FW77]   S. Fiorini and R. Wilson. *Edge-colourings of graphs*. Research Notes in Mathematics 16, Pitman, London, 1977.

[FW93]   M. Fredman and D. Willard. Surpassing the information theoretic bound with fusion trees. *J. Computer and System Sci.*, 47:424–436, 1993.

[Gab76]   H. Gabow. An efficient implementation of Edmond's algorithm for maximum matching on graphs. *J. ACM*, 23:221–234, 1976.

[Gab77]   H. Gabow. Two algorithms for generating weighted spanning trees in order. *SIAM J. Computing*, 6:139–150, 1977.

[Gal86]   Z. Galil. Efficient algorithms for finding maximum matchings in graphs. *ACM Computing Surveys*, 18:23–38, 1986.

[GBDS80]   B. Golden, L. Bodin, T. Doyle, and W. Stewart. Approximate traveling salesman algorithms. *Operations Research*, 28:694–711, 1980.

[GBY91]   G. Gonnet and R. Baeza-Yates. *Handbook of Algorithms and Data Structures*. Addison-Wesley, Wokingham, England, second edition, 1991.

[GGJ77]   M. Garey, R. Graham, and D. Johnson. The complexity of computing Steiner minimal trees. *SIAM J. Appl. Math.*, 32:835–859, 1977.

[GGJK78]   M. Garey, R. Graham, D. Johnson, and D. Knuth. Complexity results for bandwidth minimization. *SIAM J. Appl. Math.*, 34:477–495, 1978.

[GH85]   R. Graham and P. Hell. On the history of the minimum spanning tree problem. *Annals of the History of Computing*, 7:43–57, 1985.

[GHMS93]   L. J. Guibas, J. E. Hershberger, J. S. B. Mitchell, and J. S. Snoeyink. Approximating polygons and subdivisions with minimum link paths. *Internat. J. Comput. Geom. Appl.*, 3(4):383–415, December 1993.

[GHR95]   R. Greenlaw, J. Hoover, and W. Ruzzo. *Limits to Parallel Computation: P-completeness theory*. Oxford University Press, New York, 1995.

[GI89]   D. Gusfield and R. Irving. *The Stable Marriage Problem: structure and algorithms*. MIT Press, Cambridge MA, 1989.

[GI91]   Z. Galil and G. Italiano. Data structures and algorithms for disjoint set union problems. *ACM Computing Surveys*, 23:319–344, 1991.

[Gib76]   N. E. Gibbs. A hybrid profile reduction algorithm. *ACM Trans. Math. Softw.*, 2(4):378–387, December 1976.

[GJ77]   M. Garey and D. Johnson. The rectilinear Steiner tree problem is NP-complete. *SIAM J. Appl. Math.*, 32:826–834, 1977.

[GJ79]   M. R. Garey and D. S. Johnson. *Computers and Intractability: A Guide to the theory of NP-completeness*. W. H. Freeman, San Francisco, 1979.

[GJPT78]   M. Garey, D. Johnson, F. Preparata, and R. Tarjan. Triangulating a simple polygon. *Info. Proc. Letters*, 7:175–180, 1978.

[GK79]   A. Goralcikiova and V. Konbek. A reduct and closure algorithm for graphs. In *Mathematical Foundations of Computer Science*, pages 301–307. Springer Verlag, Lecture Notes in Computer Science V. 74, 1979.

[GK93]   A. Goldberg and R. Kennedy. An efficient cost scaling algorithm for the assignment problem. Technical Report 93-1481, Department of Computer Science, Stanford University, 1993.

[GKK74]   F. Glover, D. Karney, and D. Klingman. Implementation and computational comparisons of primal-dual computer codes for minimum-cost network flow problems. *Networks*, 4:191–212, 1974.

[GKP89]   R. Graham, D. Knuth, and O. Patashnik. *Concrete Mathematics*. Addison-Wesley, Reading MA, 1989.

[Glo89a]   F. Glover. Tabu search - Part 1. *ORSA Journal on Computing*, 1(3):190–206, 1989.

[Glo89b]   F. Glover. Tabu search - Part 2. *ORSA Journal on Computing*, 2(1):4–32, 1989.

[Glo90]   F. Glover. Tabu search: A tutorial. *Interfaces*, 20 (4):74–94, 1990.

[GM86]   G. Gonnet and J. I. Munro. Heaps on heaps. *SIAM J. Computing*, 15:964–971, 1986.

[Gol89]   D. E. Goldberg. *Genetic Algorithms in Search, Optimization, and Machine Learning*. Addison-Wesley, 1989.

[Gol92]   A. Goldberg. An efficient implementation of a scaling minimum-cost flow algorithm. Technical Report 92-1439, Department of Computer Science, Stanford University, 1992.

[Gol93]   L. Goldberg. *Efficient Algorithms for Listing Combinatorial Structures*. Cambridge University Press, 1993.

[GP68]   E. Gilbert and H. Pollak. Steiner minimal trees. *SIAM J. Applied Math.*, 16:1–29, 1968.

[GP79]   B. Gates and C. Papadimitriou. Bounds for sorting by prefix reversals. *Discrete Mathematics*, 27:47–57, 1979.

[GPS76]   N. Gibbs, W. Poole, and P. Stockmeyer. A comparison of several bandwidth and profile reduction algorithms. *ACM Trans. Math. Software*, 2:322–330, 1976.

[Gra53]   F. Gray. Pulse code communication. US Patent 2632058, March 17, 1953.

[Gra72]   R. Graham. An efficient algorithm for determining the convex hull of a finite planar point set. *Info. Proc. Letters*, 1:132–133, 1972.

[GS62]   D. Gale and L. Shapely. College admissions and the stability of marriages. *American Math. Monthly*, 69:9–14, 1962.

[GS78]   L. Guibas and R. Sedgewick. A dichromatic framework for balanced trees. In *Proc. 19th IEEE Symp. Foundations of Computer Science*, pages 8–21, 1978.

[GT88]   A. Goldberg and R. Tarjan. A new approach to the maximum flow problem. *J. ACM*, pages 921–940, 1988.

[GT89]   H. Gabow and R. Tarjan. Faster scaling algorithms for network problems. *SIAM J. Computing*, 18:1013–1036, 1989.

[Gup66]   R. P. Gupta. The chromatic index and the degree of a graph. *Notices of the Amer. Math. Soc.*, 13:719, 1966.

[Gus97]   D. Gusfield. *String Algorithms*. Cambridge University Press, 1997.

[GW95]   M. Goemans and D. Williamson. .878-approximation algorithms for MAX CUT and MAX 2SAT. *J. ACM*, 42:1115–1145, 1995.

[HD80]   P. Hall and G. Dowling. Approximate string matching. *ACM Computing Surveys*, 12:381–402, 1980.

[HG95]   P. Heckbert and M. Garland. Fast polygonal approximation of terrains and height fields. Technical Report CMU-CS-95-181, School of Computer Science, Carnegie-Mellon University, 1995.

[Him94]   M. Himsolt. GraphEd: A graphical platform for the implementation of graph algorithms. In *Proc. Graph Drawing (GD '94)*, volume 894, pages 182–193, 1994.

[Hir75]   D. Hirschberg. A linear-space algorithm for computing maximum common subsequences. *Communications of the ACM*, 18:341–343, 1975.

[HJS84]   R. E. Haymond, J. P. Jarvis, and D. R. Shier. Minimum spanning tree for moderate integer weights. *ACM Trans. Math. Softw.*, 10(1):108–111, March 1984.

[HK73]   J. Hopcroft and R. Karp. An $n^{5/3}$ algorithm for maximum matchings in bipartite graphs. *SIAM J. Computing*, 2:225–231, 1973.

[HK90]   D. P. Huttenlocher and K. Kedem. Computing the minimum Hausdorff distance for point sets under translation. In *Proc. 6th Annu. ACM Sympos. Comput. Geom.*, pages 340–349, 1990.

[HM83]   S. Hertel and K. Mehlhorn. Fast triangulation of simple polygons. In *Proc. 4th Internat. Conf. Found. Comput. Theory*, pages 207–218. Lecture Notes in Computer Science, Vol. 158, 1983.

[HMMN84]   S. Hertel, K. Mehlhorn, M. Mäntylä, and J. Nievergelt. Space sweep solves intersection of two convex polyhedra elegantly. *Acta Informatica*, 21:501–519, 1984.

[Hoa61]   C. A. R. Hoare. Algorithm 63 (partition) and algorithm 65 (find). *Communications of the ACM*, 4:321–322, 1961.

[Hoa62]   C. A. R. Hoare. Quicksort. *Computer Journal*, 5:10–15, 1962.

[Hoc96]   D. Hochbaum, editor. *Approximation Algorithms for NP-hard Problems*. PWS Publishing, Boston, 1996.

[Hof82]   C. Hoffmann. *Group-theoretic algorithms and graph isomorphism*, volume 136 of *Lecture Notes in Computer Science*. Springer-Verlag Inc., New York, 1982.

[Hof89]   C. Hoffman. The problem of accuracy and robustness in geometric computation. *Computer*, 22:31–42, 1989.

[Hol75]   J. Holland. *Adaptation in Natural and Artificial Systems*. University of Michigan Press, Ann Arbor, 1975.

[Hol81]   I. Holyer. The NP-completeness of edge colorings. *SIAM J. Computing*, 10:718–720, 1981.

[Hol92]   J. Holland. Genetic algorithms. *Scientific American*, 267(1):66–72, July 1992.

[Hop71]   J. Hopcroft. An $n \log n$ algorithm for minimizing the states in a finite automaton. In Z. Kohavi, editor, *The theory of machines and computations*, pages 189–196. Academic Press, New York, 1971.

[Hor80]   R. N. Horspool. Practical fast searching in strings. *Software – Practice and Experience*, 10:501–506, 1980.

[HP73]   F. Harary and E. Palmer. *Graphical enumeration*. Academic Press, New York, 1973.

[HRW92]   R. Hwang, D. Richards, and P. Winter. *The Steiner Tree Problem*, volume 53 of *Annals of Discrete Mathematics*. North Holland, Amsterdam, 1992.

[HS77]   J. Hunt and T. Szymanski. A fast algorithm for computing longest common subsequences. *Communications of the ACM*, 20:350–353, 1977.

[HS94]   J. Hershberger and J. Snoeyink. An $O(n \log n)$ implementation of the Douglas-Peucker algorithm for line simplification. In *Proc. 10th Annu. ACM Sympos. Comput. Geom.*, pages 383–384, 1994.

[HSS87]   J. Hopcroft, J. Schwartz, and M. Sharir. *Planning, geometry, and complexity of robot motion*. Ablex Publishing, Norwood NJ, 1987.

[HT73a]   J. Hopcroft and R. Tarjan. Dividing a graph into triconnected components. *SIAM J. Computing*, 2:135–158, 1973.

[HT73b]   J. Hopcroft and R. Tarjan. Efficient algorithms for graph manipulation. *Communications of the ACM*, 16:372–378, 1973.

[HT74]   J. Hopcroft and R. Tarjan. Efficient planarity testing. *J. ACM*, 21:549–568, 1974.

[HT84]   D. Harel and R. E. Tarjan. Fast algorithms for finding nearest common ancestors. *SIAM J. Comput.*, 13:338–355, 1984.

[Hua92]   X. Huang. A contig assembly program based on sensitive detection of fragment overlaps. *Genomics*, 1992.

[Hua94]   X. Huang. On global sequence alignment. *CABIOS*, 10:227–235, 1994.

[Huf52]   D. Huffman. A method for the construction of minimum-redundancy codes. *Proc. of the IRE*, 40:1098–1101, 1952.

[HW74]   J. E. Hopcroft and J. K. Wong. Linear time algorithm for isomorphism of planar graphs. In *Proc. Sixth Annual ACM Symposium on Theory of Computing*, pages 172–184, 1974.

[HWK94]   T. He, S. Wang, and A. Kaufman. Wavelet-based volume morphing. In *Proc. IEEE Visualization '94*, pages 85–92, 1994.

[IK75]   O. Ibarra and C. Kim. Fast approximation algorithms for knapsack and sum of subset problems. *J. ACM*, 22:463–468, 1975.

[Ita78]   A. Itai. Two commodity flow. *J. ACM*, 25:596–611, 1978.

[JAMS91]   D. Johnson, C. Aragon, C. McGeoch, and D. Schevon. Optimization by simulated annealing: an experimental evaluation; part II, graph coloring and number partitioning. In *Operations Research*, volume 39, pages 378–406, 1991.

[Jan76]   W. Janko. A list insertion sort for keys with arbitrary key distribution. *ACM Trans. Math. Softw.*, 2(2):204–206, June 1976.

[Jar73]   R. A. Jarvis. On the identification of the convex hull of a finite set of points in the plane. *Info. Proc. Letters*, 2:18–21, 1973.

[JD88]   A. Jain and R. Dubes. *Algorithms for Clustering Data*. Prentice-Hall, Englewood Cliffs NJ, 1988.

[JM93]   D. Johnson and C. McGeoch, editors. *Network Flows and Matching: First DIMACS Implementation Challenge*, volume 12. American Mathematics Society, Providence RI, 1993.

[Joh63]   S. M. Johnson. Generation of permutations by adjacent transpositions. *Math. Computation*, 17:282–285, 1963.

[Joh74]   D. Johnson. Approximation algorithms for combinatorial problems. *J. Computer and System Sciences*, 9:256–278, 1974.

[Joh90]   D. Johnson. A catalog of complexity classes. In J. van Leeuwen, editor, *Handbook of Theoretical Computer Science: Algorithms and Complexity*, volume A, pages 67–162. MIT Press, 1990.

[Jon86]   D. W. Jones. An empirical comparison of priority-queue and event-set implementations. *Communications of the ACM*, 29:300–311, 1986.

[Kah67]   D. Kahn. *The Code breakers: the story of secret writing*. Macmillan, New York, 1967.

[Kah80]   D. A. Kahaner. Fortran implementation of heap programs for efficient table maintenance. *ACM Trans. Math. Softw.*, 6(3):444–449, September 1980.

[Kar72]   R. M. Karp. Reducibility among combinatorial problems. In R. Miller and J. Thatcher, editors, *Complexity of Computer Computations*, pages 85–103. Plenum Press, 1972.

[Kar84]   N. Karmarkar. A new polynomial-time algorithm for linear programming. *Combinatorica*, 4:373–395, 1984.

[Kar96a]   D. Karger. Minimum cuts in near-linear time. In *Proc. 28th ACM Symp. Theory of Computing*, pages 56–63, 1996.

[Kar96b]   H. Karloff. How good is the Goemans-Williamson MAX CUT algorithm? In *Proc. Twenty-Eighth Annual ACM Symposium on Theory of Computing*, pages 427–434, 1996.

[Kei85]   M. Keil. Decomposing a polygon into simpler components. *SIAM J. Computing*, 14:799–817, 1985.

[KGV83]   S. Kirkpatrick, C. D. Gelatt, Jr., and M. P. Vecchi. Optimization by simulated annealing. *Science*, 220:671–680, 1983.

[KH80]   J. Kennington and R. Helgason. *Algorithms for Network Programming*. Wiley-Interscience, New York, 1980.

[Kha79]   L. Khachian. A polynomial algorithm in linear programming. *Soviet Math. Dokl.*, 20:191–194, 1979.

[Kir79]   D. Kirkpatrick. Efficient computation of continuous skeletons. In *Proc. 20th IEEE Symp. Foundations of Computing*, pages 28–35, 1979.

[Kir83]   D. Kirkpatrick. Optimal search in planar subdivisions. *SIAM J. Computing*, 12:28–35, 1983.

[KKT95]   D. Karger, P. Klein, and R. Tarjan. A randomized linear-time algorithm to find minimum spanning trees. *J. ACM*, 42:321–328, 1995.

[KL70]   B. W. Kernighan and S. Lin. An efficient heuristic procedure for partitioning graphs. *The Bell System Technical Journal*, pages 291–307, 1970.

[KM72]   V. Klee and G. Minty. How good is the simplex algorithm. In *Inequalities III*, pages 159–172, New York, 1972. Academic Press.

[KM95]   J. D. Kececioglu and E. W. Myers. Combinatorial algorithms for DNA sequence assembly. *Algorithmica*, 13(1/2):7–51, January 1995.

[KMP77]   D. Knuth, J. Morris, and V. Pratt. Fast pattern matching in strings. *SIAM J. Computing*, 6:323–350, 1977.

[Kno79]   H. D. Knoble. An efficient one-way enciphering algorithm. *ACM Trans. Math. Softw.*, 5(1):108–111, March 1979.

[Knu73a]   D. E. Knuth. *The Art of Computer Programming, Volume 1: Fundamental Algorithms*. Addison-Wesley, Reading MA, second edition, 1973.

[Knu73b]   D. E. Knuth. *The Art of Computer Programming, Volume 3: Sorting and Searching*. Addison-Wesley, Reading MA, 1973.

[Knu81]   D. E. Knuth. *The Art of Computer Programming, Volume 2: Seminumerical Algorithms*. Addison-Wesley, Reading MA, second edition, 1981.

[Knu94]   D. E. Knuth. *The Stanford GraphBase: a platform for combinatorial computing*. ACM Press, New York, 1994.

[KO63]   A. Karatsuba and Yu. Ofman. Multiplication of multi-digit numbers on automata. *Sov. Phys. Dokl.*, 7:595–596, 1963.

[KOS91]   A. Kaul, M. A. O'Connor, and V. Srinivasan. Computing Minkowski sums of regular polygons. In *Proc. 3rd Canad. Conf. Comput. Geom.*, pages 74–77, 1991.

[Koz92]   J. R. Koza. *Genetic Programming*. MIT Press, Cambridge, MA, 1992.

[KPS89]   B. Korte, H. Prömel, and A. Steger. Steiner trees in VLSI-layout. In B. Korte, L. Lovasz, H. Prömel, and A. Schrijver, editors, *Paths, Flows, and VLSI-layout*, pages 185–214. Springer-Verlag, Berlin, 1989.

[KR87]   R. Karp and M. Rabin. Efficient randomized pattern-matching algorithms. *IBM J. Research and Development*, 31:249–260, 1987.

[KR91]   A. Kanevsky and V. Ramachandran. Improved algorithms for graph four-connectivity. *J. Comp. Sys. Sci.*, 42:288–306, 1991.

[Kru56]   J. B. Kruskal. On the shortest spanning subtree of a graph and the traveling salesman problem. *Proc. of the American Mathematical Society*, 7:48–50, 1956.

[KS85]   M. Keil and J. R. Sack. *Computational Geometry*, chapter Minimum decomposition of geometric objects, pages 197–216. North-Holland, 1985.

[KS86]   D. Kirkpatrick and R. Siedel. The ultimate planar convex hull algorithm? *SIAM J. Computing*, 15:287–299, 1986.

[KS90]   K. Kedem and M. Sharir. An efficient motion planning algorithm for a convex rigid polygonal object in 2-dimensional polygonal space. *Discrete and Computational Geometry*, 5:43–75, 1990.

[Kuh75]   H. W. Kuhn. Steiner's problem revisited. In G. Dantzig and B. Eaves, editors, *Studies in Optimization*, pages 53–70. Mathematical Association of America, 1975.

[Kur30]   K. Kuratowski. Sur le problème des courbes gauches en topologie. *Fund. Math.*, 15:217–283, 1930.

[Kwa62]   M. Kwan. Graphic programming using odd and even points. *Chinese Math.*, 1:273–277, 1962.

[Lat91]   J.-C. Latombe. *Robot Motion Planning*. Kluwer Academic Publishers, Boston, 1991.

[Law76]   E. Lawler. *Combinatorial Optimization: Networks and Matroids*. Holt, Rinehart, and Winston, Fort Worth TX, 1976.

[Lee82]   D. T. Lee. Medial axis transformation of a planar shape. *IEEE Trans. Pattern Analysis and Machine Intelligence*, PAMI-4:363–369, 1982.

[Lee94]   H. Leeb. pLab—a system for testing random numbers. In *Proc. Int. Workshop Parallel Numerics*, pages 89–99. Slovak Academy of Science, Institute for Informatics, Slovakia, 1994. Also available via ftp from http://random.mat.sbg.ac.at/ftp/pub/publications/leeb/smolenice/.

[Len87a] T. Lengauer. Efficient algorithms for finding minimum spanning forests of hierarchically defined graphs. *J. Algorithms*, 8, 1987.

[Len87b] H. W. Lenstra. Factoring integers with elliptic curves. *Annals of Mathematics*, 126:649–673, 1987.

[Len89]   T. Lengauer. Hierarchical planarity testing algorithms. *J. ACM*, 36(3):474–509, July 1989.

[Len90]   T. Lengauer. *Combinatorial Algorithms for Integrated Circuit Layout*. Wiley, Chichester, England, 1990.

[Lev92]   J. L. Leva. A normal random number generator. *ACM Trans. Math. Softw.*, 18(4):454–455, December 1992.

[Lew82]   J. G. Lewis. The Gibbs-Poole-Stockmeyer and Gibbs-King algorithms for reordering sparse matrices. *ACM Trans. Math. Softw.*, 8(2):190–194, June 1982.

[Lin65]   S. Lin. Computer solutions of the traveling salesman problem. *Bell System Tech. J.*, 44:2245–2269, 1965.

[LK73]   S. Lin and B. Kernighan. An effective heuristic algorithm for the traveling salesman problem. *Operations Research*, 21:498–516, 1973.

[LLK83]   J. K. Lenstra, E. L. Lawler, and A. Rinnooy Kan. *Theory of Sequencing and Scheduling*. Wiley, New York, 1983.

[LLKS85] E. Lawler, J. Lenstra, A. Rinnooy Kan, and D. Shmoys. *The Traveling Salesman Problem*. John Wiley, 1985.

[LLS92]   L. Lam, S.-W. Lee, and C. Suen. Thinning methodologies – a comprehensive survey. *IEEE Trans. Pattern Analysis and Machine Intelligence*, 14:869–885, 1992.

[LP86]   L. Lovász and M. Plummer. *Matching Theory*. North-Holland, Amsterdam, 1986.

[LPW79]   T. Lozano-Perez and M. Wesley. An algorithm for planning collision-free paths among polygonal obstacles. *Comm. ACM*, 22:560–570, 1979.

[LR93]   K. Lang and S. Rao. Finding near-optimal cuts: An empirical evaluation. In *Proc. 4th Annual ACM-SIAM Symposium on Discrete Algorithms (SODA '93)*, pages 212–221, 1993.

[LS87]   V. Lumelski and A. Stepanov. Path planning strategies for a point mobile automaton moving amidst unknown obstacles of arbitrary shape. *Algorithmica*, 3:403–430, 1987.

[LS95]   Y.-L. Lin and S. Skiena. Algorithms for square roots of graphs. *SIAM J. Discrete Mathematics*, 8:99–118, 1995.

[LT79]   R. Lipton and R. Tarjan. A separator theorem for planar graphs. *SIAM Journal on Applied Mathematics*, 36:346–358, 1979.

454   Bibliography

[LT80]   R. Lipton and R. Tarjan. Applications of a planar separator theorem. *SIAM J. Computing*, 9:615–626, 1980.

[Luc91]   E. Lucas. *Récréations Mathématiques*. Gauthier-Villares, Paris, 1891.

[Luk80]   E. M. Luks. Isomorphism of bounded valence can be tested in polynomial time. In *Proc. of the 21st Annual Symposium on Foundations of Computing*, pages 42–49. IEEE, 1980.

[LV93]   M. Li and P. Vitányi. *An introduction to Kolmogorov complexity and its applications*. Springer-Verlag, New York, 1993.

[LW77]   D. T. Lee and C. K. Wong. Worst-case analysis for region and partial region searches in multidimensional binary search trees and balanced quad trees. *Acta Informatica*, 9:23–29, 1977.

[LW88]   T. Lengauer and E. Wanke. Efficient solution of connectivity problems on hierarchically defined graphs. *SIAM J. Computing*, 17:1063–1080, 1988.

[LY93]   C. Lund and M. Yannakakis. On the hardness of approximating minimization problems. In *Proc. 25th ACM Symp. Theory of Computing (STOC)*, pages 286–293, 1993.

[Man89]   U. Manber. *Introduction to Algorithms*. Addison-Wesley, Reading MA, 1989.

[Mar83]   S. Martello. An enumerative algorithm for finding Hamiltonian circuits in a directed graph. *ACM Trans. Math. Softw.*, 9(1):131–138, March 1983.

[Mat87]   D. W. Matula. Determining edge connectivity in $O(nm)$. In *28th Ann. Symp. Foundations of Computer Science*, pages 249–251. IEEE, 1987.

[McC76]   E. McCreight. A space-economical suffix tree construction algorithm. *J. ACM*, 23:262–272, 1976.

[McH90]   J. McHugh. *Algorithmic Graph Theory*. Prentice-Hall, Englewood Cliffs NJ, 1990.

[McN83]   J. M. McNamee. A sparse matrix package – part II: Special cases. *ACM Trans. Math. Softw.*, 9(3):344–345, September 1983.

[Meg83]   N. Megiddo. Linear time algorithm for linear programming in r^3 and related problems. *SIAM J. Computing*, 12:759–776, 1983.

[Meh84]   K. Mehlhorn. *Data structures and algorithms*, volume 1-3. Springer-Verlag, Berlin, 1984.

[Men27]   K. Menger. Zur allgemeinen Kurventheorie. *Fund. Math.*, 10:96–115, 1927.

[MGH81]   J. J. Moré, B. S. Garbow, and K. E. Hillstrom. Fortran subroutines for testing unconstrained optimization software. *ACM Trans. Math. Softw.*, 7(1):136–140, March 1981.

[MH78]   R. Merkle and M. Hellman. Hiding and signatures in trapdoor knapsacks. *IEEE Trans. Information Theory*, 24:525–530, 1978.

[Mic92]   Z. Michalewicz. *Genetic Algorithms + Data Structures = Evolution Programs*. Springer, Berlin, 1992.

[Mie58]   W. Miehle. Link-minimization in networks. *Operations Research*, 6:232–243, 1958.

[Mil76]   G. Miller. Riemann's hypothesis and tests for primality. *J. Computer and System Sciences*, 13:300–317, 1976.

[Mil89]   V. Milenkovic. Double precision geometry: a general technique for calculating line and segment intersections using rounded arithmetic. In *Proc. 30th IEEE Symp. Foundations of Computer Science (FOCS)*, pages 500–505, 1989.

[Mil97]   V. Milenkovic. Multiple translational containment. part II: exact algorithms. *Algorithmica*, 1997.

[Min78]   H. Minc. *Permanents*, volume 6 of *Encyclopedia of Mathematics and its Applications*. Addison-Wesley, Reading MA, 1978.

[Mit96]   J. Mitchell. Guillotine subdivisions approximate polygonal subdivisions: Part II – A simple polynomial-time approximation scheme for geometric k-MST, TSP, and related problems. University at Stony Brook, Part I appears in *SODA'96*, pp. 402-408, 1996.

[MM90]   U. Manber and G. Myers. Suffix arrays: A new method for on–line string searches. In *Proceedings First ACM–SIAM Symposium on Discrete Algorithms*, pages 319–327. SIAM, 1990.

[MMI72]   D. Matula, G. Marble, and J. Isaacson. Graph coloring algorithms. In R. C. Read, editor, *Graph Theory and Computing*, pages 109–122. Academic Press, 1972.

[MN95]   K. Mehlhorn and S. Näher. LEDA, a platform for combinatorial and geometric computing. *Communications of the ACM*, 38:96–102, 1995.

[MO63]   L. E. Moses and R. V. Oakford. *Tables of Random Permutations*. Stanford University Press, Stanford, Calif., 1963.

[Moo59]   E. F. Moore. The shortest path in a maze. In *Proc. International Symp. Switching Theory*, pages 285–292. Harvard University Press, 1959.

[MP78]   D. Muller and F. Preparata. Finding the intersection of two convex polyhedra. *Theoretical Computer Science*, 7:217–236, 1978.

[MP80]   W. Masek and M. Paterson. A faster algorithm for computing string edit distances. *J. Computer and System Sciences*, 20:18–31, 1980.

[MR95]   R. Motwani and P. Raghavan. *Randomized Algorithms*. Cambridge University Press, New York, 1995.

[MRRT53]   N. Metropolis, A. W. Rosenbluth, M. N. Rosenbluth, and A. H. Teller. Equation of state calculations by fast computing machines. *Journal of Chemical Physics*, 21(6):1087–1092, June 1953.

[MS91]   B. Moret and H. Shapiro. *Algorithm from P to NP: Design and Efficiency*. Benjamin/Cummings, Redwood City, CA, 1991.

[MS93]   M. Murphy and S. Skiena. Ranger: A tool for nearest neighbor search in high dimensions. In *Proc. Ninth ACM Symposium on Computational Geometry*, pages 403–404, 1993.

[MS95a]   D. Margaritis and S. Skiena. Reconstructing strings from substrings in rounds. Proc. 36th IEEE Symp. Foundations of Computer Science (FOCS), 1995.

[MS95b]   J. S. B. Mitchell and S. Suri. Separation and approximation of polyhedral objects. *Comput. Geom. Theory Appl.*, 5:95–114, 1995.

[MT85]   S. Martello and P. Toth. A program for the *0-1* multiple knapsack problem. *ACM Trans. Math. Softw.*, 11(2):135–140, June 1985.

[MT87]   S. Martello and P. Toth. Algorithms for knapsack problems. In S. Martello, editor, *Surveys in Combinatorial Optimization*, volume 31 of *Annals of Discrete Mathematics*, pages 213–258. North-Holland, 1987.

[MT90a]   S. Martello and P. Toth. *Knapsack problems: algorithms and computer implementations*. Wiley, New York, 1990.

[MT90b]   K. Mehlhorn and A. Tsakalidis. Data structures. In J. van Leeuwen, editor, *Handbook of Theoretical Computer Science: Algorithms and Complexity*, volume A, pages 301–341. MIT Press, 1990.

[Mul94]   K. Mulmuley. *Computational Geometry: an introduction through randomized algorithms*. Prentice-Hall, New York, 1994.

[MV80]   S. Micali and V. Vazirani. An $O(\sqrt{|V|}|e|)$ algorithm for finding maximum matchings in general graphs. In *Proc. 21st. Symp. Foundations of Computing*, pages 17–27, 1980.

[MW93]   J. More and S. Wright. *Optimization Software Guide*. SIAM, Philadelphia PA, 1993.

[Mye86]   E. Myers. An $O(nd)$ difference algorithm and its variations. *Algorithmica*, 1:514–534, 1986.

[NC88]   T. Nishizeki and N. Chiba. *Planar Graphs: Theory and Algorithms*. North-Holland, Amsterdam, 1988.

[Neu63]   J. Von Neumann. Various techniques used in connection with random digits. In A. H. Traub, editor, *John von Neumann, Collected Works*, volume 5. Macmillan, 1963.

[NH93]   J. Nievergelt and K. Hinrichs. *Algorithms and Data Structures with applications to graphics and geometry*. Prentice Hall, Englewood Cliffs NJ, 1993.

[NU95]   S. Näher and C. Uhrig. The LEDA user manual, version r3.2. Available by ftp from ftp.mpi-sb.mpg.de in directory /pub/LEDA, 1995.

[NW78]   A. Nijenhuis and H. Wilf. *Combinatorial Algorithms for Computers and Calculators*. Academic Press, Orlando FL, second edition, 1978.

[NW86]   J. C. Nash and R. L. C. Wang. Subroutines for testing programs that compute the generalized inverse of a matrix. *ACM Trans. Math. Softw.*, 12(3):274–277, September 1986.

[NZ80]   I. Niven and H. Zuckerman. *An Introduction to the Theory of Numbers*. Wiley, New York, fourth edition, 1980.

[Ogn93]   R. Ogniewicz. *Discrete Voronoi Skeletons*. Hartung-Gorre Verlag, Konstanz, Germany, 1993.

[O'R87]   J. O'Rourke. *Art Gallery Theorems and Algorithms*. Oxford University Press, Oxford, 1987.

[O'R94]   J. O'Rourke. *Computational Geometry in C*. Cambridge University Press, New York, 1994.

[Ort88]   J. Ortega. *Introduction to Parallel and Vector Solution of Linear Systems*. Plenum, New York, 1988.

[OvL81]   M. Overmars and J. van Leeuwen. Maintenance of configurations in the plane. *J. Computer and System Sciences*, 23:166–204, 1981.

[OW85]   J. O'Rourke and R. Washington. Curve similarity via signatures. In G. T. Toussaint, editor, *Computational Geometry*, pages 295–317. North-Holland, Amsterdam, Netherlands, 1985.

[P57]   G. Pólya. *How to Solve It*. Princeton University Press, Princeton NJ, second edition, 1957.

[Pag74]   R. L. Page. A minimal spanning tree clustering method. *Commun. ACM*, 17(6):321–323, June 1974.

[Pal85]   E. M. Palmer. *Graphical Evolution: An Introduction to the Theory of Random Graphs*. Wiley-Interscience, New York, 1985.

[Pap76a]   C. Papadimitriou. The complexity of edge traversing. *J. ACM*, 23:544–554, 1976.

[Pap76b]   C. Papadimitriou. The NP-completeness of the bandwidth minimization problem. *Computing*, 16:263–270, 1976.

[Pap80]   U. Pape. Shortest path lengths. *ACM Trans. Math. Softw.*, 6(3):450–455, September 1980.

[Par90]   G. Parker. A better phonetic search. *C Gazette*, 5-4, June/July 1990.

[Pav82]   T. Pavlidis. *Algorithms for Graphics and Image Processing*. Computer Science Press, Rockville MD, 1982.

[PFTV86]   W. Press, B. Flannery, S. Teukolsky, and W. T. Vetterling. *Numerical Recipes: the art of scientific computing*. Cambridge University Press, 1986.

[PGD82]   D. Phillips and A. Garcia-Diaz. *Fundamentals of Network Analysis*. Prentice-Hall, Englewood Cliffs NJ, 1982.

[PH80]   M. Padberg and S. Hong. On the symmetric traveling salesman problem: a computational study. *Math. Programming Studies*, 12:78–107, 1980.

[PL94]   P. A. Pevzner and R. J. Lipshutz. Towards DNA sequencing chips. In *19th Int. Conf. Mathematical Foundations of Computer Science*, volume 841, pages 143–158, Lecture Notes in Computer Science, 1994.

[PM88]   S. Park and K. Miller. Random number generators: Good ones are hard to find. *Communications of the ACM*, 31:1192–1201, 1988.

[Pom84]   C. Pomerance. The quadratic sieve factoring algorithm. In T. Beth, N. Cot, and I. Ingemarrson, editors, *Advances in Cryptology*, volume 209, pages 169–182. Lecture Notes in Computer Science, Springer-Verlag, 1984.

[PR79]   S. Peleg and A. Rosenfield. Breaking substitution ciphers using a relaxation algorithm. *Comm. ACM*, 22:598–605, 1979.

[Pra75]   V. Pratt. Every prime has a succinct certificate. *SIAM J. Computing*, 4:214–220, 1975.

[Pri57]   R. C. Prim. Shortest connection networks and some generalizations. *Bell System Technical Journal*, 36:1389–1401, 1957.

[Prü18]   H. Prüfer. Neuer Beweis eines Satzes über Permutationen. *Arch. Math. Phys.*, 27:742–744, 1918.

[PS82]   C. Papadimitriou and K. Steiglitz. *Combinatorial Optimization: Algorithms and Complexity*. Prentice-Hall, Englewood Cliffs, NJ, 1982.

[PS85]   F. Preparata and M. Shamos. *Computational Geometry*. Springer-Verlag, New York, 1985.

[PSW92]   T. Pavlides, J. Swartz, and Y. Wang. Information encoding with two-dimensional bar-codes. *IEEE Computer*, 25:18–28, 1992.

[Pug90]   W. Pugh. Skip lists: A probabilistic alternative to balanced trees. *Communications of the ACM*, 33:668–676, 1990.

[PW83]   S. Pizer and V. Wallace. *To compute numerically: concepts and strategies*. Little, Brown, Boston, 1983.

[Rab80]   M. Rabin. Probabilistic algorithm for testing primality. *J. Number Theory*, 12:128–138, 1980.

[Rab95]   F. M. Rabinowitz. A stochastic algorithm for global optimization with constraints. *ACM Trans. Math. Softw.*, 21(2):194–213, June 1995.

[Raw92]   G. Rawlins. *Compared to What?* Computer Science Press, New York, 1992.

[RC55]   Rand-Corporation. *A million random digits with 100,000 normal deviates*. The Free Press, Glencoe, IL, 1955.

[RDC93]   E. Reingold, N. Dershowitz, and S. Clamen. Calendrical calculations II: Three historical calendars. *Software – Practice and Experience*, 22:383–404, 1993.

[Rei72]   E. Reingold. On the optimality of some set algorithms. *J. ACM*, 19:649–659, 1972.

[Rei91]   G. Reinelt. TSPLIB – a traveling salesman problem library. *ORSA J. Computing*, 3:376–384, 1991.

[Rei94]   G. Reinelt. The traveling salesman problem: Computational solutions for TSP applications. In *Lecture Notes in Computer Science 840*, pages 172–186. Springer-Verlag, Berlin, 1994.

[Ren84]   R. J. Renka. Triangulation and interpolation at arbitrarily distributed points in the plane. *ACM Trans. Math. Softw.*, 10(4):440–442, December 1984.

[RHS89]   A. Robison, B. Hafner, and S. Skiena. Eight pieces cannot cover a chessboard. *Computer Journal*, 32:567–570, 1989.

[Riv77]   R. Rivest. On the worst-case behavior of string-searching algorithms. *SIAM J. Computing*, 6:669–674, 1977.

[Riv92]   R. Rivest. The MD5 message digest algorithm. RFC 1321, 1992.

[RND77]   E. Reingold, J. Nievergelt, and N. Deo. *Combinatorial algorithms: theory and practice*. Prentice-Hall, Englewood Cliffs NJ, 1977.

[RS96]   H. Rau and S. Skiena. Dialing for documents: an experiment in information theory. *Journal of Visual Languages and Computing*, pages 79–95, 1996.

[RSA78]   R. Rivest, A. Shamir, and L. Adleman. A method for obtaining digital signatures and public-key cryptosystems. *Communications of the ACM*, 21:120–126, 1978.

[RSL77]   D. Rosenkrantz, R. Stearns, and P. M. Lewis. An analysis of several heuristics for the traveling salesman problem. *SIAM J. Computing*, 6:563–581, 1977.

[RSST96]   N. Robertson, D. Sanders, P. Seymour, and R. Thomas. Efficiently four-coloring planar graphs. In *Proc. 28th ACM Symp. Theory of Computing*, pages 571–575, 1996.

[RT81]   E. Reingold and J. Tilford. Tidier drawings of trees. *IEEE Trans. Software Engineering*, 7:223–228, 1981.

[Rus97]   F. Ruskey. *Combinatorial Generation*. in preparation, 1997.

[Ryt85]   W. Rytter. Fast recognition of pushdown automata and context-free languages. *Information and Control*, 67:12–22, 1985.

[SA95]   M. Sharir and P. Agarwal. *Davenport-Schinzel sequences and their geometric applications*. Cambridge University Press, New York, 1995.

[Sam90a]   H. Samet. *Applications of spatial data structures*. Addison-Wesley, Reading MA, 1990.

[Sam90b]   H. Samet. *The design and analysis of spatial data structures*. Addison-Wesley, Reading MA, 1990.

[Sax80]   J. B. Saxe. Dynamic programming algorithms for recognizing small-bandwidth graphs in polynomial time. *SIAM J. Algebraic and Discrete Methods*, 1:363–369, 1980.

[Sch94]   B. Schneier. *Applied Cryptography*. Wiley, New York, 1994.

[SD75]   M. Syslo and J. Dzikiewicz. Computational experiences with some transitive closure algorithms. *Computing*, 15:33–39, 1975.

[SD76]   D. C. Schmidt and L. E. Druffel. A fast backtracking algorithm to test directed graphs for isomorphism using distance matrices. *J. ACM*, 23:433–445, 1976.

[SDK83]   M. Syslo, N. Deo, and J. Kowalik. *Discrete Optimization Algorithms with Pascal Programs*. Prentice Hall, Englewood Cliffs NJ, 1983.

[Sed77]   R. Sedgewick. Permutation generation methods. *Computing Surveys*, 9:137–164, 1977.

[Sed78]   R. Sedgewick. Implementing quicksort programs. *Communications of the ACM*, 21:847–857, 1978.

[Sed92]   R. Sedgewick. *Algorithms in C++*. Addison-Wesley, Reading MA, 1992.

[SF92]   T. Schlick and A. Fogelson. TNPACK—a truncated Newton minimization package for large-scale problems: I. algorithm and usage. *ACM Trans. Math. Softw.*, 18(1):46–70, March 1992.

[SFG82]   M. Shore, L. Foulds, and P. Gibbons. An algorithm for the Steiner problem in graphs. *Networks*, 12:323–333, 1982.

[SH75]   M. Shamos and D. Hoey. Closest point problems. In *Proc. Sixteenth IEEE Symp. Foundations of Computer Science*, pages 151–162, 1975.

[SH76]   M. Shamos and D. Hoey. Geometric intersection problems. In *Proc. 17th IEEE Symp. Foundations of Computer Science*, pages 208–215, 1976.

[Sha78]   M. Shamos. *Computational Geometry*. PhD thesis, Yale University, UMI #7819047, 1978.

[Sha87]   M. Sharir. Efficient algorithms for planning purely translational collision-free motion in two and three dimensions. In *Proc. IEEE Internat. Conf. Robot. Autom.*, pages 1326–1331, 1987.

[Sha93]   R. Sharda. *Linear and Discrete Optimization and Modeling Software: A Resource Handbook*. Lionheart Publishing Inc., 1993.

[She78]   A. H. Sherman. NSPIV: a Fortran subroutine for sparse Gaussian elimination with partial pivoting. *ACM Trans. Math. Softw.*, 4(4):391–398, December 1978.

[She96]   J. R. Shewchuk. Robust adaptive floating-point geometric predicates. In *Proc. 12th ACM Computational Geometry*, pages 141–150, 1996.

[SK83]   D. Sankoff and J. Kruskal. *Time Warps, String Edits, and Macromolecules: the theory and practice of sequence comparison*. Addison-Wesley, Reading MA, 1983.

[SK86]   T. Saaty and P. Kainen. *The Four-Color Problem*. Dover, New York, 1986.

[SK93]   R. Skeel and J. Keiper. *Elementary Numerical computing with Mathematica*. McGraw-Hill, New York, 1993.

[Ski88]   S. Skiena. Encroaching lists as a measure of presortedness. *BIT*, 28:775–784, 1988.

[Ski90]   S. Skiena. *Implementing Discrete Mathematics*. Addison-Wesley, Redwood City, CA, 1990.

[Sla61]   P. Slater. Inconsistencies in a schedule of paired comparisons. *Biometrika*, 48:303–312, 1961.

[SM73]   L. Stockmeyer and A. Meyer. Word problems requiring exponential time. In *Proc. Fifth ACM Symp. Theory of Computing*, pages 1–9, 1973.

[Smi91]   D. M. Smith. A Fortran package for floating-point multiple-precision arithmetic. *ACM Trans. Math. Softw.*, 17(2):273–283, June 1991.

[Sou96]   E. Southern. DNA chips: analysing sequence by hybridization to oligonucleotides on a large scale. *Trends in Genetics*, 12:110–115, 1996.

[SPP76]   A. Schönhage, M. Paterson, and N. Pippenger. Finding the median. *J. Computer and System Sciences*, 13:184–199, 1976.

[SR83]   K. Supowit and E. Reingold. The complexity of drawing trees nicely. *Acta Informatica*, 18:377–392, 1983.

[SR95]   R. Sharda and G. Rampal. Algebraic modeling languages on PCs. *OR/MS Today*, 22-3:58–63, 1995.

[SS71]   A. Schönhage and V. Strassen. Schnelle Multiplikation grosser Zahlen. *Computing*, 7:281–292, 1971.

[SS90]   J. T. Schwartz and M. Sharir. Algorithmic motion planning. In J. van Leeuwen, editor, *Handbook of Theoretical Computer Science: Algorithms and Complexity*, volume A, pages 391–430. MIT Press, 1990.

[SSS74]   I. Sutherland, R. Sproull, and R. Shumacker. A characterization of ten hidden surface algorithms. *ACM Computing Surveys*, 6:1–55, 1974.

[ST85]   D. Sleator and R. Tarjan. Self-adjusting binary search trees. *J. ACM*, 32:652–686, 1985.

[Sta90]   J. Stasko. Tango: A framework and system for algorithm animation. *Computer*, 23:27–39, 1990.

[Sta95]   W. Stallings. *Protect your Privacy: a guide for PGP Users*. Prentice Hall PTR, Englewood Cliffs NJ, 1995.

[Ste94]   G. A. Stephen. *String Searching Algorithms*. Lecture-Notes-Series-on-Computing. World-Scientific-Publishing, October 1994.

[Sto88]   J. Storer. *Data compression: methods and theory*. Computer Science Press, Rockville MD, 1988.

[Str69]   V. Strassen. Gaussian elimination is not optimal. *Numerische Mathematik*, 14:354–356, 1969.

[SV87]   J. Stasko and J. Vitter. Pairing heaps: Experiments and analysis. *Communications of the ACM*, 30(3):234–249, 1987.

[SV88]   B. Schieber and U. Vishkin. On finding lowest common ancestors: simplification and parallelization. *SIAM J. Comput.*, 17(6):1253–1262, December 1988.

[SW86]   D. Stanton and D. White. *Constructive Combinatorics*. Springer-Verlag, New York, 1986.

[SW94]   M. Stoer and F. Wagner. A simple min cut algorithm. In *European Symp. Algorithms (ESA)*, volume 855, pages 141–147. Springer Verlag Lecture Notes in Computer Science, 1994.

[SW95]   J. Salowe and D. Warme. Thirty-five-point rectilinear Steiner minimal trees in a day. *Networks*: 25, 1995.

[SWM95]   J. Shallit, H. Williams, and F. Moraine. Discovery of a lost factoring machine. *The Mathematical Intelligencer*, 17-3:41–47, Summer 1995.

[Tar95]   G. Tarry. Le problème de labyrinthes. *Nouvelles Ann. de Math.*, 14:187, 1895.

[Tar72]   R. Tarjan. Depth-first search and linear graph algorithms. *SIAM J. Computing*, 1:146–160, 1972.

[Tar75]   R. Tarjan. Efficiency of a good but not linear set union algorithm. *J. ACM*, 22:215–225, 1975.

[Tar79]   R. Tarjan. A class of algorithms which require non-linear time to maintain disjoint sets. *J. Computer and System Sciences*, 18:110–127, 1979.

[Tar83]   R. Tarjan. *Data Structures and Network Algorithms*. Society for Industrial and Applied Mathematics, Philadelphia, 1983.

[Tho68]   K. Thompson. Regular expression search algorithm. *Communications of the ACM*, 11:419–422, 1968.

[Tho96]   Mikkel Thorup. On RAM priority queues. In *Proc. Seventh Annual ACM-SIAM Symposium on Discrete Algorithms*, pages 59–67, 1996.

[Tin90]   G. Tinhofer. Generating graphs uniformly at random. *Computing*, 7:235–255, 1990.

[Tro62]   H. F. Trotter. Perm (algorithm 115). *Comm. ACM*, 5:434–435, 1962.

[Tur88]   J. Turner. Almost all k-colorable graphs are easy to color. *J. Algorithms*, 9:63–82, 1988.

[TW88]   R. Tarjan and C. Van Wyk. An $O(n \lg \lg n)$ algorithm for triangulating a simple polygon. *SIAM J. Computing*, 17:143–178, 1988.

[Ukk92]   E. Ukkonen. Constructing suffix trees on-line in linear time. In *Intern. Federation of Information Processing (IFIP '92)*, pages 484–492, 1992.

[Vai88]   P. Vaidya. Geometry helps in matching. In *Proc. 20th ACM Symp. Theory of Computing*, pages 422–425, 1988.

[Val79]   L. Valiant. The complexity of computing the permanent. *Theoretical Computer Science*, 8:189–201, 1979.

[Var91]   I. Vardi. *Computational Recreations in Mathematica*. Addison-Wesley, Redwood City CA, 1991.

[Vau80]   J. Vaucher. Pretty printing of trees. *Software Practice and Experience*, 10:553–561, 1980.

[vEBKZ77] P. van Emde Boas, R. Kaas, and E. Zulstra. Design and implementation of an efficient priority queue. *Math. Systems Theory*, 10:99–127, 1977.

[Veg90]   G. Vegter. The visibility diagram: a data structure for visibility problems and motion planning. In *Proc. second Scand. Workshop on Algorithm Theory (SWAT)*, volume 447, pages 97–110. Springer Verlag Lecture Notes in Computer Science, 1990.

[Vit89]   J. S. Vitter. Dynamic Huffman coding. *ACM Trans. Math. Softw.*, 15(2):158–167, June 1989.

[Viz64]   V. G. Vizing. On an estimate of the chromatic class of a p-graph (in Russian). *Diskret. Analiz*, 3:23–30, 1964.

[vL90a]   J. van Leeuwen. Graph algorithms. In J. van Leeuwen, editor, *Handbook of Theoretical Computer Science: Algorithms and Complexity*, volume A, pages 525–631. MIT Press, 1990.

[vL90b]   J. van Leeuwen, editor. *Handbook of Theoretical Computer Science: Algorithms and Complexity*, volume A. MIT Press, 1990.

[Vos92]   S. Voss. Steiner's problem in graphs: heuristic methods. *Discrete Applied Mathematics*, 40, 1992.

[War62]   S. Warshall. A theorem on boolean matrices. *J. ACM*, 9:11–12, 1962.

[Wat95]   M. Waterman. *Introduction to Computational Biology*. Chapman Hall, London, 1995.

[WBCS77]   J. Weglarz, J. Blazewicz, W. Cellary, and R. Slowinski. An automatic revised simplex method for constrained resource network scheduling. *ACM Trans. Math. Softw.*, 3(3):295–300, September 1977.

[Wei73]   P. Weiner. Linear pattern-matching algorithms. In *Proc. 14th IEEE Symp. on Switching and Automata Theory*, pages 1–11, 1973.

[Wei92]   M. Weiss. *Data Structures and Algorithm Analysis*. Benjamin/Cummings, Redwood City, CA, 1992.

[Wel84]   T. Welch. A technique for high-performance data compression. *IEEE Computer*, 17-6:8–19, 1984.

[Wes83]   D. H. West. Approximate solution of the quadratic assignment problem. *ACM Trans. Math. Softw.*, 9(4):461–466, December 1983.

[WF74]   R. A. Wagner and M. J. Fischer. The string-to-string correction problem. *J. ACM*, 21:168–173, 1974.

[Whi32]   H. Whitney. Congruent graphs and the connectivity of graphs. *American J. Mathematics*, 54:150–168, 1932.

[Wig83]   A. Wigerson. Improving the performance guarantee for approximate graph coloring. *J. ACM*, 30:729–735, 1983.

[Wil64]   J. W. J. Williams. Algorithm 232 (heapsort). *Communications of the ACM*, 7:347–348, 1964.

[Wil84]   H. Wilf. Backtrack: An $O(1)$ expected time algorithm for graph coloring. *Info. Proc. Letters*, 18:119–121, 1984.

[Wil85]   D. E. Willard. New data structures for orthogonal range queries. *SIAM J. Computing*, 14:232–253, 1985.

[Wil89]   H. Wilf. *Combinatorial Algorithms: an update*. SIAM, Philadelphia PA, 1989.

[Win68]   S. Winograd. A new algorithm for inner product. *IEEE Trans. Computers*, C-17:693–694, 1968.

[Win80]   S. Winograd. *Arithmetic Complexity of Computations*. SIAM, Philadelphia, 1980.

[WM92a]   S. Wu and U. Manber. Agrep – a fast approximate pattern-matching tool. In *Usenix Winter 1992 Technical Conference*, pages 153–162, 1992.

[WM92b]   S. Wu and U. Manber. Fast text searching allowing errors. *Comm. ACM*, 35:83–91, 1992.

[Wol79]   T. Wolfe. *The Right Stuff*. Bantam Books, Toronto, 1979.

[Woo93]   D. Wood. *Data Structures, Algorithms, and Performance*. Addison-Wesley, Reading MA, 1993.

[WS79]   C. Wetherell and A. Shannon. Tidy drawing of trees. *IEEE Trans. Software Engineering*, 5:514–520, 1979.

[WW95]   F. Wagner and A. Wolff. Map labeling heuristics: provably good and practically useful. In *Proc. 11th ACM Symp. Computational Geometry*, pages 109–118, 1995.

[Yao81]   A. C. Yao. A lower bound to finding convex hulls. *J. ACM*, 28:780–787, 1981.

[Yao90]   F. Yao. Computational geometry. In J. van Leeuwen, editor, *Handbook of Theoretical Computer Science: Algorithms and Complexity*, volume A, pages 343–389. MIT Press, 1990.

[YS96]   F. Younas and S. Skiena. Randomized algorithms for identifying minimal lottery ticket sets. *Journal of Undergraduate Research*, 2-2:88–97, 1996.

[YY76]   A. Yao and F. Yao. The complexity of searching in an ordered random table. In *Proc. 17th IEEE Symp. Foundations of Computer Science*, pages 173–177, 1976.

[ZL78]   J. Ziv and A. Lempel. A universal algorithm for sequential data compression. *IEEE Trans. Information Theory*, IT-23:337–343, 1978.

Index